T0331146

An Introduction to Electrochemical Impedance Spectroscopy

An Introduction to Electrochemical Impedance Spectroscopy

Dr. Ramanathan Srinivasan and Dr. Fathima Fasmin

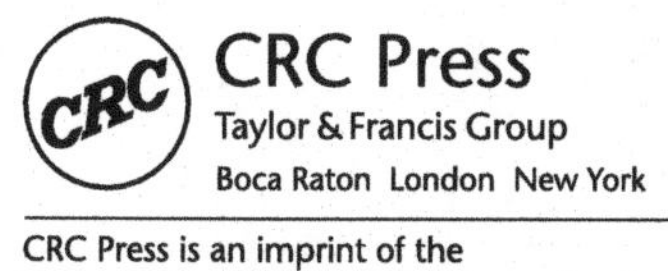

CRC Press is an imprint of the
Taylor & Francis Group, an informa business

MATLAB® is a trademark of The MathWorks, Inc. and is used with permission. The MathWorks does not warrant the accuracy of the text or exercises in this book. This book's use or discussion of MATLAB® software or related products does not constitute endorsement or sponsorship by The MathWorks of a particular pedagogical approach or particular use of the MATLAB® software.

First edition published 2021
by CRC Press
6000 Broken Sound Parkway NW, Suite 300, Boca Raton, FL 33487-2742

and by CRC Press
2 Park Square, Milton Park, Abingdon, Oxon, OX14 4RN

© 2021 Ramanathan Srinivasan and Fathima Fasmin

CRC Press is an imprint of Taylor & Francis Group, LLC

The right of Ramanathan Srinivasan and Fathima Fasmin to be identified as authors of this work has been asserted by them in accordance with sections 77 and 78 of the Copyright, Designs and Patents Act 1988.

Reasonable efforts have been made to publish reliable data and information, but the author and publisher cannot assume responsibility for the validity of all materials or the consequences of their use. The authors and publishers have attempted to trace the copyright holders of all material reproduced in this publication and apologize to copyright holders if permission to publish in this form has not been obtained. If any copyrighted material has not been acknowledged, please write and let us know so we may rectify it in any future reprint.

Except as permitted under U.S. Copyright Law, no part of this book may be reprinted, reproduced, transmitted, or utilized in any form by any electronic, mechanical, or other means, now known or hereafter invented, including photocopying, microfilming, and recording, or in any information storage or retrieval system, without written permission from the publishers.

For permission to photocopy or use material electronically from this work, access www.copyright.com or contact the Copyright Clearance Center, Inc. (CCC), 222 Rosewood Drive, Danvers, MA 01923, 978-750-8400. For works that are not available on CCC, please contact mpkbookspermissions@tandf.co.uk

Trademark notice: Product or corporate names may be trademarks or registered trademarks and are used only for identification and explanation without intent to infringe.

Library of Congress Cataloging-in-Publication Data
Names: Srinivasan, Ramanathan, author. | Fasmin, Fathima, author.
Title: Introduction to electrochemical impedance spectroscopy /
Dr. Ramanathan Srinivasan and Dr. Fathima Fasmin.
Description: First edition. | Boca Raton : CRC Press, 2021. |
Includes bibliographical references and index.
Identifiers: LCCN 2020049887 (print) | LCCN 2020049888 (ebook) |
ISBN 9780367651176 (hardback) | ISBN 9781003127932 (ebook)
Subjects: LCSH: Impedance spectroscopy. | Electrochemical analysis–Data processing. | Spectrum analysis.
Classification: LCC QD116.I57 S65 2021 (print) | LCC QD116.I57 (ebook) |
DDC 543/.6–dc23
LC record available at https://lccn.loc.gov/2020049887
LC ebook record available at https://lccn.loc.gov/2020049888

ISBN: 978-0-367-65117-6 (hbk)
ISBN: 978-0-367-65122-0 (pbk)
ISBN: 978-1-003-12793-2 (ebk)

Typeset in Times
by codeMantra

Access the [companion website/Support Material]: www.Routledge.com/9780367651176.

Ramanathan Srinivasan

dedicates this book to his parents

K R Srinivasan and Lalitha Srinivasan

Contents

Preface

Electrochemical impedance spectroscopy (EIS) is one of the many techniques used to characterize an electrochemical system. In this era of relatively more generous funding and less expensive instruments, it has become easier to collect impedance spectra. To analyze the data, the usual tendency is to look at publications on similar systems, pick up an electrical circuit model from these publications, and apply the same model to the data. The easy availability of canned software for fitting the electrical circuit models has aided this trend.

Unfortunately, this can lead to interpretations of questionable value. On the other hand, many journal articles illustrate better methods of interpretation of the EIS data, especially for electrochemical reactions. However, no book explains, at an introductory level, the relationship between the reactions and EIS data or describes in detail the subtleties involved in analyzing the data. This has kept most of the researchers, particularly the graduate students, away from employing such reliable methods to analyze the data. We hope that this book will help to bridge the gap.

Our target audiences are graduate students working in the general area of electrochemistry. They may have an undergraduate degree in engineering; or in physics or chemistry, but with some additional mathematical background. Working with EIS data involves working with complex numbers and elementary calculus such as differentiation and integration. We assume that the reader will be comfortable with these, and there is no real introduction to these topics in this book. There is only a very short refresher for complex numbers and differential equations in the appendix. Similarly, the students are expected to have a basic knowledge of electrochemistry. The first chapter is a brief introduction to electrochemistry. Still, if the reader is completely new to this field, it will probably be necessary to gain sufficient background in electrochemistry before the real benefits of this book can be reaped.

This book is written with a focus on electrochemical reactions, and it explains in detail the use of reaction mechanism analysis to interpret the spectra. Most of the books already available in the market spare at the most a few pages to this topic and the presentation is terse, which makes it difficult for the beginners to learn the details. Nonlinear EIS (NLEIS) is an emerging technique but is only briefly discussed in the books available, and this book devotes an entire chapter to NLEIS. In a sense, a significant part of the material presented in this book is either complementary to what is discussed in other EIS books, or it is discussed in much more detail than is available elsewhere.

This book is organized as follows. The first chapter is an introduction to electrochemistry and the idea of impedance. It explains the concept of a double-layer and the need for a three-electrode cell in most electrochemical experiments. The idea of impedance as a generalized resistance is explained with the help of the response of ideal electrical circuit elements such as resistors and capacitors to sinusoidal potential. The commonly used graphical representations of impedance are also presented. The second chapter deals with the instrumentation and various techniques, such as single sine and multi-sine techniques employed to measure EIS. A few lab experiments that one can perform to familiarize oneself with the practical aspects of EIS data acquisition are also described.

The third chapter treats the important subject of data validation, using Kramers–Kronig transforms or an equivalent method. The fourth chapter introduces the analysis of EIS data using circuit analogy. The fifth chapter shows, in detail, how the impedance of several electrochemical reactions can be calculated. A distributed element, called a constant phase element (CPE), which is often necessary to model the real-life experimental data, is described in the sixth chapter. In addition, procedures to account for diffusion effects, the porosity of electrodes, and film formation and to calculate the impedance response are described in the sixth chapter. The seventh chapter illustrates the application of EIS to characterize a few electrochemical systems, i.e., corrosion, biosensing, and battery technology. The eighth and final chapter describes the latest developments, particularly the simulations, in EIS at large amplitude perturbations, viz., nonlinear EIS. At the end of each chapter, a list of exercise problems is given. Many of them will require the use of software like Microsoft Excel® or a programming environment like MATLAB®. These are designed to help the reader understand the ideas described in the chapters better and to be familiar with the simulation and analysis of EIS data and the challenges involved.

The work on the mechanistic analysis of EIS data started in the lab of Ramanathan, with Jeevan Maddala, Krishnaraj S, and Vinodkumar V, in 2008 and has continued ever since. Several research scholars in our lab, viz., Nagendraprasad Y, Karthik P, Sruthi S, Prasanna V, Noyel Victoria S, Fathima Fasmin (who is also an author of this book), Praveen BVS, Amrutha MS, Tirumal Rao M, Ranjith PM, Rajesh P, Twinkle Paul, and Saibi R employed mechanistic analysis of EIS data as part of their research work, and our understanding of the power of this technique and the subtleties involved in analyzing the data grew with each application. We have attempted to distill the learnings and present them here with the hope that others can also use EIS and appreciate it as much as we do, if not more.

In Jan–Apr 2019, one of the authors (Ramanathan Srinivasan) conducted a 30 hours online video course called "Introduction to Electrochemical Impedance Spectroscopy," under the auspices of the National Program on Technology Enhanced Learning (NPTEL), Ministry of Human Resources and Development, India, and the videos are freely available at http://www.che.iitm.ac.in/~srinivar/EIS-Book.html. The contents of this book closely follow the material presented in that course. Although we have checked the contents of this book thoroughly, we are aware that mistakes and errors would have slipped in. We will be grateful to receive corrections and suggestions for improvements. You can reach us at learnimpedance@gmail.com.

Ramanathan Srinivasan
Chennai, India
2020
Fathima Fasmin
Calicut, India
2020.

MATLAB® is a registered trademark of The MathWorks, Inc. For product information, please contact:

The MathWorks, Inc.
3 Apple Hill Drive
Natick, MA 01760-2098 USA
Tel: 508-647-7000
Fax: 508-647-7001
E-mail: info@mathworks.com
Web: www.mathworks.com

Access the Support Material: https://www.routledge.com/9780367651176.

Authors

Dr. Ramanathan Srinivasan obtained his B.Tech. in Chemical Engineering from A.C. Tech, Anna University in 1993. He completed his Masters. (Chem. Engg.) in 1996 and Ph.D. (Chem. Engg.) in 2000 from Clarkson University and joined PDF Solutions, USA as a Senior Consulting Engineer. He joined IIT Madras as faculty in 2003 and is currently working as a Professor in the Department of Chemical Engineering, IIT Madras. His main field of specialization is in the application of electrochemical impedance spectroscopy to obtain mechanistic information of electrochemical reactions. He has created freely available 30 hours of video lectures on electrochemical impedance spectroscopy under the auspices of NPTEL, India.

Dr. Fathima Fasmin obtained her B.Tech. degree in Chemical Engineering from Thangal Kunju Musaliar College of Engineering, Kerala in 2005. She completed her Ph.D. (Chemical Engineering) in 2016 from the Indian Institute of Technology Madras and joined Qatar Environment & Energy Research Institute, Qatar as a postdoctoral researcher. Currently, she is an Assistant Professor in the Department of Chemical Engineering, National Institute of Technology, Calicut. She pursues research in areas such as corrosion, fuel cells, and batteries; particularly in understanding EIS and nonlinear EIS data using physics-based models.

1 Introduction

To doubt everything and to believe everything are two equally convenient solutions; each saves us from thinking.

—Henri Poincare

In this chapter, first, a brief introduction to the electrode–electrolyte interface is given. Then, the formation of a double-layer is presented, followed by the kinetics of electrochemical reactions. The energy barrier to electrochemical reactions and the effect of potential on the reaction rate are discussed. The difficulty in measuring the absolute potential across a single electrode–electrolyte interface is discussed, and the need for a three-electrode cell to measure the changes in potential of an interface is explained. The idea of impedance as generalized resistance is introduced, followed by a simulation of the impedance spectra of a simple electrical circuit. Finally, we touch upon a few other techniques that are frequently used to characterize general electrochemical systems.

1.1 ELECTRODE–ELECTROLYTE INTERFACE

Metal corrosion is an example of an electrochemical reaction. Here, chemical reactions take place, and electrons are either produced or consumed at the electrodes. Any metal immersed in an aqueous solution, even if it is not corroding, will immediately develop a potential (e.g., gold in saltwater). Although a metal–salt solution is used as an example here, electrochemistry encompasses nonmetallic electrodes and nonaqueous solutions as well as molten salts. However, the majority of real-world cases includes metal electrodes in aqueous solutions. Therefore, we restrict our discussion to a metal–aqueous solution case. A metal–liquid interface is referred to as an electrode–electrolyte interface. In electrochemistry, the electrode–electrolyte interface plays a critical role. It is usually more important than the individual electrode phase or the electrolyte phase, and hence the discussion is focused on the interface. A typical electrolyte consists of a salt, or an acid, or a base in water. An electrolyte contains positively charged ions (cations) and negatively charged ions (anions). Usually, water molecules surround the cations and are tightly bound to them, and these are called 'solvated' cations. Water molecules also surround the anions, but they can be removed relatively easily, with a few exceptions.

If a small dc potential is applied, then the electrode with a positive charge will attract anions, and the electrode with a negative charge will attract the cations. If the electrode potential is large enough, the water molecules surrounding the ions can be removed, and the ions can adsorb on the electrode, as shown in Figure 1.1. The water molecules are not shown for the sake of simplicity. In this figure, two dotted lines mark the inner Helmholtz plane (IHP) and the outer Helmholtz plane (OHP). The IHP is the imaginary plane that goes through the center of ions adsorbed on

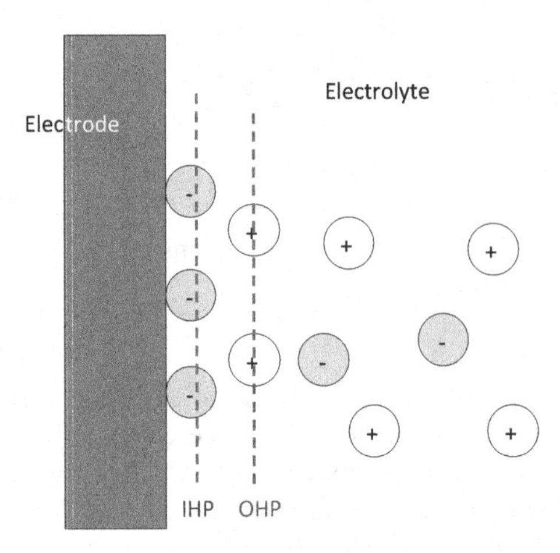

FIGURE 1.1 Schematic of a metal in water, with cations and anions present in the solution.

the electrode. These are the ions without solvation. The OHP is the imaginary plane that goes through the solvated ions, which are right next to the electrode, but not as close as the ions without solvation. A thorough introduction to basic electrochemistry is accessible in many classical texts, e.g., Bockris, Reddy, and Gamboa-Aldeco (2002).

1.2 ELECTROCHEMICAL REACTION

For the cations to gain an electron from the metal electrode, the electrons have to pass through an energy barrier. Similarly, for the anions to give electrons to the metal electrode, the electrons have to pass through an energy barrier. The process of gaining or giving up an electron is called charge-transfer. When a suitably large dc potential is applied, this energy barrier can be crossed, and charge-transfer can occur. The minimum energy required for a reaction to occur can be calculated using thermodynamics. However, just because a reaction *can* occur, it does not mean that the reaction *will* occur at a measurable rate. The reaction rate may be fast, slow, or medium. The kinetics tells us the rate at which the reaction occurs *in practice*. The energy barrier calculated using thermodynamics corresponds to a potential. For example, water can split into hydrogen and oxygen, and the gases can evolve if the dc potential is 1.23 V or larger. If a smaller dc potential is applied, then there will not be any reaction, and the current will be zero.

In practice, the type of electrode used also makes a difference to this dc potential value. That is, the minimum potential at which the 'charge-transfer' occurs depends on the electrode. So, at many of the electrodes, water will not split if a potential of 1.23 V is applied. If the electrode is a 'good catalyst,' a somewhat higher dc potential is sufficient to split water. If it is a 'poor catalyst,' then a very large dc potential is needed to split water. In this example, platinum serves as a good electrode, whereas mercury is a poor electrode. In other words, Pt is a good catalyst, and Hg is a poor catalyst for water splitting. Other examples of electrochemical reactions are hydrogen oxidation in fuel cells and corrosion of metals, such as Zn in dilute acid. Zinc dissolution in sulfuric acid can be represented as $Zn \rightarrow Zn^{2+} + 2e^-$. These electrons must be taken up by some entity, and in this case, it is the hydrogen ion, which gets reduced as $2H^+ + 2e^- \rightarrow H_2$. The overall reaction is $Zn + H_2SO_4 \rightarrow ZnSO_4 + H_2$. Since electrons participate in this reaction, we can expect that the reaction rate can be altered by controlling the potential. Applying a positive potential to the Zn electrode will increase the rate of Zn ionization, and hence Zn dissolution. A reaction in which the charges are transferred is often called a *faradaic process*, in recognition of the pioneering electrochemist Michael Faraday.

1.2.1 Electrode–Electrolyte Interface – The Double-Layer

When a metal is immersed in an electrolyte, the metal is charged, and the ions line up near the electrode, as shown in Figure 1.1. If the energy barrier to cross the interface is large, then there is no reaction between the electrode and the electrolyte. The interface appears like a capacitor where a small gap separates two plates having opposite charges. This simple model of an electrode–electrolyte interface is called the Helmholtz model. Consider an example where a metal electrode is immersed in NaCl solution. In the solution, NaCl is dissociated into positively charged Na^+ cations and negatively charged Cl^- anions. The Na^+ and Cl^- ions are usually hydrated. However, the Na^+ ion has a smaller radius and hence a larger charge density than the Cl^- ions. Therefore, the water molecules, which exhibit a significant dipole moment, are more tightly bound to the Na^+ ions. The Cl^- anions are also hydrated, but usually, the water molecules in the hydration sheath of anions can be removed relatively easily. Often, for the sake of simplicity, the hydration sheath around the ions is not drawn in the schematics.

Note that in the schematic, the electrode is negative and more of Na^+ ions are near the electrode compared to the number of Cl^- ions. Some Cl^- ions are also present, but the net charge on the solutions side of the interface is positive because an excess of Na^+ ions would be present near the electrode. This is often referred to as a *double-layer* structure. In this Helmholtz model, the entire potential difference occurs across a distance of less than 1 nm. Thus, even if a relatively small potential of 1 V is applied, the electric field at the interface would be very large (10^9 V m^{-1}).

A different model for the electrode–electrolyte interface is proposed by Gouy and Chapman, where the charges on the solution side are spread out a bit more, i.e., they are present at a distance of ~1–10 nm from the electrode. The distance up to which the excess ions are present depends on the concentration of the electrolyte and other variables such as the solvent and the electrode potential. For example, if the salt concentration is 1 M, then the distance is of the order of 1 nm, whereas if the salt concentration is 1 mM, then the distance is of the order of 10 nm. The Stern model, a combination of the Helmholtz and Gouy–Chapman models, gives a better description of the interface. A major part of the potential change occurs in the Helmholtz layer, and the remaining part occurs in the diffuse layer. For many common metal–salt solution systems, the double-layer capacitance is in the range of 10–30 μF cm^{-2}. The unit of capacitance is Farad (F), and in the case of electrochemical systems, capacitance is always expressed per unit electrode area.

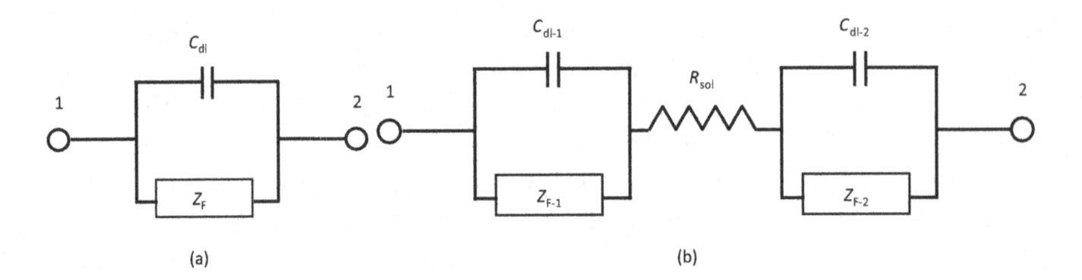

FIGURE 1.2 Electrical circuit representation of (a) one electrode–electrolyte interface and (b) two electrodes with an electrolyte.

1.2.2 Electrical Circuit Model of the Interface

The above discussion assumes that there is no transfer of electrons across the metal–solution interface. In this case, the interface can be modeled as a capacitor. If there is a reaction, the interface can be modeled as a capacitor in parallel with another element, which accounts for the reaction. This is often called the *faradaic impedance* and is represented as Z_F in Figure 1.2a. A capacitor, denoted by C_{dl}, models the double-layer.

Since at least two electrodes are present in an electrochemical cell, two such pairs must be combined to represent the complete cell. In addition, if the solution resistance (R_{sol}) is significant, then the complete cell would be given by the circuit shown in Figure 1.2b. In most cases, the impedance of the second electrode can be made negligible, and the solution resistance effect is also made small so that the entire cell impedance is approximately given by the single electrode impedance (Figure 1.2a).

1.2.3 Effect of Potential on the Rate Constant

The following discussion is adopted from *Electrode Kinetics* by John Albery (Albery 1975). Consider the reduction/oxidation reaction $Fe^{3+}_{sol} + e^- \rightleftharpoons Fe^{2+}_{sol}$ occurring on an inert electrode like Pt. The forward reaction is the reduction of Fe^{3+}_{sol} and the reverse reaction is the oxidation of Fe^{2+}_{sol}. Together, it is referred to as a *redox* reaction. The free-energy curve corresponding to this reaction is given in Figure 1.3.

To change the state from Fe^{3+}_{sol} to Fe^{2+}_{sol} (or vice versa), a free-energy barrier has to be crossed. In Figure 1.3, for the species to move from O to Q, it has to cross the point T. The species corresponding to the point T is called the transition state. Here, the maximum in free energy occurs. The species at location O has to 'climb' the mountain at location T to move to the location Q. In normal chemical reactions, the temperature can be increased so that the species gain energy and cross the barrier more easily. In electrochemistry, there is an additional knob, viz., the potential. Any decrease in the barrier height with respect to the reactant (product) corresponds to an increase in the rate constant for the forward (reverse) reaction because the species has to climb a relatively shorter 'mountain.'

When the potential changes, the free energies of Fe^{3+}_{sol} and Fe^{2+}_{sol} also change. In addition, the free energy of the transition state also changes. The extent of the changes need not be the same for Fe^{3+}_{sol}, Fe^{2+}_{sol} and the transition state species. An order of magnitude analysis shows (Albery 1975) that the rate constant can be changed by

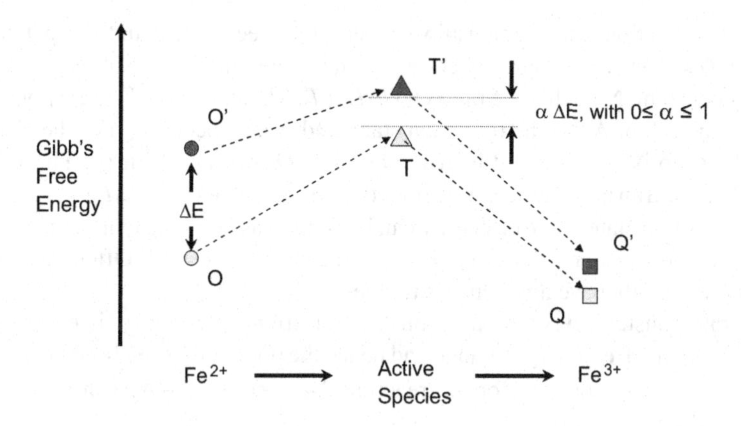

FIGURE 1.3 Free-energy diagram, with chemical species in the abscissa. Application of potential (ΔE) changes the free-energy diagram.

a factor of 10 by changing the potential by 118 mV. The free energy changes linearly with the potential. The rate constant depends exponentially on the free energy, and hence the potential. This is why the electrode potential has such an extraordinary influence on the reaction rate constant. Note that although we call it as rate constant, it actually varies with the temperature and the potential; i.e., it is a constant *when the potential and the temperature are fixed.*

Now let us consider a second example, the reaction $Ag^{+}_{sol} + e^{-} \rightleftharpoons Ag$ occurring on a silver electrode. Imagine that the electrode is grounded and that we can alter the potential of the solution at will. This example is constructed so that the idea of the charge-transfer coefficient can be understood easily. When the potential of the solution is changed, the free energy (G) of Ag^{+}_{sol} is also changed. On the other hand, the potential of the Ag electrode remains the same since it is always grounded. This is shown in Figure 1.4.

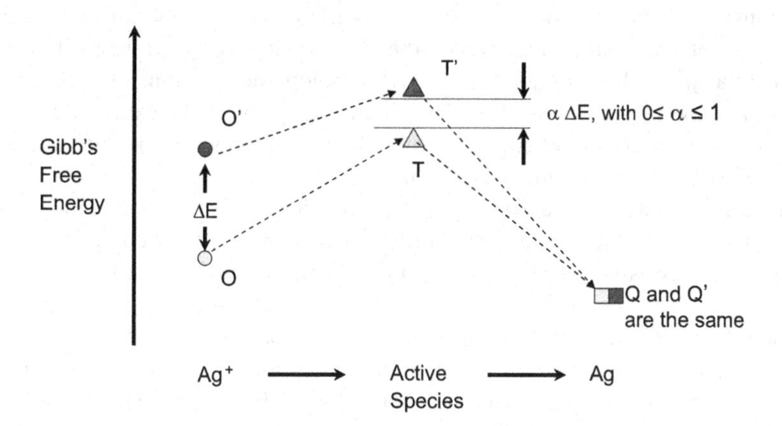

FIGURE 1.4 Free-energy diagram, with the product grounded. Application of potential (ΔE) changes the free-energy diagram.

For this reaction, the relationship between the free energy and the potential is given by $G = \text{constant} + FE_{sol}$, where E_{sol} is the potential of the solution and F is the Faraday constant. Note that the ratio $OO'/\Delta E$, $TT'/\Delta E$, and $QQ'/\Delta E$ correspond to the value of $\Delta G/\Delta E$ of Ag^+_{sol}, the transition state, and Ag, respectively. For the electrode (Ag), the ratio is zero since it is grounded, i.e., Q overlaps with Q'. The value of $\Delta G/\Delta E$ of the transition state is less than that of the value of $\Delta G/\Delta E$ of Ag^+_{sol}. Thus, for the transition state, the free energy can be written as $G = \text{constant} + \alpha F E_{sol}$. Here α is the charge-transfer coefficient, with a value between 0 and 1. Often it is taken as 0.5 since we do not have any better estimate.

The rate constant, written based on the transition state theory, is exponentially dependent on the free-energy change and hence the potential. It can be shown (Albery 1975) that the rate constant for the reaction $Ag^+_{sol} + e^- \xrightarrow{k_1} Ag$ can be written as $k_1 = k_{10}e^{\frac{(1-\alpha)FE_{sol}}{RT}}$ and that the rate constant for the reverse reaction $Ag \xrightarrow{k_{-1}} Ag^+_{sol} + e^-$ can be written as $k_{-1} = k_{-10}e^{\frac{-\alpha FE_{sol}}{RT}}$. Now, if the metal is not grounded and if the metal electrode potential is also allowed to vary, then the effective potential across the interface is $(E_{sol} - E_m)$ rather than E_{sol}. Here, E_m is the metal electrode potential. This allows us to rewrite the rate constants as $k_1 = k_{10}e^{\frac{-(1-\alpha)F(E_m-E_{sol})}{RT}}$ and $k_{-1} = k_{-10}e^{\frac{\alpha F(E_m-E_{sol})}{RT}}$

1.2.4 DC Current *vs.* Potential in an Electrochemical Reaction

We want to develop the equations relating the current to the potential in an electrochemical reaction. In the previous section, we saw that the rate constant depends exponentially on the potential. For a simple electron transfer reaction, under certain conditions, the reaction rate is proportional to the rate constant. Independently, based on experimental observations, Julius Tafel proposed a relationship between the current and the potential in many electrochemical reactions as $i = \text{constant} \times e^{bE}$ and this is also referred to as the *Tafel equation*. Here, b is a parameter that depends on the electrode material and the electrolytes.

For more general cases with multiple steps in the reaction, there are multiple rate constants. Then the net reaction rate is a complex expression involving the rate constants and other terms, but the individual rate constants still depend exponentially on the potential. Next, we develop the relationship between the current and the rate of the reaction. By combining this with the expression relating the rate to the potential, we can write the relationship between the current and the potential in an electrochemical reaction.

Michael Faraday was the first to propose a relationship between the total quantity of a material deposited by electrochemical reaction and the total charge consumed. The total charge is the product of the current and the time. For a reaction with a single electron transfer such as $M^+_{sol} + e^- \rightarrow M$, the relationship is $i = F \times$ rate of the reaction. Here, F is the Faraday constant, which is 96,485 C mol^{-1}, which is usually written in Farad, a derived unit. Note that the proportionality constant F is written in italics while the unit Farad is written in normal style (F). Since the rate varies exponentially with the potential, the dc current *vs.* potential relationship is also nonlinear. The current due to the reaction is called the *faradaic* current.

However, for most electrochemical reactions, if we measure the dc current as a function of dc potential, the current will increase exponentially with the potential only up to some potential. Beyond that, the current will tend to saturate. A qualitative representation of dc current *vs.* potential is given in Figure 1.5.

Let us consider the case of the deposition of Cu from the $CuSO_4$ solution. The Cu^{2+} ions present in the solution will deposit on the electrode by taking up two electrons. If the reaction is simple, then the rate of the reaction will be given by $k\left[Cu^{2+}\right]$, where k is the rate constant and $\left[Cu^{2+}\right]$ is the concentration of the Cu^{2+} ions at the interface. If the solution is not stirred during the deposition, then at high potentials, the concentration of Cu^{2+} ions at the interface will decrease due to the rapid deposition. Because of this, (a) Cu^{2+} ions from the bulk of the solution will diffuse to the interface and (b) there will be a reduction in the rate of the reaction and hence the current. The steady-state (dc) current will be limited by the rate of diffusion of Cu^{2+} ions. Thus, the steady-state current will not increase exponentially with the potential but will rather be independent of potential. This is referred to as the 'mass transfer limited' regime or 'diffusion limited' regime. This is shown in Figure 1.5. Stirring the solution will increase the net mass transfer rate and hence the saturation current value. If the dc potential is small, then the rate of deposition of Cu will be low. Then the concentration of the Cu^{2+} ions in the solution will be more or less uniform, and the current is limited by the reaction rate only. This is called the 'kinetic controlled' or 'reaction limited' regime. At intermediate potentials, both mass transfer and kinetics play equally important roles and this is known as 'mixed control regime'. The relevant mass transfer and kinetics equations are described in more detail in Section 6.2.

In most cases, the overall reaction is the result of many elementary reactions. For example, the dissolution of copper may occur in two steps as $Cu^{2+} + e^- \rightarrow Cu^+$ followed by $Cu^+ + e^- \rightarrow Cu$, and the rates of each step may be different. The set of elementary reactions is called the reaction mechanism. Even when the overall reaction is not elementary, many a time, the dc current will depend exponentially on the potential. Therefore, it is not easy to distinguish between two different mechanisms

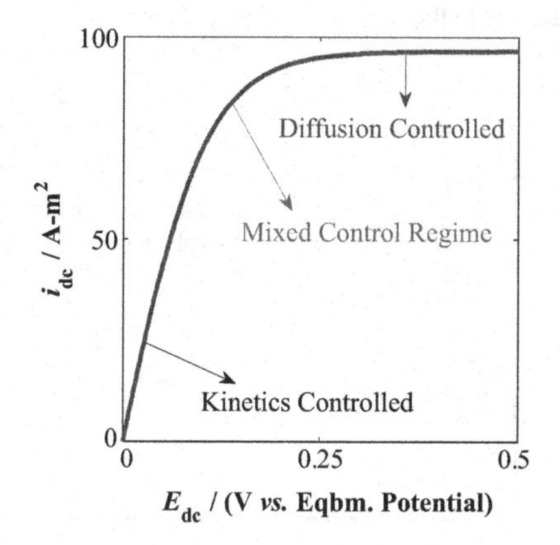

FIGURE 1.5 Plot of dc current *vs.* potential, for an electrochemical reaction occurring under kinetic and diffusion-controlled regimes.

with knowledge of only the dc current *vs.* potential for the two reactions. Impedance spectroscopy is a technique that can give more information about the mechanism than the "dc current *vs.* potential data."

1.3 THREE-ELECTRODE CELL

The absolute potential change across a metal–liquid interface cannot be measured experimentally. This is because two connections are needed to measure the potential, which means another metal has to be introduced into the solution. That will cause a second potential change. Thus, only the sum of the two potential changes can be measured. The difficulty in measuring the potential across a single electrode–electrolyte interface is explained well in *Modern Electrochemistry* by Bockris, Reddy, and Gamboa-Aldeco (2002).

Let the electrode be at some potential x with respect to the electrolyte. To measure the potential drop across the electrode–electrolyte interface, we have to connect it to a voltmeter. A voltmeter has two probes. One probe can be connected to the electrode, but when the second probe is introduced into the liquid, it also develops a potential y, which is unknown, across the probe–electrolyte interface. What we measure is the difference $x-y$ rather than the value of x alone. This is given in Figure 1.6.

We, therefore, see that it is impossible to measure the potential drop across one electrode–electrolyte interface. If we introduce two electrodes into an electrolyte and apply a potential between them (e.g., 1 V), we will not know the individual potential drops at the electrode–electrolyte interface. Consider the electrochemical system given in Figure 1.7. Let us denote the electrode on the left as M_1 and that on

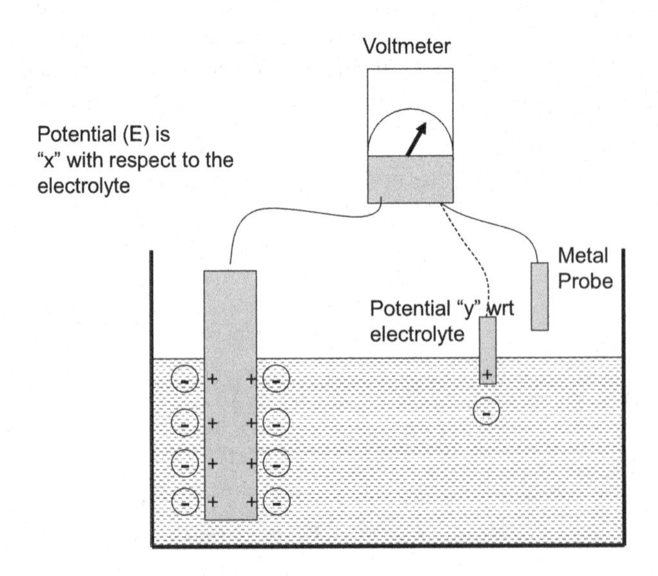

FIGURE 1.6 Schematic to illustrate the challenges in measuring the electrode–electrolyte potential difference.

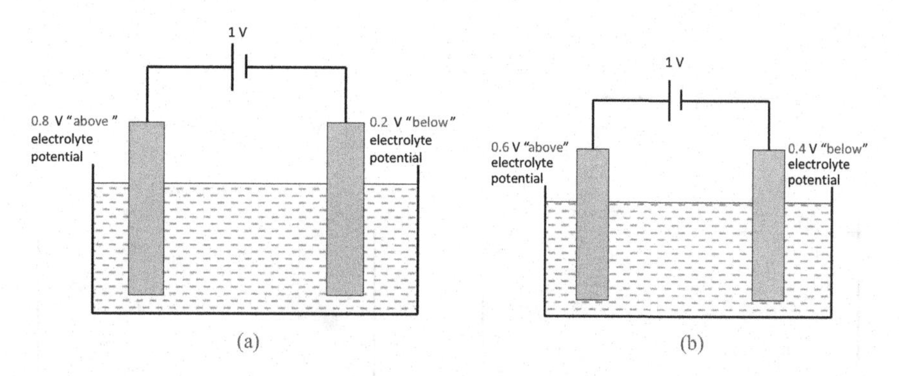

FIGURE 1.7 Schematic to illustrate the challenges in estimating the potential drop in a two-electrode cell. Across the two electrodes, the potential drop can be (a) 0.8 and −0.2 V or (b) 0.6 and −0.4 V. Note that other combinations are also possible.

the right as M_2. While the potential difference between M_1 and M_2 is known to be 1 V, M_1 may be +0.8 V *wrt* the electrolyte, and M_2 is −0.2 V *wrt* the electrolyte as given in Figure 1.7a. It is quite possible that M_1 is 0.6 V *wrt* the electrolyte, and M_2 is −0.4 V *wrt* the electrolyte (Figure 1.7b). Any other combination is also possible, and it depends on the electrode material, the electrolyte composition, and the temperature. In summary, although in theory, we relate the potential across the interface to the electrochemical reaction rate, and hence the current, experimentally we cannot measure the potential across one electrode–electrolyte interface. We can measure only the sum of the potential changes across two interfaces. Relating this quantity to the reaction rate or the current becomes a challenge.

As if this is not challenging enough, we encounter the following problem. Now, let us assume that when we apply 1 V across M_1 and M_2, the reality is described in Figure 1.7b, i.e., M_1 and M_2 are +0.6 and −0.4 V respectively *wrt* the electrolyte. Now, if we increase the potential to 1.5 V across M_1 and M_2, we cannot say where the additional potential drop will occur; i.e., the new potential drops may be described between two electrodes by Figure 1.8a or b or c or some other combination of potentials, which adds up to 1.5 V. In other words, we cannot determine if the additional potential change occurs completely at the M_1–electrolyte interface, or at the M_2–electrolyte interface or if it is split in some ratio between the two interfaces.

When the potential across the interface is changed, the extent of electrochemical reaction can change. Experimentally, we change the potential and observe a corresponding change in the current and the reaction rates. However, we will not be able to tell how much potential change occurred in which electrode. In other words, we find that even relating the *change* in current or reaction rate to the change in potential becomes a challenge.

1.4 USE OF REFERENCE ELECTRODES

There is a solution to this problem. In some electrodes, the potential across the electrodes remains constant. Although we cannot measure that potential, we can

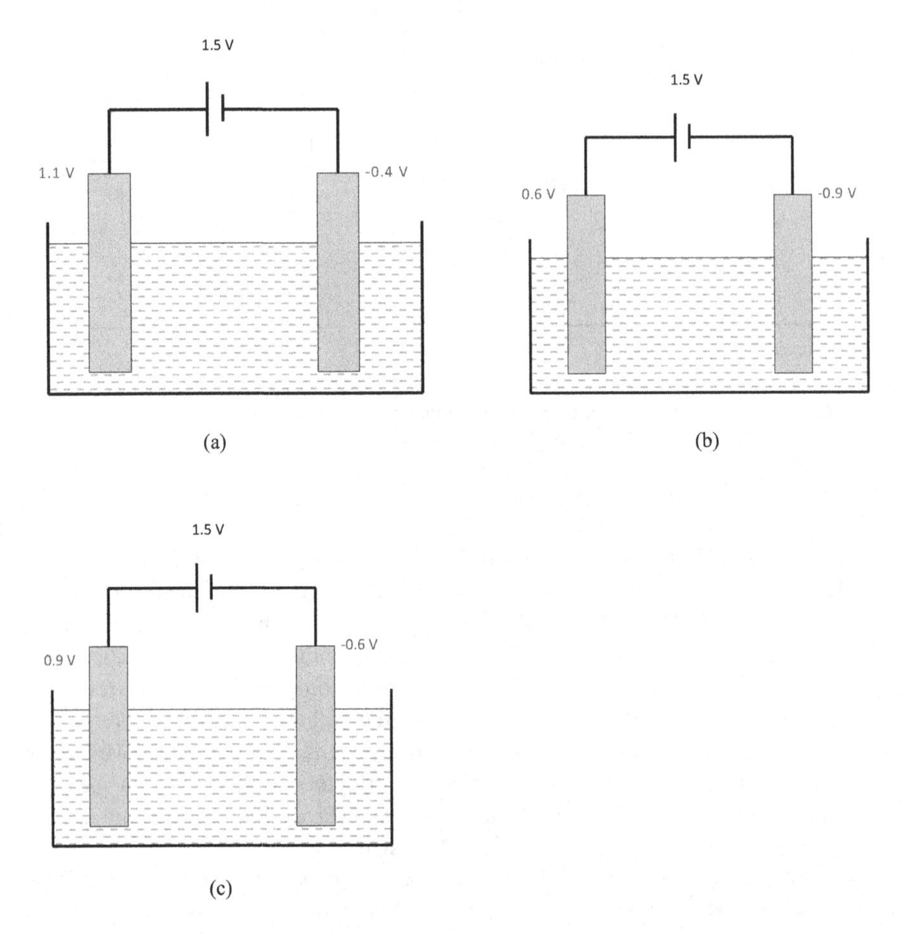

FIGURE 1.8 Cartoon to illustrate the challenges in measuring even the 'change in potential across an electrode–electrolyte interface' in a two-electrode system. When the potential across the electrodes is increased by 0.5 V, the change can occur (a) only at the anode or (b) only at the cathode or (c) partially at the anode and partially at the cathode.

be confident that the potential across the electrode–electrolyte interface does not change for these electrodes. These are known as ideally *nonpolarizable* electrodes and are also called *reference* electrodes. Commonly used reference electrodes are silver/silver chloride (Ag/AgCl) and saturated calomel electrode (SCE). Not so common reference electrodes are standard hydrogen electrode (SHE), mercury/mercury sulfate electrode, and so on.

Now, if one of the two electrodes used in our imaginary experiment is a reference electrode, then we still would not know how much potential drop occurs in case (a) when we apply 1 V as in Figure 1.9a and case (b) when we apply 1.5 V (Figure 1.9b). However, since we know that the potential drop across the reference electrode is a constant, we can be sure that compared to case (a), in case (b), an *additional* potential drop of 0.5 V has occurred across the *other, nonreference* electrode. Usually, this is the electrode we want to study and is called the *working electrode.* Therefore, by

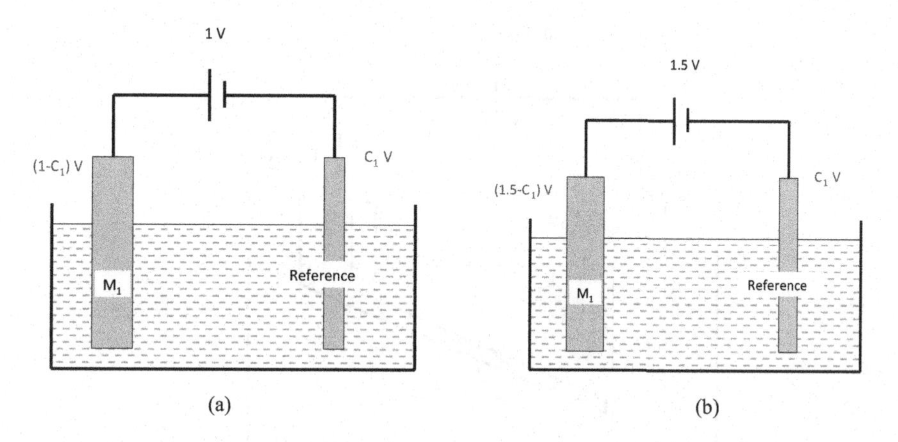

FIGURE 1.9 Schematic to illustrate the advantage of using a reference electrode. (a) The potential across the M_1–electrolyte is not known. (b) When the cell potential is increased from 1 to 1.5 V, the change is impressed only on the M_1–electrolyte interface.

using a reference electrode as one electrode, we can be confident that any *change* in potential that we impose on the system occurs on the working electrode.

When two different types of reference electrodes such as Ag/AgCl and SHE are immersed in a solution and the potential across them is measured, it is always a constant; i.e., there is an offset between the reference electrode potentials. Conventionally, the potential across SHE is arbitrarily taken as zero. In any case, when the potential across the working electrode is reported, it is always reported with respect to a particular reference electrode. It does not make much sense to report the potential applied without mentioning the reference electrode.

Sometimes the reference electrode is large, and it may not be physically possible to bring it close to the working electrode without making a complicated setup. Then the reference electrode itself is not brought close to the working electrode. Instead, a capillary tube filled with a salt solution is used as a bridge between the reference electrode and the location close to the working electrode. This is often referred to as the Luggin capillary or Luggin probe. On one side, the Luggin tube is large so that the reference electrode is accommodated while at the other end, it is a small tube, as shown in Figure 1.10.

If the current through the working electrode is small, then, two electrodes, i.e., the working electrode and the reference electrode, are sufficient to conduct the experiment. If the current through the working electrode is large, then the current will flow through the reference electrode. Unfortunately, when a large current flows through the reference electrode, the assumption that *only a constant potential change occurs across the electrode–electrolyte interface* becomes invalid, i.e., it is no longer a proper reference electrode.

The solution is to either (a) make the working electrode very small, i.e., microelectrode or ultra-micro electrode, or (b) bring another, a third, electrode and draw all the current through that electrode. This third electrode is called a *counter* electrode or *auxiliary* electrode.

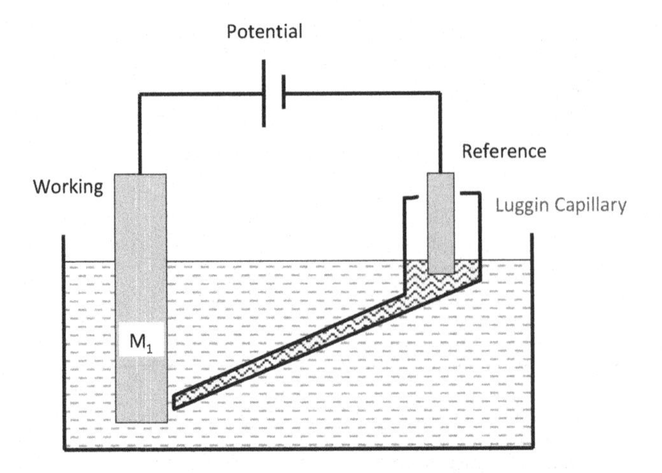

FIGURE 1.10 Schematic to illustrate the utilization of a Luggin capillary to minimize solution resistance effects.

How do we ensure that all the current flowing out of (into) the working electrode comes to (from) the counter electrode? The counter electrode is also polarizable. The potential of this electrode is controlled using a feedback circuit, and it is altered until all the current flowing out of (into) the working electrode comes to (from) this counter electrode. Thus, the electrochemical instrument is built to maintain a given potential between the working and the reference electrode by suitably altering the potential of the counter electrode. Since a significant quantity of current passes through the counter electrode, reactions occur at this electrode also, and the products will be released there.

The counter electrode must be inert. Otherwise, it can easily be corroded or passivated by these reactions. Usually, a Pt wire or mesh is used as a counter electrode. The instrument used in electrochemical experiments is called a **potentiostat**. By definition, it should be able to maintain a given potential between the working and the reference electrode and monitor the current. However, commercial potentiostats can do much more than that. They can apply a potential sweep at a specified rate or apply a potential step. In most cases, they can also maintain a constant current from the working electrode and monitor the potential, i.e., galvanostatic mode. If used in conjunction with a function generator and a frequency response analyzer, they can apply a sinusoidal potential and monitor the current. This is the preferred method of obtaining the impedance. If two-electrode mode is used (e.g., we use a microelectrode and hence do not need a counter electrode, or in cases like batteries where we cannot introduce a reference electrode into the 'cell'), then the reference and counter electrode leads are combined into one lead and applied to one terminal. The working electrode lead is connected to the second terminal.

1.5 WHAT IS EIS?

We now introduce the idea of impedance. Impedance is a generalized form of electrical resistance. When a potential E is applied across a simple resistance R, current I flows through it. The relationship between the three quantities is given by $R = E/I$.

When we apply an ac potential, usually a sinusoidal potential wave of the form E_{ac0} sin(ωt) across the resistor (R), then the current will also be sinusoidal and is given by I_{ac0}sin(ωt), where a similar relationship among R, E_{ac0}, and I_{ac0} holds; i.e., $R = E_{ac0}/I_{ac0}$. Here E_{ac0} is the potential amplitude and I_{ac0} is the current amplitude. The potential across a resistor and the ensuing current are plotted in Figure 1.11a and b. In this example, two different frequencies are employed.

However, when a similar ac potential is applied across a capacitor (C), the current is given by

$$I_{ac} = C\frac{dE}{dt} = C\frac{d\{E_{ac0}\sin(\omega t)\}}{dt} = C\omega E_{ac0}\cos(\omega t) \tag{1.1}$$

This can be written as $I_{ac} = C\omega E_{ac0}\sin(\omega t + 90°)$. The current *leads* the potential by 90° or π/2 radians. This equation is sometimes referred to as the *time-domain* equation. The results are presented in Figure 1.12a and b. The ratio of potential to

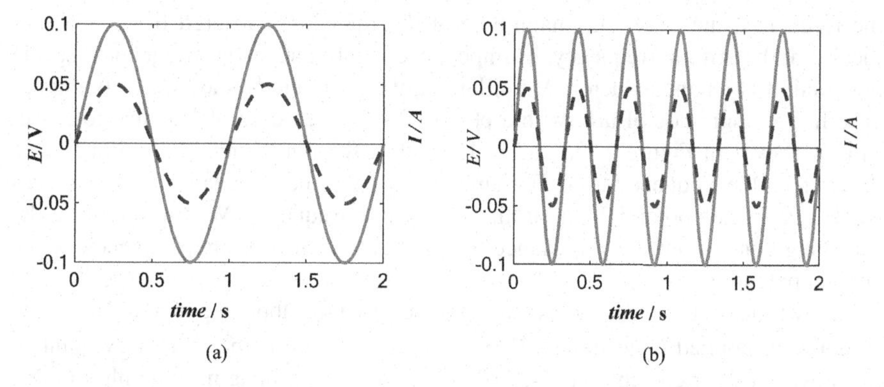

FIGURE 1.11 Potential *vs.* time and current *vs.* time for a resistor at (a) 1 Hz and (b) 3 Hz. Note that the magnitude of the current is independent of frequency.

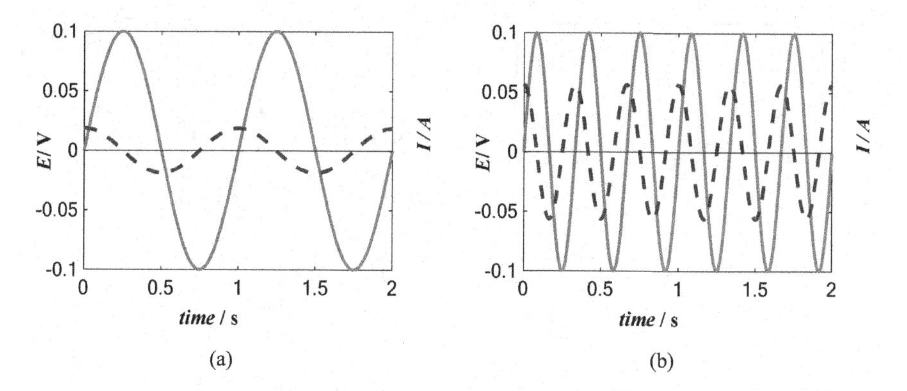

FIGURE 1.12 Potential *vs.* time and current *vs.* time for a capacitor at (a) 1 Hz and (b) 3 Hz. Note that the magnitude of the current increases with frequency.

current now has a magnitude given by $1/(\omega C)$. It also has a phase of −90°. If the current *leads* the potential, we can also say that potential *lags* the current by 90°. The ratio of potential to current can be thought of as a generalized form of resistance. It has a magnitude and a phase and can be represented easily as a complex number. A complex number has a magnitude and phase in polar coordinates. It can also be written as a pair of numbers, one corresponding to the real part and another corresponding to the imaginary part, in the Cartesian coordinates. Note that impedance is represented as a complex number with real and imaginary parts only for our convenience in mathematical manipulations. There is nothing *imaginary* or unreal about the physical quantity impedance.

1.5.1 Phase and Magnitude

Incidentally, for a simple resistor the phase difference is zero. The current does not *lag* or *lead* the potential. It is exactly in phase with the potential. Now let us consider Figure 1.12a and b. Note, first, that unlike the simple resistor the magnitude of the impedance of a capacitor depends on the frequency of the applied wave. The higher the frequency, the lower the magnitude of the impedance offered by a capacitor. Ideally, at the infinite frequency, its impedance is zero and at zero frequency, i.e., dc potential, it is a nonconductor. A similar analysis of inductor shows that its impedance has a magnitude of ωL, with a phase of +90°, i.e., current *lags* the potential. This is shown in Figure 1.13a and b. Here also, the magnitude of the impedance depends on the frequency used. For an ideal inductor the impedance is zero at zero frequency, i.e., dc potential, and infinity at infinite frequency. We thus arrive at the following conclusion: the impedance, in general, depends on the frequency of the applied potential.

Many times, we use complex notation to describe the applied potential. Thus, the ac potential applied is written as $E_{ac0}e^{j\omega t}$, where $j = \sqrt{-1}$ is the imaginary number. Since we use i for current density, j is used to denote an imaginary number in this book. Incidentally, the software MATLAB® uses both i and j as imaginary numbers.

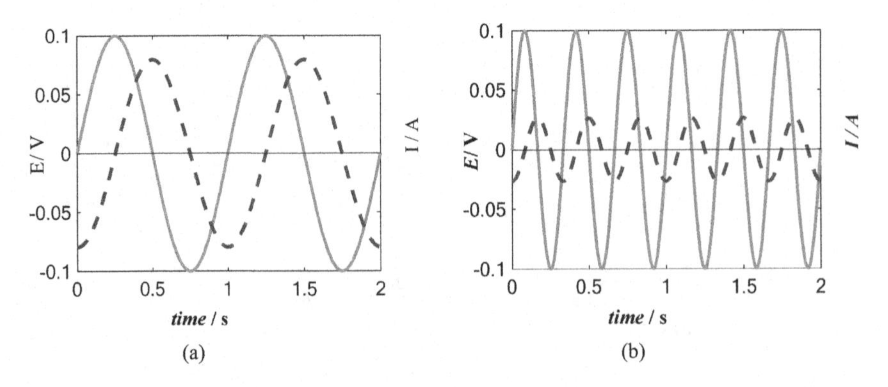

FIGURE 1.13 Plots of potential *vs.* time and current *vs.* time for an inductor at (a) 1 Hz and (b) 3 Hz. Note that the magnitude of current decreases with frequency.

Note that $e^{j\omega t} = \cos(\omega t) + j\sin(\omega t)$. Usually, Z is used to denote impedance, and a subscript is used to denote the component. Thus, the impedance of a resistor $Z_R = R$, that of a capacitor $Z_C = \frac{1}{(j\omega C)}$, and that of an inductor $Z_L = j\omega L$. This complex number representation takes care of the phase and magnitude correctly. These are sometimes referred to as *frequency domain* equations.

In electrochemistry, we tend to use the current density (i) more frequently than the current (I). Current density is the current value per unit electrode surface area, and the usual unit employed is A cm^{-2} or in multiples of it such as mA cm^{-2}, μA cm^{-2}, nA cm^{-2}. Then the impedance is also expressed after area-based normalization, and the usual unit employed is Ω cm^2. In the rest of this book, when we are dealing with electrodes, we will use Z to indicate normalized impedance in Ω cm^2 or multiples of it, such as kΩ cm^2 or MΩ cm^2. The resistance and capacitance will also be expressed in terms of Ω cm^2 and F cm^{-2}, respectively. However, when we analyze electrical circuits that are not used to model electrochemical systems, there is no physically meaningful area to consider. Hence, we will not normalize the values with the area.

1.5.2 DC *vs.* AC Potential

It is very important to note that electrochemical impedance spectroscopy (EIS) is essentially an ac technique, and most of the other electrochemical techniques are dc techniques. In EIS, however, we do not always apply a "pure ac" without any dc bias. Instead, we may apply an ac potential superimposed on a dc potential, i.e., there is an offset or dc bias. This is illustrated in Figure 1.14.

In a typical electrochemical cell, if a dc potential is applied, then a steady current, i.e., a dc current will flow only if a reaction occurs. Depending on the electrode material and the solution composition, there may be a wide potential range where no reaction occurs. In that case, the dc current *vs.* potential curve will show a wide region of zero current. However, even when there is no reaction, an ac potential will

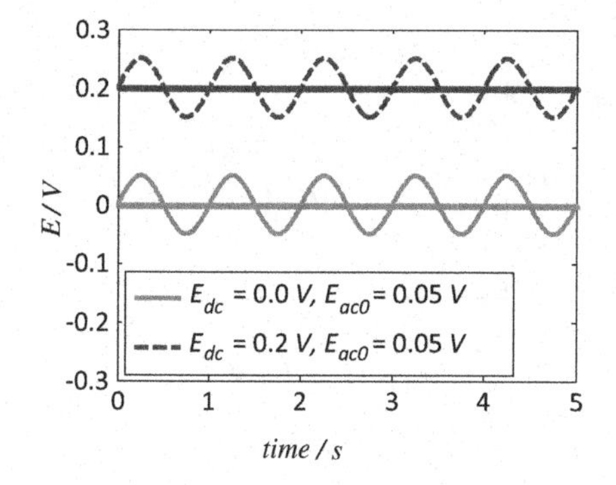

FIGURE 1.14 A plot of potential *vs.* time for a 'pure ac' given as a continuous line and 'ac superimposed on dc' given as a dashed line.

always give rise to an ac current, as long as the frequency is not very low. Note that if the frequency is very low, then the ac potential is very similar to a dc potential, and at zero frequency, it is identical to dc.

1.5.3 Differential Impedance

When a dc potential is applied along with an ac potential, the total potential becomes $E = E_{dc} + E_{ac}$. A generalized form of resistance can then be calculated as the ratio E/i. However, we will use only the ratio E_{ac}/i_{ac} to calculate impedance. There is a subtle difference between using these two ratios to represent impedance. The latter quantity is what we refer to as *impedance* in electrochemical literature, but the more precise term is *differential impedance.* In electrochemistry, we do *not* use the ratio $E/i = (E_{dc} + E_{ac})/(i_{dc} + i_{ac})$ to calculate the impedance. For example, consider an example plot of current *vs.* potential as shown in Figure 1.15, without bothering what circuit or reaction can give rise to this behavior. At the point marked as X in the figure, the overall impedance is E/i and it will be a positive quantity. In this particular case, it is 1.5 V/ 0.75 A = 2 Ω. The slope of the curve is given by the derivative di/dE and is a negative quantity. In this case, it is –3 Ω^{-1}. The inverse of the slope is dE/di and is also a negative quantity. This is the differential impedance measured by applying a small ac potential and by measuring the phase and magnitude of the ensuing ac current to calculate the ratio E_{ac}/i_{ac}. Here, in this example, the differential impedance (*at low frequency*, more on this later) is a negative quantity, whereas the overall impedance is still a positive quantity.

Another variable that is sometimes used in the analysis is *admittance.* It is usually denoted by Y and is the inverse of impedance. Admittance can also be thought of as a generalized form of conductance. Again, in the context of electrochemistry, differential admittance is normally employed, but the word *differential* is often left out for

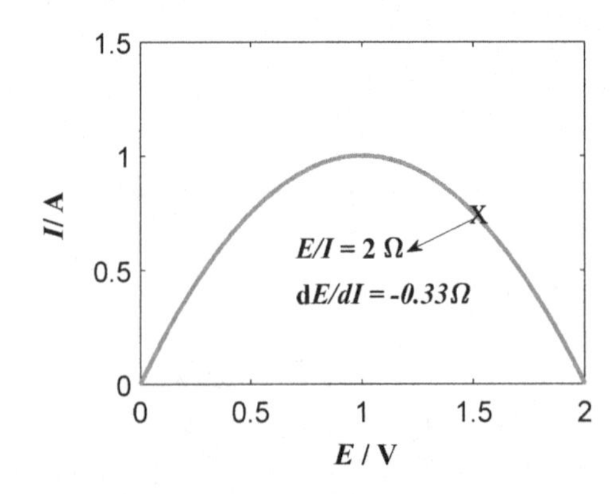

FIGURE 1.15 Plot of current *vs.* potential for an arbitrary system. Note that the 'impedance' is given by the ratio of potential to current, while the 'differential impedance' is given by the inverse of the slope (dE/dI). Electrochemical impedance spectroscopy refers to the differential impedance.

convenience. Sometimes, the word *immittance* is used to refer to either admittance or impedance, but that usage is not very common now.

1.6 SERIES AND PARALLEL CONNECTIONS

In late 1800, **Oliver Heaviside** introduced operational calculus, where he applied Laplace transform to solve differential equations arising from the theory of electrical circuits and laid the foundation of impedance.

When several resistors (or capacitors or inductors or any other electrical elements) are connected in series, as shown in Figure 1.16a, then the total impedance is the sum of the individual impedances. When the resistors are connected in parallel, as shown in Figure 1.16b, then the total admittance is the sum of the individual admittances. These rules are simplified versions of Kirchhoff's laws, but these are sufficient to analyze many systems that we encounter in electrochemistry.

The electrochemical impedance spectrum is the impedance data of an electrochemical system, which are acquired over many frequencies. Science and engineering students would be familiar with UV-Visible spectroscopy, where the absorption of light by a liquid sample is measured as a function of wavelength or frequency. By analyzing the UV-Vis spectrum, we can hope to estimate perhaps the concentration of a particular chemical in the sample. Similarly, by measuring the impedance of the electrochemical system as a function of frequency and by analyzing the spectrum, we can hope to understand the electrochemical system.

1.6.1 Example Circuit – 1

In the case of a resistor (R_1) in series with a capacitor (C_1), the total impedance can be written as $Z_T = R_1 + (j\omega C_1)^{-1}$, and at a given frequency, the impedance can be calculated. For example, in the circuit shown in Figure 1.17a, if $R_1 = 1\ \Omega$, $C_1 = 0.01$ F, and a sinusoidal potential of 10 Hz frequency is applied, the impedance offered is $(+1 - 1.59j)$. In polar coordinates, the magnitude is 1.88 and the phase is $-58°$.

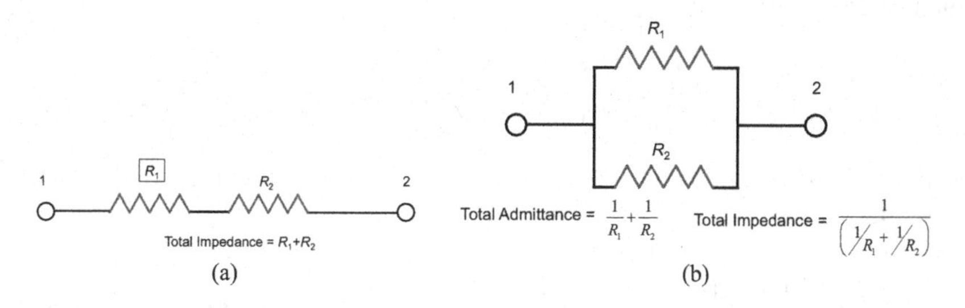

FIGURE 1.16 Simple schematic to show the calculation of total impedance in a (a) series connection and (b) parallel connection.

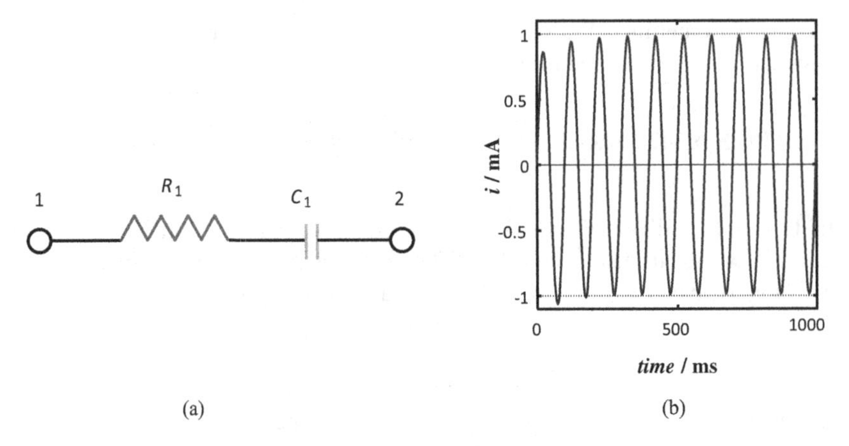

FIGURE 1.17 (a) Example of an electrical circuit that can represent some simple electrochemical systems. (b) Time domain current response of this system to a sinusoidal potential. $R_1 = 10\ \Omega$, $C_1 = 0.01$ F, $E_{ac0} = 10\,\text{mV}$, and $f = 10\,\text{Hz}$.

We can write the time domain equation as follows. The potential across the circuit can be visualized as the sum of the potential across the resistor (E_1) and the potential across the capacitor (E_2). The current passing through the resistor and capacitor are the same. In other words,

$$E = E_1 + E_2 \tag{1.2}$$

$$i_T = i_1 = i_2 \tag{1.3}$$

For the resistor, the relationship between the current and potential is given by $E_1 = i_1 R_1$ and for the capacitor, it is given by $i_2 = C\left(\frac{dE_2}{dt}\right)$. When the applied potential is a sine wave of amplitude E_{ac0} and angular frequency ω, we can write Eqs. 1.2 and 1.3 as

$$\frac{E_{ac0}\sin(\omega t) - E_2}{R_1} = C_1\frac{dE_2}{dt} \tag{1.4}$$

which can be re-arranged as

$$\frac{dE_2}{dt} + \frac{1}{R_1 C_1}E_2 = \frac{1}{R_1 C_1}E_{ac0}\sin(\omega t) \tag{1.5}$$

With the initial condition that at $t = 0$, $E = E_1 = E_2 = 0$, we can show that the solution is given by

$$E_2 = \frac{E_{ac0}}{\tau}\left\{\left[\frac{1/\tau}{1/\tau^2 + \omega^2}\sin(\omega t) - \frac{\omega}{1/\tau^2 + \omega^2}\cos(\omega t)\right] + \frac{\omega}{1/\tau^2 + \omega^2} \times e^{-t/\tau}\right\} \tag{1.6}$$

The current is given by

$$i_T = i_1 = \frac{\left(E_{ac0}\sin(\omega t) - \frac{E_{ac0}}{\tau}\left\{ \begin{bmatrix} \left[\frac{1/\tau}{1/\tau^2+\omega^2}\sin(\omega t) - \frac{\omega}{1/\tau^2+\omega^2}\cos(\omega t) \right] \\ + \frac{\omega}{1/\tau^2+\omega^2} \times e^{-t/\tau} \end{bmatrix} \right\} \right)}{R_1} \tag{1.7}$$

where $\tau = R_1 \times C_1$.

In Eq. 1.7, the last term exhibits an exponential decay, i.e., after a long time, the last term will vanish. The time constant (τ) gives an idea of how long it will take for this transient to become very small. Now, for $R_1 = 10\ \Omega$ and $C_1 = 0.01$ F, we can plot the current given in Eq. 1.7 as a function of time. The potential amplitude (E_{ac0}) is 0.01 V, and the frequency is 10 Hz. The results are shown in Figure 1.17b. It is seen that initially, the current starts at zero value at time $t = 0$. The first current peak is smaller than the subsequent peak values. In this example, the response stabilizes within three cycles. The next few cycles exhibit steady periodic results.

1.6.2 Example Circuit – 2

Consider the electrical circuit shown in Figure 1.18a, made of a resistor (R_2) in parallel with a capacitor (C_2), and the combination is in series with a resistor (R_1). This is one of the simplest circuits used to represent some electrochemical systems, and hence we start with this circuit. Getting familiar with this would help one with learning the EIS response of simple electrochemical systems. Assume that the values are $R_1 = 20\ \Omega$, $R_2 = 100\ \Omega$, and $C_2 = 10^{-5}$ F. These values are chosen such that they would represent a realistic electrochemical system.

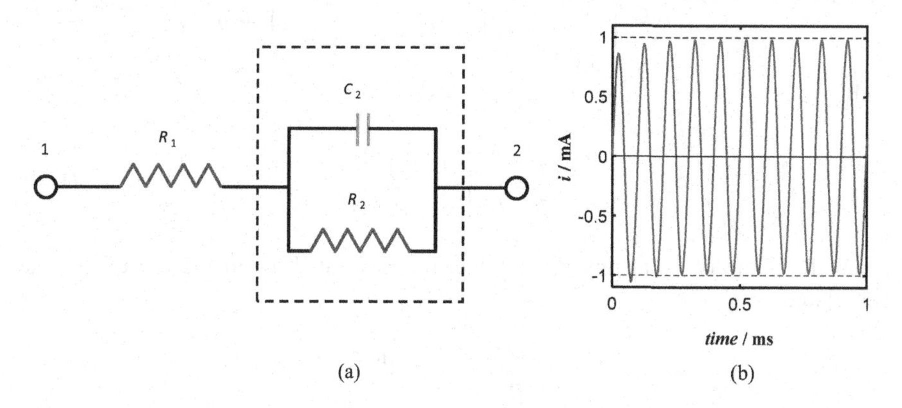

FIGURE 1.18 (a) Example of an electrical circuit that can represent some simple electrochemical systems. (b) Time-domain current response of this system to a sinusoidal potential perturbation. $R_1 = 10\ \Omega$, $R_2 = 100\ \Omega$, $C_2 = 10^{-5}$ F, $E_{ac0} = 10$ mV, and $f = 10$ kHz.

When a potential is applied, the current passing through R_1 (i_1) is split into two components. One of them (i_{2a}) flows through the capacitor C_2 while the other (i_{2b}) flows through the resistance R_2. The algebraic sum of the currents through R_2 and C_2 is equal to the current flowing through R_1. At the same time, the potential drop across the resistor R_2 and capacitor C_2 is the same and can be denoted as E_2.

The current flowing through R_1 can be written as

$$i_1 = \frac{E - E_2}{R_1} = \frac{E_{\text{ac0}} \sin(\omega t) - E_2}{R_1} \tag{1.8}$$

The current flowing through C_2 can be written as

$$i_{2a} = C_2 \frac{\mathrm{d}E_2}{\mathrm{d}t} \tag{1.9}$$

and that flowing through R_2 can be written as

$$i_{2b} = \frac{E_2}{R_2} \tag{1.10}$$

Therefore, the time-domain equation can be written as

$$i_1 = \frac{E - E_2}{R_1} = \frac{E_{\text{ac0}} \sin(\omega t) - E_2}{R_1} = i_{2a} + i_{2b} = C \frac{\mathrm{d}E_2}{\mathrm{d}t} + \frac{E_2}{R_2} \tag{1.11}$$

and the solution of this equation, with the initial condition that at $t = 0$, $E = E_2 = 0$, is given by

$$E_2 = \frac{E_{\text{ac0}}}{R_1 C_2} \left[\frac{\left(\frac{1}{\tau}\right) \sin(\omega t) - \omega \cos(\omega t)}{\left(\frac{1}{\tau}\right)^2 + \omega^2} + \frac{\omega e^{\frac{-t}{\tau}}}{\left(\frac{1}{\tau}\right)^2 + \omega^2} \right] \tag{1.12}$$

with

$$\frac{1}{\tau} = \frac{1}{C_2} \left[\frac{1}{R_1} + \frac{1}{R_2} \right] \tag{1.13}$$

From the value of E_2 at any given time, the current through the circuit can be calculated as

$$i = \frac{E_1}{R_1} = \frac{E - E_2}{R_1} = \frac{E_{\text{ac0}} \sin(\omega t) - E_2}{R_1} \tag{1.14}$$

For a particular set of element values and perturbation, the solution (Eq. 1.14) is shown in Figure 1.18b. The result shows that the solution has a steady periodic component and a transient component. At longer times (in this case, after 0.5 ms), the transient term will tend to zero and can be ignored.

In the frequency domain, the impedance is calculated as follows. For the part where the capacitor C_2 is in parallel with the resistor R_2, the admittance is given by

$$Y_2 = Y_{C_2} + Y_{R_2} = j\omega C_2 + {}^{1}\!/_{R_2} \tag{1.15}$$

The impedance of that part is given by $Z_2 = {}^{1}\!/_{Y_2}$. The total impedance is given by $Z_T = R_1 + Z_2$.

$$\text{i.e., } Z_T = R_1 + \frac{1}{\frac{1}{R_2} + j\omega C_2} = R_1 + \frac{\frac{1}{R_2} - j\omega C_2}{\frac{1}{R_2^2} + \omega^2 C_2^2} = R_1 + \frac{R_2 - j\omega R_2 C_2}{1 + \omega^2 R_2^2 C_2^2} \tag{1.16}$$

Using these steps, the impedance at any frequency can be calculated and plotted.

1.7 DATA VISUALIZATION

For the circuit given in Figure 1.18a, if the impedance values are measured in a frequency range of 1 mHz–100 kHz, the impedance, which consists of the real and imaginary parts, at each frequency can be tabulated. Thus, there are two responses, Z_{Re} and Z_{Im}, at one frequency. They are not exactly independent responses, and we will see later that they are coupled. Now, there are many ways of graphically presenting the data. One is to plot $-Z_{Im}$ *vs.* Z_{Re}, as shown in Figure 1.19a. We plot $-Z_{Im}$ and *not* $+Z_{Im}$ in the ordinate. This is because for this particular circuit, and most electrochemical systems, the imaginary component of the impedance is negative. By plotting $-Z_{Im}$ instead of $+Z_{Im}$, we bring the spectrum to the first quadrant in the graph. The spectrum appears to be a semicircle. This is because of the circuit we have employed and the values of parameters and frequencies that are chosen. Here, the frequency is not explicitly shown. At best, a few data points can be marked,

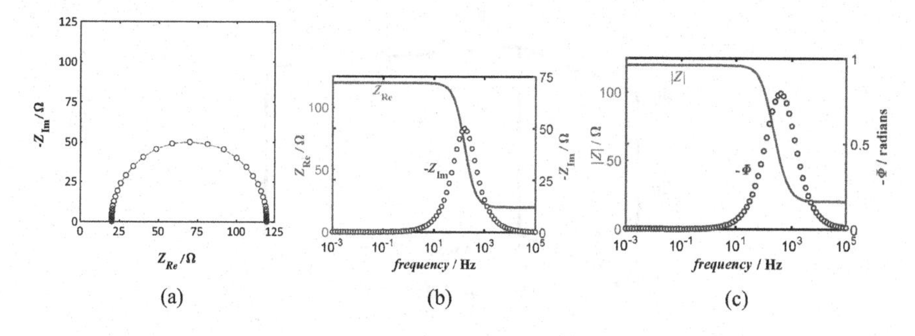

FIGURE 1.19 (a) A complex plane plot and (b) Bode plot of Z_{Re} and $-Z_{Im}$ *vs.* frequency. (c) Bode plots of |Z| and phase *vs.* frequency. The data are generated from the circuit given in Figure 1.18a, with $R_1 = 20\ \Omega$, $C_2 = 10\ \mu F$, and $R_2 = 100\ \Omega$.

and the frequency values can be written next to them. This is also referred to as the **Nyquist** plot or the **complex plane** plot. Note that the abscissa and ordinate scales are equal in Figure 1.19a. If the capacitor used is nonideal, the current through the capacitor will be described by a different equation, and the plot would be distorted, as explained in the chapter on constant phase elements. The distortion of semicircle will be obvious only if the complex plane plots are shown with equal scaling of the axes, and hence it is essential to follow this practice.

The second method is to plot the real part Z_{Re} *vs.* frequency, with the frequency expressed in the logarithmic scale (Figure 1.19b). The imaginary part, which is usually $-Z_{Im}$ instead of Z_{Im}, can also be shown in the same figure or a separate figure. The third method is to plot the magnitude of the impedance $|Z|$ *vs.* frequency, with frequency on a log scale (Figure 1.19c). The phase, usually presented as $-\phi$ instead of $+\phi$, can be plotted either in the same figure or separately. The second and third types of plots are often referred to as **Bode** plots. In this particular example, the results of the second and third methods appear to be very similar, but they will appear different for more complicated circuits. Some of the variations include plotting the ordinate (Z_{Re}, $-Z_{Im}$, and $|Z|$) also in the log scale in Bode plots. A three-dimensional plot of frequency, Z_{Re}, and $-Z_{Im}$, with projections on the planes, can also be used to give all the information compactly, as shown in Figure 1.20. A video showing this 3D plot from different perspectives is available at http://www.che.iitm.ac.in/~srinivar/EIS-Book.html.We recommend plotting the data in at least the three basic forms to analyze and understand the results.

Now, which is a *better* form of representation? For any given system, whichever representation brings forth the important phenomena clearly, is the better form, and it should be used in the final presentation.

1.8 CIRCUIT PARAMETER VALUE EXTRACTION FROM DATA

Now, we ask an important question. Let us assume that we somehow *know* that the circuit in Figure 1.18a is suitable to represent the system, but we do not know the values of R_1, R_2, and C_2. Can we estimate R_1, R_2, and C_2 from the impedance data? The answer is, "Certainly, yes." If we have three or more data points, i.e., impedance values at three or more frequencies, we can estimate these three parameter values.

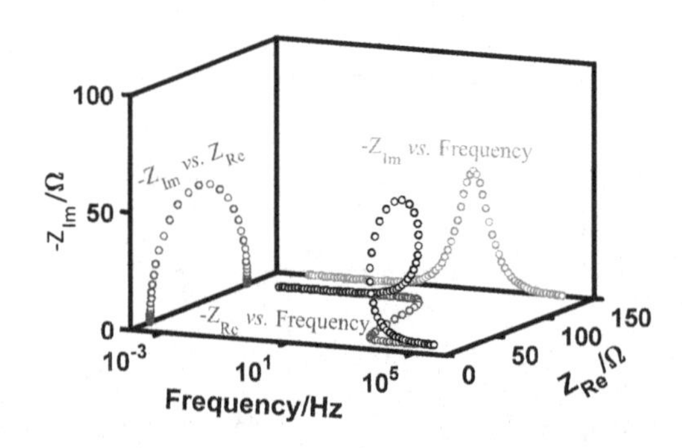

FIGURE 1.20 Three-dimensional plot of frequency, Z_{Re}, and $-Z_{Im}$ of the data presented in Figure 1.18a.

Of course, it is better to have many data points to get confidence in our estimates of R_1, R_2, and C_2. Usually, we have 40 or more impedance data points for this type of analysis. Note that here one *data point* means one frequency and the complex impedance value at that frequency $[f, Z(f)]$.

Software applications that can estimate the best-fit values for the electrical circuit representation are available from many commercial and a few noncommercial sources. They use algorithms such as Levenberg–Marquardt (Moré 1978) or the downhill simplex (Olsson and Nelson 1975) method to obtain the best fit. Commercial software applications are fast and very easy to use. For example, using the software Zsimpwin®, which employs the downhill simplex method, we can estimate the resistances and capacitance of the circuit elements from the data within a few seconds. An advantage of this method is that it can converge to the correct result in most cases, even if the initial guess is not very good. A few other sources are listed below. They are available as of 2020. The LEVM program by Ross J Macdonald is available to download and use in MS-DOS and Windows® operating systems. ZView®, from Scribner, is a commercial program. A trial version of a software called MEISP® can be downloaded and used. Another option is to use a freeware program called EIS Spectrum Analyzer®. A MATLAB® program called Zfit® and a GUI version called ZFitGUI® are available in MATLAB file exchange and are free to use under the BSD license.

1.9 REACTION MECHANISM ANALYSIS

Consider the impedance measurement of a gold electrode in a solution containing 100 mM $NaNO_3$, 5 mM $K_3[Fe(CN)_6]$, and 5 mM $K_4[Fe(CN)_6]$. Note: while cyanides are very toxic, $K_3[Fe(CN)_6]$ and $K_4[Fe(CN)_6]$ are of relatively low toxicity and are therefore commonly used in electrochemical experiments as a redox couple. Under strongly acidic conditions, they release the highly toxic HCN gas. As long as they are not exposed to acidic conditions, they can be handled easily. In this solution, $NaNO_3$ is used as a supporting electrolyte. The overall reaction can be simplified as $Fe^{2+} \rightleftharpoons Fe^{3+} + e^-$.

If we measure the impedance spectrum of this cell, what would be the spectrum? A more important question from the practical viewpoint is this: if we measure the spectrum of this cell, (a) can we analyze the data and *conclude* that the reaction under study is a one-step reversible reaction and (b) can we *estimate* the values of the forward and reverse reaction rate constants? This book is an attempt to show how to answer these systematically. To answer these properly, it is important to acquire certain skills, viz., one should know how to obtain the data experimentally, how to validate the data, how to develop analytical expressions for the impedance for a given reaction, how to perform the numerical calculations in computes, and how to use optimization programs to estimate the parameters.

1.10 OTHER ELECTROCHEMICAL TECHNIQUES

Here, we touch upon a few of the many other techniques that are used in electrochemistry. For detailed information on these and many more such techniques, the reader is directed to *Electrochemical Methods: Fundamentals and Applications* by Bard and Faulkner (1980).

1.10.1 Open-Circuit Potential *vs.* Time

Here, the electrode potential is measured over time without supplying any current, i.e., no external current is applied, and the net current is zero because the cathodic and anodic currents in the working electrode cancel each other. Under these conditions, called *open-circuit* conditions, the potential is measured over time. This is called open-circuit potential or OCP. If the electrode–electrolyte interface undergoes significant change over time, then the potential will vary over time. Ideally, EIS must be acquired when the potential is stable for a long time and the variation is within 5 mV h^{-1}.

1.10.2 Potentiodynamic Polarization

In this method, the electrode potential is held at one value and the current is measured. Then the potential is changed to another value, and once the current is stabilized, the value is recorded. This is repeated in a potential range, usually from 250 mV below the OCP to 250 mV above the OCP. In practical terms, the electrode potential is swept from −250 mV (*vs.* OCP) to +250 mV (*vs.* OCP) at a slow scan rate of 0.1 or 1 mV s^{-1}. The electrode is usually rotated, and/or the electrolyte is stirred so that the concentration of all the dissolved species is uniform throughout the cell. When the potential is scanned, the current passes through the double-layer capacitance as well as through the faradaic process, i.e., reaction with electron transfer, and the total current is the sum of the two currents. However, at slow scan rates, the current through the double-layer capacitance will be negligible, and we can approximate the faradaic current by the total current. Since the concentrations of all species are uniform within the electrolyte, there are no mass transfer effects on the current. In general, the scan starts at the cathodic potential (−250 mV *vs.* OCP) and ends at the anodic potential. This is because the electrode tends to get oxidized at an anodic potential (+250 mV *vs.* OCP). If the scan starts at the anodic potential, then the oxidation of the electrode quickly modifies the electrode–electrolyte interface. Effectively, the system is changing during the experiment. The data collected is not obtained from the original metal–electrolyte interface that we intended to probe; instead, it is collected from the oxidized metal in contact with the electrolyte. The potential scan is started at the cathodic potential to avoid this problem.

1.10.3 Voltammetry

In this analysis, the potential of the electrode is changed linearly at a particular scan rate (e.g., 10 or 50 mV s^{-1}), and the current is monitored (Compton and Banks 2011). The electrode is held stationary, and the electrolyte is not stirred. This means that the concentration of the solution species can vary within the cell. Frequently, the electrode potential is increased up to a particular value (E_{high}) and then decreased at the same scan rate to a particular value (E_{low}) and then increased again. This is called cyclic voltammetry. The current is plotted against the potential. The technique is very similar to potentiodynamic polarization. However, the sweep rates are generally high, i.e., $\gg$1 mV s^{-1}, and the charging current of the double-layer capacitance cannot be neglected here. Similarly, diffusion can become the overall rate-limiting step if the kinetics of the electrochemical reaction is fast. In comparison, in the analysis of potentiodynamic polarization data, it is normally assumed that the double-layer charging current is negligible and that the mass transfer limitations are not present.

1.10.4 Potential Step

In this method, a potential step is applied onto the electrode, i.e., the electrode is initially held at some constant potential, and suddenly the potential is changed to a higher or lower value (Bard and Faulkner 1980). The current is measured as a function of time. Here also the double-layer charging current and the diffusion limitations must be taken into consideration during the analysis of the data. Monitoring the current as a function of time is referred to as chronoamperometry, and monitoring the charge as a function of time is referred to as chronocoulometry.

1.10.5 Electrochemical Quartz Crystal Microbalance (EQCM)

Electrochemical quartz crystal microbalance analysis enables simultaneous measurement of the electrical charge used and the mass deposited in an electrochemical reaction (Buttry and Ward 1992). A very small quantity of mass deposited per unit area, i.e., <1 μg cm^{-2}, can be measured. This is possible by monitoring the resonance frequency of a quartz crystal on which the electrode is coated. If there is a change in the mass, e.g., due to the deposition of a material, then the resonance frequency would change, and this can be detected accurately. Thus, the effect of any material adsorbing onto or desorbing from the electrode, or depositing onto or dissolving from the electrode, at a given potential can be measured.

1.10.6 Scanning Electrochemical Microscopy (SECM)

A very fine metal electrode tip, called an ultra-micro electrode, is used in this technique. The tip is positioned very close to the surface under study. The tip and the sample surface function as two electrodes in the electrochemical system. An electrochemical reaction, usually the Fe^{2+}/Fe^{3+} redox couple, is conducted. By recording the current at a given potential and by moving the tip in a controlled manner, information on the surface reactivity can be gathered. A reference electrode is also used in SECM (Bard and Mirkin 2012). In contrast, a related method called electrochemical atomic force microscopy (AFM) uses two electrodes, and the two-electrode nature of the system limits the extent of information one can gather.

1.11 EXERCISE – ELECTROCHEMISTRY BASICS AND CIRCUIT-BASED IMPEDANCE

Q1.1 An electrochemical reaction rate constant is exponentially related to the potential, but the reaction rate may or may not be exponentially related to the potential. Explain.

Q1.2 In an electrochemical cell, why should we use a reference electrode? What is the purpose of a counter electrode? What is the need to use a large area counter electrode?

Q1.3 Select a publication in the last 5 years where impedance and current potential data are shown. Digitize the data. Estimate the low-frequency impedance from the impedance data. Also, estimate the values of E/I as well as dE/di at the potential where EIS was acquired. Compare these three values and comment.

Q1.4 Using the expressions for current in Eq. 1.1, calculate the current for a capacitor of $C = 10^{-5}$ F, when a sine wave potential of 10 mV amplitude is applied. Plot the results as a function of time. The frequency is (a) 1 kHz and (b) 0.01 Hz.

Q1.5 From a suitable reference, find the expression relating current and potential for an ideal inductor. Plot potential *vs.* time and current *vs.* time, when a potential of 10 mV is applied across an inductor of 0.1 H inductance. The frequency is (a) 1 kHz and (b) 0.01 Hz.

Q1.6 Analyze the plots generated in the above problem and determine the phase difference between the current and potential for an inductor.

Q1.7 Generate a set of frequencies (a) in the range 1 Hz–100 kHz, in the log space (i.e., geometric series) with 7 frequencies per decade. (b) In the same range, in linear space (i.e., arithmetic series) with a total of 50 points. (c) Using the circuit given in Figure 1.18a, calculate the impedance when $R_1 = 10\ \Omega$, $R_2 = 10$, and $C_2 = 15\ \mu F$ in the frequency range generated.

Q1.8 In continuation of the above question, plot the impedance spectra in (a) complex plane plot, (b) $|Z|$ and Φ *vs.* frequency, and (c) Z_{Re} and $-Z_{Im}$ *vs.* frequency for both frequency sets. In the case of data with log space frequencies, use the log scale in abscissa of Bode plots.

Q1.9 Evaluate the effect of changing the values of (a) resistance R_1, (b) capacitance C_2, and (c) resistance R_2 in Q1.7 and generate the plots. What inferences can we draw? In particular, when the capacitance value is changed, what are the changes seen in a complex plane plot and a Bode plot?

Q1.10 In question Q1.7, calculate the time domain response when a sine wave of 1 kHz frequency and 5 mV amplitude is applied until a steady periodic response is obtained. Plot the current response as a function of time. How long does it take to obtain steady periodic results?

Q1.11 In the above question, repeat the calculations for a sine wave of 100 kHz frequency and 10 mV amplitude.

Q1.12 Consider the circuit given in Figure 1.17a. If $R_1 = 20\ \Omega$ and $C_1 = 0.0001$ F, calculate the time domain response when a sine wave of 100 Hz frequency and 5 mV amplitude is used until a steady periodic response is obtained. Plot the current response as a function of time. How long does it take to obtain steady periodic results?

Q1.13 In the above question, repeat the calculations for a sine wave of 1 kHz frequency and 10 mV amplitude.

Q1.14 Consider the circuits given in Figures 1.17a and 1.18a. Derive the expressions for the time domain current response, when a linear potential ramp is applied to these circuits.

Q1.15 Select a publication in the last 5 years, where an electrochemical technique that is not mentioned in this chapter is used. Digitize the published data, learn the basics of that technique, and interpret the data.

2 Experimental Aspects

The language of experiment is more authoritative than any reasoning: facts can destroy our ratiocination—not vice versa.

—Count Alessandro Giuseppe Antonio Anastasio Volta

In theory, there is no difference between theory and practice. In practice, there is

—Yogi Berra

We start with the description of instruments commonly used to acquire the impedance spectra of electrochemical systems. Then experimental details, such as the use of supporting electrolytes and placement of reference electrodes, are discussed. Impedance can be acquired using single sine or multi-sine waves. Similarly, they can be used in pseudo-potentiostatic or pseudo-galvanostatic mode, and these details are also presented.

2.1 INSTRUMENTATION

In the case of electrical circuits, the impedance spectrum can be measured to characterize them. However, the equipment used in those cases connects to the two terminals of the circuit. Their frequency range is also typically from milli Hz (mHz) to many mega Hz (MHz). The equipment needed for *electrochemical* impedance spectroscopy should connect to at least three terminals since three-electrode cells are common; i.e., the equipment should be capable of measuring the impedance between the working electrode and the reference electrode, while making sure that very little current passes through the reference electrode, and use the counter electrode to draw the current and complete the circuit. The frequency range is also different here, viz. 10 μHz–1 MHz. As of 2016, a few of the electrochemical impedance spectrum (EIS) equipment manufacturers are bringing forth equipment with an upper-frequency limit of 32 MHz, but that is not relevant for typical electrochemical systems in an aqueous medium. It may be useful for dielectric spectroscopy or solid electrolytes. For the common electrolytes in an aqueous medium, an upper limit of 300 kHz is sufficient. The instruments can superimpose an ac potential over a dc potential and run the experiment. This is called a *pseudo*- or *quasi*-potentiostatic mode. Usually, they also can run the experiment in a *pseudo*- or *quasi*-galvanostatic mode where an ac current is superimposed on dc current, and the potential is measured so that the impedance can be calculated.

In the last few decades, frequency response analyzers (FRAs) are the most commonly used instruments to obtain EIS data. Previously, lock-in amplifiers were also used, but they are not suitable for acquiring very-low-frequency data. A few decades ago, oscilloscopes were used but are practically out of use now for EIS measurements. For most electrochemical reactions, the important information on the kinetics is present mainly in the low-frequency data, and hence oscilloscopes are not very useful. FRAs used to be more expensive; however, like all the digital electronic devices, they have become cheaper over time. Several manufacturers around the world supply FRAs. FRAs are used in other fields as well. For electrochemical applications, the FRAs are usually combined with a potentiostat.

2.2 POTENTIOSTATIC *VS.* GALVANOSTATIC MODES

While investigating an electrochemical system with EIS, we would normally apply a dc bias onto the working electrode and superimpose an *ac* perturbation on it. For example, the electrode may be held at +200 mV *vs.* the Ag/AgCl reference electrode, and an ac potential of 10 mV peak-to-peak may be superimposed on this. This mode is called a *pseudo*-potentiostatic mode since the potential is *more or less* +200 mV *vs.* the Ag/AgCl reference electrode. The experiment can be done at other dc potentials, e.g., 250, 300, and 400 mV *vs.* the Ag/AgCl reference electrode, and the spectra at all these dc potentials can be modeled using a reaction mechanism. The measured current will have a dc component and an ac component, and the equipment hardware and software would separate these components, measure the magnitude and phase of the current, and record the impedance values at each frequency. Note: In the literature, the dc potential at which the electrode is held is reported either with respect to a reference electrode such as Ag/AgCl or with respect to the open-circuit potential (OCP).

It is also possible to conduct the experiments by holding the electrode at a constant dc current and superimposing an ac current. This mode is called the *pseudo*-galvanostatic mode. For example, a system may be held at +200 mA dc, and an ac current of 10 mA may be superimposed on this. The dc and ac potential applied to maintain this current would be used by the equipment hardware and software to calculate the impedance at each frequency. In many corrosion and electrodeposition experiments, pseudo-potentiostatic is the mode in which experiments are usually conducted. However, a pseudo-galvanostatic mode is frequently employed while acquiring EIS data of fuel cells or batteries. For the sake of brevity, we will drop the word *pseudo* in the rest of the book. The development of equations under potentiostatic conditions is discussed in Chapter 5. With regards to the simulations, there are certain differences between the development of equations for potentiostatic and galvanostatic EIS, and these are discussed in detail in the final chapter.

2.3 SUPPORTING ELECTROLYTE

If the resistance of the solution is very small, it would be ideal because the potential drop across the cell will be primarily across the double-layer of the electrodes. However, in many cases, the system under study may have high solution-resistance. For example, one may want to study the deposition of Cu onto a gold electrode from a solution containing 0.1 mM concentration of $CuSO_4$ in water. The conductivity of that solution will be poor. Now, it is possible to increase the conductivity by adding a salt such as $KClO_4$. Here, $KClO_4$ is called the supporting electrolyte. The supporting electrolyte is chosen such that it will not participate in the reaction that is studied, and its only purpose is to increase the conductivity without undergoing any reaction.

Because of the presence of the ions of the supporting electrolyte, the solution resistance is low, and the electric field is more or less the same throughout the solution. The field change occurs mainly across the double-layer, in the electrode–electrolyte interface. Besides, perchlorate anions are large, and hence the charge density, i.e., the charge per unit volume, will be small. Therefore, it will not adsorb strongly onto the electrode. Potassium ion will have hydration sheath around it, and hence it also effectively has a low charge density and thus will not adsorb onto the electrode. In some cases, such as the test of biological fluids, adding salt to the electrolyte will cause unwanted effects such as precipitation of colloidal particles and change in the structure of the proteins in the electrolytes. In those cases, the unsupported system should be investigated, and the results must be analyzed after taking the solution resistance into account, as described in the final chapter.

2.4 LOCATION OF THE REFERENCE ELECTRODE

In a typical electrochemical cell employing three electrodes, the reference electrode is usually brought close to the working electrode to reduce the solution resistance effects. If the reference electrode is too far away from the working electrode, then a significant part of the applied potential drop would occur across the solution. This would lead to an incorrect interpretation of the results, as shown in the final chapter. However, if the reference electrode is brought too close to the working electrode, then the current distribution across the working electrode is altered, and an inductive loop would be introduced in the high-frequency regime (Figure 2.1a). This would unnecessarily complicate the analyses of the results. This can be modeled using a circuit shown in Figure 2.1b. To minimize these effects, it is recommended to keep the reference electrode at a distance of two to three times the dimension of the working electrode; i.e., if the working electrode is 5 mm in diameter, then the reference electrode should be placed approximately 10–15 mm from the working electrode.

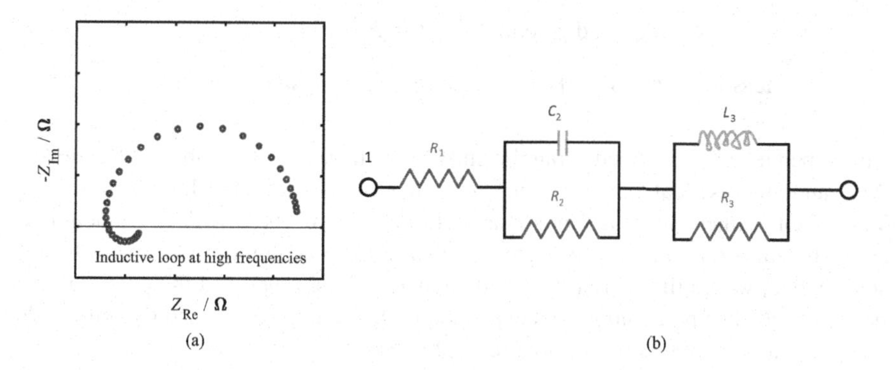

FIGURE 2.1 (a) Complex plane plot of impedance spectra showing an inducting loop at high frequencies. It can arise if the reference electrode is brought too close to the working electrode. (Adapted from *Electrochimica Acta*, 55, C. Blanc, M.E. Orazem, N. Pébère, B. Tribollet, V.Vivier, S. Wu, The origin of the complex character of the Ohmic impedance, 6313–6321, Copyright (2010), with permission from Elsevier.) (b) A circuit that can be used to model this data.

2.5 SHIELDED CABLES AND FARADAY CAGE

The wires which connect the cell to the electrochemical instrument must be shielded. That is, the lead metal wire is covered by an insulator; a wire mesh surrounds the insulator, and a final insulating layer covers the wire mesh. Under ambient conditions, electromagnetic waves present in the atmosphere will introduce noise in the measured spectrum. If the lead wires are not shielded, they can act as antennas and pick up the noise.

Similarly, the electrode and other electrically connected components of the cell can pick up the noise. The cell and most of the connections can be placed inside a metal box to minimize these effects. This is referred to as a Faraday cage. The metal box will absorb most of the electromagnetic waves from outside and shield the cell and the leads. Ideally, the metal box should be completely closed with a small opening for the lead wires to enter. In our lab, we have observed a clear reduction in the noise of the spectra, especially in the mid and low-frequency regime, when the wires are shielded, and a Faraday cage is used. If it is difficult to enclose the system in a metal box for whatever reason, a metal mesh can be placed around the system, as a substitute. A metal mesh will absorb most of the waves whose wavelength is larger than the gap in the mesh. Since the electromagnetic waves that contaminate the results have wavelengths of the order of cm or m, a mesh with a few mm gap will be suitable.

2.6 SINGLE SINE *vs.* MULTI-SINE

2.6.1 Single Sine Input

FRAs use the idea of Fourier transform to calculate the impedance. For example, one can apply a sinusoidal potential and measure the current. We can write the current in the Fourier series and find the amplitude and phase at the applied frequency and also its higher harmonics.

$$\text{e.g., Applied potential} = E(t) = E_{dc} + E_{ac0}\sin(\omega t) \tag{2.1}$$

$$\text{Measured current} = i(t) = i_0 + i_1\sin(\omega t + \phi_1) + i_2\sin(2\omega t + \phi_2) + \cdots \tag{2.2}$$

The impedance is calculated using the amplitude and phase at the applied frequency. Any random noise that may *contaminate* the reading is rejected well by this method. This is sometimes called a single-sine technique. Here, the EIS data are acquired under potentiostatic mode. Likewise, the data can be acquired under the galvanostatic mode, where the current is specified as $i_{dc} + i_{ac0}\sin(\omega t)$, and the potential can be measured. The potential is then expanded in the Fourier series, and the amplitude and phase at the applied frequency are used to calculate the impedance.

2.6.2 Multi-Sine

In FRAs, it is also possible to apply many sine waves (with different frequencies, amplitudes, and phases) simultaneously and measure the current and resolve the current using Fourier transform. Fourier transform of a periodic signal allows us to resolve the signal into its Fourier series. This is sometimes called as a 'multi-sine' technique. It is not possible to use a multi-sine technique with the lock-in amplifier equipment. The advantage of simultaneously applying many waves is that the total run time of the experiment can be reduced. The disadvantage is that the contribution from 'noise' to the data is higher compared to the single-sine case. If the amplitude (E_{ac0}) is increased to a large level to obtain a higher signal-to-noise ratio, then the nonlinear effects are likely to manifest in the data. More details on multi-sine are given in Appendix 3.

While using multi-sine, the following points are to be noted. In theory, the sampling frequency has to be at least twice the highest frequency employed. In practice, it is usually at least ten times the highest frequency employed. For a single-sine wave, the period is the inverse of the frequency, i.e., $T = 1/f$. The period of a multi-sine wave is at least one period based on the wave with the lowest frequency; i.e., if the multi-sine comprises sine waves of 5, 15, and 25 Hz, then the sampling frequency is 250 Hz, and the period is 1/5 = 0.2 seconds. Thus, when waves of many frequencies are added, then the period is long, and the sampling frequency is high. This means that a very large amount of data has to be processed and transferred to the computer for recording. In addition, if too many waves are packed into one multi-sine, then the peak amplitude will be large. Even with optimization of the phases, it will not be possible to reduce the crest factor (the ratio of peak amplitude to the *rms* value) below a limit. Then the nonlinear effects of the electrochemical system will certainly manifest. These factors limit the number and range of frequencies used to synthesize the multi-sine wave.

Although a few publications show a multi-sine with a wide range of frequencies (Breugelmans et al. 2012, Breugelmans, Tourwé, Jorcin, et al. 2010, Breugelmans, Tourwé, Van Ingelgem, et al. 2010, Van Ingelgem et al. 2011), it is more common to apply a multi-sine with one or two decades of frequencies at the most, i.e., a multi-sine wave comprising 15 or so frequencies, in the range of 100–1 kHz, will be synthesized and applied onto the system. The resulting current is then subjected to Fourier analysis to extract the amplitude and phase. The impedance will be calculated at all those frequencies. Then another multi-sine comprising sine waves in the frequency range of 1 kHz–10 Hz will be synthesized and applied onto the system, and the impedance will be calculated in this frequency range. Then a third multi-sine wave having waves of frequencies 10–0.1 Hz will be applied, and so on. Thus, a complete spectrum in the frequency range of 100 kHz to 0.1 Hz will be obtained by using three multi-sine waves, each spanning two decades of frequencies.

In some cases, a 'random noise' potential can be applied to the cell. The input potential wave and the output current wave are subjected to fast Fourier transform (FFT), and the impedance spectrum is obtained. This is similar to the multi-sine method, except that the input is not deliberately synthesized with a series of pure sine waves. While it is tempting to use multi-sine because it appears to be elegant and a time saver, one has to note that multi-sine experiments are difficult to debug. If something goes wrong and there are problems in the data such as discontinuity or noise or reproducibility issues, then it is hard to salvage the remaining part of the data. In single-sine data, on the other hand, each of the data is independently acquired. If a few data points appear to be 'noisy,' it is easier to identify them and remove them from the analysis. There are other issues with multi-sine, as explained in the final chapter. We recommend using a single-sine technique for research purposes in electrochemical cells.

For pure electrical circuits with only linear, passive elements such as ideal resistances, capacitances, and inductances, multi-sine does work well. If an EIS experiment is conducted routinely and the results are very reproducible, then the multi-sine technique can be employed. For example, let us consider a metal coated with a protective layer and immersed in a corrosive liquid medium. If pinholes develop in the protective coating, then corrosion will occur in those regions, and pits will be formed on the metal. The EIS response of a pin-hole free system and that of the defective system will be different, and also the data would be very reproducible. EIS can be used as a routine test to check for the formation of pinholes in the coating, and in this case, a multi-sine input would be preferred because it is faster. But as a general rule, for studying a new system, the single-sine method is usually better.

2.7 REPEATABILITY, LINEARITY, AND STABILITY

2.7.1 Repeatability

Any experiment must be repeated at least twice to check for reproducibility. The experiment must be repeated several times to obtain statistically meaningful data. Thus, a spectra must be obtained at the same E_{dc} and E_{ac0} at least twice, and preferably a few more times, to ensure reproducibility.

2.7.2 Linearity

Most of the analyses assume that the system response is linear, and this can be ensured if E_{ac0}, (I_{ac0} in galvanostatic mode) is *small*. Now, how small is small enough? One should acquire the spectra at two different E_{ac0}, e.g., 5 and 10 mV, and if the two results match each other, within noise limits, then the E_{ac0} is small enough, and the system response can be considered as linear. Note that one should plot all the data in both the complex plane and Bode representation and compare. If one uses only one representation, some of the differences may be missed.

2.7.3 Stability

Generally, it is assumed that the system has remained unchanged during the experiment. It takes 15 minutes or more to conduct a typical EIS experiment. If the system has changed significantly *during* the experiment, then the data would include the effect of those changes, and the standard analysis would lead to incorrect conclusions. To ensure that the system has not changed significantly, at the end of the first experiment, one should start the repeat experiment, i.e., acquire EIS data with the same setting once more. If both data are identical, within noise limits, then the system has remained stable. Otherwise, the system was not stable, and the reason for instability has to be identified and addressed.

2.7.4 Data Validation

There is an elegant test available to validate the EIS data, and it is called Kramers Kronig transform (KKT) (Macdonald and Urquidi-Macdonald 1985, Urquidi-Macdonald, Real, and Macdonald 1986). We can subject the acquired EIS data to this test. Very briefly, using the frequency and Z_{Re} data, it is possible to estimate Z_{Im} if the system were linear, causal, and stable. A system is linear if the output signal is proportional to the input signal. When we apply a small perturbation to the system and then remove the perturbation, if the system comes back to its original state, the system is stable. If the signal we get is caused only by the input perturbation we applied, then the system is causal. This also means that the noise or external disturbances are negligible. We can compare the estimated Z_{Im} with actually measured Z_{Im}, and if both are the same, then we can be confident that the data were from a linear, causal, and stable system. If they do not match, then at least one of the three conditions was not met during the experiment, and the data cannot be interpreted easily. Similarly, using frequency and Z_{Im} data, we can estimate Z_{Re} data and compare those with the actual data. In certain cases, the admittance (Y_{Re} and Y_{Im}) must be used instead of the impedance data in the KKT validation. The details and limitations of KKT and similar data validation techniques are explained in Chapter 3.

Earlier it was believed that the data acquired through multi-sine input are automatically KKT compliant (Orazem and Tribollet 2011), and hence KKT cannot be used as a validation tool for those data. However, simulation results (Srinivasan, Ramani, and Santhanam 2013) show that even data from multi-sine perturbation can be validated using KKT. The details are also presented in the final chapter.

2.8 LAB EXPERIMENTS

2.8.1 Experimental Variables and Data Acquisition Software

Here we describe several variables that the user can select to perform EIS experiments. The actual choices depend on the equipment manufacturer, i.e., not all the variables described here are always available for the user to select, but a few key variables should be specified in any EIS experiment. Most of the modern potentiostats have an in-built FRA. Some of the older models do not have an FRA and require a separate FRA, connected to the potentiostat by appropriate cables, to conduct EIS experiments. Here we assume that a potentiostat with a built-in FRA is used.

2.8.1.1 Single-Channel *vs.* Multi-Channel Potentiostats

Furthermore, commercial potentiostats can be classified as single-channel or multi-channel models. Single-channel models are common and denote a single potentiostat. In multi-channel potentiostats, more than one, independent, potentiostats are enclosed in a single physical frame and connected to the computer using a single cable. Here, each potentiostat is referred to as a *channel*. It is even possible to connect more than one potentiostat (from the same or different manufacturers) to the computer using a separate cable for each piece of equipment. The software interface to control the equipment and acquire the data is proprietary and is matched with the hardware. Therefore, if more than one single-channel potentiostat manufactured by different companies is connected to a computer, then each potentiostat can be controlled by the corresponding software from the manufacturer. If more than one single-channel potentiostat manufactured by the same company is connected to a computer, then in the software, the user has to select the potentiostat to use. Only after that, the user should proceed with the experiment. Likewise, if a multi-channel potentiostat is employed, the user has to first select the channel to be used.

2.8.1.2 Equipment to Cell Connections

The potentiostat hardware should be connected to the cell to perform the experiments. At the minimum, the hardware should provide three connections, one to the working electrode, one to the reference electrode, and one to the counter or auxiliary electrode. In addition, a ground connection may be provided. Many modern potentiostats provide a sense electrode connection. If a three-electrode cell is used, the sense and working electrode leads are combined and connected to the working electrode. If a two-electrode cell is used, then the sense and working electrode leads are combined and connected to the working electrode. In addition, the reference and counter electrode leads are combined and connected to the other electrode. The user should refer to the hardware manual and make suitable connections. It is very important to follow the instructions carefully since an incorrect connection can destroy the equipment.

2.8.1.3 Type of EIS Experiment

After the appropriate software is opened and the hardware is connected, the user should choose the electrochemical technique required to perform the experiment. Two techniques are common in impedance – (pseudo)potentiostatic and (pseudo)galvanostatic EIS. These also may be referred to as *constant E* and *constant I*, respectively. In potentiostatic EIS, the dc potential is held constant, and an ac potential is superimposed. The current is measured to calculate impedance. In galvanostatic EIS, the dc current is held constant, and an ac current is superimposed. The potential is measured to calculate the impedance.

2.8.1.4 DC Bias

In potentiostatic EIS, the dc bias potential, perturbation amplitude, and frequencies need to be specified. The dc bias potential may be described as *starting E*. The potential is measured against the reference electrode. In some cases, the user may get a choice of specifying the potential *vs.* the OCP. Still, even in that case, the equipment will measure the OCP at the beginning of the experiment and then calculate the appropriate dc bias *vs.* the reference, to meet the specification, and then proceed to apply the potential *vs.* the reference. If the OCP changes with time, then this can lead to incorrect values. For example, let us assume that the user specifies that the measurement should be performed at the OCP, i.e., 0 V *vs.* the OCP. Let the OCP at the beginning of the experiment is –240 mV *vs.* the reference. The instrument will acquire impedance by holding the electrode at –240 mV *vs.* the reference and superimposing sine waves of various frequencies, throughout the experiment. If the OCP changes with time, and after a minute, it is –270 mV *vs.* the reference, then the initial part of the impedance spectrum is acquired at the OCP, but the latter part is measured at a dc bias of +30 mV *vs.* the OCP. To ensure that the data are acquired as intended, it is necessary to monitor the OCP and ensure that it is stabilized before commencing the EIS experiment. Likewise, if EIS with a dc bias is required, then it is a good practice to change the electrode potential to the required bias, wait for the dc current to stabilize, and then commence the EIS experiment. The time required for the response current to stabilize can be determined by conducting a potential step experiment (*chronoamperometry*), holding the electrode at the desired potential and monitoring the current *vs.* time. Corrosion EIS studies are usually performed at the OCP, whereas mechanistic investigations are usually performed at several dc bias potentials.

Under galvanostatic EIS (*constant I*), the dc current bias can be specified as zero or a positive or a negative value. In this, it is important to check the convention employed by the equipment manufacturer. A cathodic potential may be denoted as positive or negative, and the user may even have a choice of selecting the convention. Next, it is important to allow the dc potential to stabilize, and the time required for the stabilization can be obtained using *chronopotentiometry* or current step experiments. If the experiment is to be conducted at zero current, then it can be started after the OCP stabilizes. Galvanostatic EIS is common in investigations of energy devices such as batteries and fuel cells.

2.8.1.5 Frequencies

EIS is acquired at multiple frequencies and can be performed using single-sine or multi-sine waves. Single-sine waves are recommended since they yield a better signal-to-noise ratio, even if it takes a longer time compared to multi-sine experiments. The starting and ending frequencies should be specified. Usually, the starting frequency is the highest, and the ending frequency is the lowest. The intermediate frequencies are decided by choosing either the total number of frequencies between the starting and ending frequencies (e.g., 30 or 50) or by specifying the number of frequencies per decade (e.g., 7 or 10). Usually, the frequencies are logarithmically spaced, i.e., they are in a geometric series. Some manufacturers also give an option of using linearly spaced frequencies, i.e., in arithmetic series. In a few cases, it is possible to specify individual frequencies in custom or manual mode. The actual frequency range and the number of frequencies required will depend on the system under investigation and the information desired. However, for most users working on metal electrodes in aqueous electrolytes, a high-frequency limit of 100 kHz and a low-frequency limit of 0.1 Hz, with seven frequencies per decade logarithmically spaced, would be a good starting point. If the noise in the data is low, then the low-frequency limit can be reduced further.

2.8.1.6 Amplitude

The potential amplitude has to be specified, and it can be specified as the magnitude or peak-to-peak or root-mean-square (*rms*) value. We use E_{ac0} to denote the magnitude, and it is equal to half of the peak-to-peak value. The rms value is given by $E_{ac0}/\sqrt{2}$. Thus, a specification of 10 mV (rms) is equal to 14.14 mV magnitude or 28.28 mV peak-to-peak. A large-amplitude can lead to nonlinear contributions to the EIS data, while a small amplitude can lead to a poor signal-to-noise ratio. A choice of 10 mV magnitude is a good starting point for most potentiostatic EIS experiments.

In galvanostatic EIS, the amplitude has to be specified as current (A), and 10 mA magnitude is a good starting point in most cases. Depending on the expected range of impedance (in Ω and not in Ω cm^2), the range of peak potentials can be estimated. The specified amplitude is applied to all the sine waves in a given experiment. Some manufacturers even allow a custom mode where the user can specify a separate amplitude at each frequency.

If the user chooses a multi-sine wave input, then the *rms* value may be specified. Often, a custom algorithm calculates the amplitude of the individual sine waves and their phase offset values. The user may not be presented with these values, and only the final impedance values are recorded.

2.8.1.7 Current/Potential Range

In potentiostatic EIS, the user may have to choose the current range as either automatic or manual. In stand-alone EIS experiments, auto-ranging can be chosen by default. If the user intends to record the potential and current waves for additional processing, then fixing the current range to a specific value (e.g., 10 mA) will be appropriate. This is because, under potentiostatic EIS, if the impedance varies over several orders of magnitude in the frequency range employed, then the measured current will also vary over several orders of magnitude. At different current values, the measurement is performed by different circuits. By specifying a fixed current range, the user can ensure that the same circuit component is used for all the measurements in a given experiment. In galvanostatic EIS, often, the user is not required to select a potential range, and the entire potential range of the potentiostat would be available.

2.8.1.8 Current/Potential Overload

In potentiostatic mode, if the current required to apply a sine wave of a particular frequency is more than the current capability of the instrument, then the instrument will be overloaded. It may either stop the experiment or skip that particular measurement and proceed to the next. Likewise, in galvanostatic mode, if the potential required is too high, the instrument will overload. These conditions must be avoided by reducing the magnitude of the applied potential/current.

2.8.1.9 Other Options

Several other variables are described below, and all of them may not be available in all the software interfaces. There is no universally accepted terminology across equipment manufacturers, and the users should consult the respective user manual to select the appropriate choice.

Stability vs. Speed: Some manufacturers give users a choice of high speed or high stability operation. The high-speed option can be used if the system is stable; if the reference electrode impedance is high, or if the working electrode capacitance is high, then the high-speed operation may result in potential oscillations. In that case, the high-stability option should be chosen.

Conditioning: Some software interfaces allow for electrochemical cleaning of the electrode before the EIS experiment, i.e., by repeatedly scanning the potential between two values. The number of scans may be specified, or the scan rate and duration can be specified.

Initial OCP Delay: After conditioning, the cell may be allowed to settle at the OCP. Alternatively, the user can specify that OCP drift should be less than a certain value, and the software will wait for the OCP drift to reduce to the specified value before commencing the experiment.

Equilibration: If a dc bias other than the OCP is used, the user can specify that the electrode must be held at the dc bias for a certain time before sine waves are applied. This variable may be denoted as *equilibration time or induction period.* The best way to select this value is to perform chronoamperometry or chronopotentiometry at the desired dc bias. From the results, the time required for obtaining a stable current/potential response can be determined and input into this field.

Measurement Delay: When a sine wave potential of a particular frequency is applied, the initial current response may not be steady periodic, and the system may require some time to yield a steady periodic response. This delay could be due to the nature of the electrochemical processes or nonidealities in the measurement system or both. Nevertheless, the user can stipulate that the response current should be recorded after a certain time, known as *measurement delay.* The impedance at that frequency is calculated based on the data recorded after the measurement delay. The measurement delay may also be specified as the number of cycles (of the applied sine wave). At high frequencies, several cycles correspond to a short time, while at low frequencies, even a few cycles will require a long delay. In some cases, the user can specify the minimum number of cycles as well as the time, and the software will use the higher of the two specifications at each frequency.

Signal Monitoring: Some manufacturers give the ability to monitor the applied potential (current) and measured current (potential), either as a function of time or as a Lissajous plot. Some manufacturers also allow the user to tap the potential and current directly as analog output, so that the users can process the signal in any way they want. This is particularly useful if the user wants to perform further signal processing, such as nonlinear EIS measurement, to record higher harmonics or total harmonic distortion (THD). In these cases, it is desirable to use a fixed current range, rather than an auto-current range. In these cases, the equipment can be set to *issue a pulse* when the sine wave is applied, so that the pulse can serve as a trigger in the external signal processing equipment. A few manufacturers also record the extent of nonlinearity or THD in each measurement. In all these cases, the users should carefully read the user manual to understand how the manufacturer defines these terms.

Remarks: Comments or remarks can be saved as text in a space reserved. Furthermore, the user may be required to specify the electrode area, material density, and equivalent weight. The reference electrode should also be specified. Strictly speaking, an incorrect input of these variables will not affect the measured response in any way. These are useful in bookkeeping and in retrieving the correct data at a later stage. In some interfaces, the user needs to specify the filename even before starting the experiment, and the results will automatically be saved as the impedance at each frequency is measured. If the experiment is stopped before completion due to any reason, then the data acquired up to the stoppage can be recovered.

Automation: Apart from the options described above, a few more options may be available to the user. In a specification named *potential scan*, the user is given the option of conducting several potentiostatic EIS experiments in sequence. The dc bias values are changed for each experiment, and these are specified by the starting potential, step potential, and the end potential. The potential is not actually scanned during a given EIS experiment; instead, multiple EIS experiments, each with a unique dc bias value, are conducted. A few manufacturers also permit another level of automation, where a sequence of electrochemical techniques can be performed. For example, first, the OCP can be measured for a certain time, followed by an EIS measurement; a potential step may be conducted next to deposit a material on the electrode for a specific time; EIS can be performed next, and finally, a stripping voltammogram can be recorded. These tasks can be achieved using a proprietary script language.

Next, we describe a few experiments that the reader can practice. These experiments are listed in an increasing order of complexity, both in terms of the effort required to perform the experiments to acquire good quality data, and in terms of the type of spectrum that is obtained. Appropriate safety precautions must be taken while working with electrical/electronic instruments and chemicals. Here we present the results and observations, but not the analysis. Initially, a simple electrical circuit can be used. Next, a classical electrochemical redox reaction is characterized. In the third example, copper electrodeposition is studied, where dc bias is applied during EIS data acquisition. In the fourth example, Zr corrosion in HF is investigated, and a low-frequency inductive loop manifests. In the fifth and final example, Ti anodic dissolution in HF is investigated. With the application of a suitable dc bias, at low frequencies, the real part of the impedance exhibits negative values.

2.8.2 Circuit Impedance Measurements

In the place of an electrochemical cell, a simple electrical circuit can be used to familiarize oneself with the operation of the instrument, the connections, and the software interface. A circuit in the configuration shown in Figure 1.18 is commonly used, although other circuits with resistors, capacitors, and inductors can be used as well. These are referred to as *dummy cells*. Here, only two terminals are available. The counter and reference electrode leads from the potentiostat are tied together and connected to one terminal of the circuit. The working electrode (and sense electrode, if present) should be connected to the other terminal. The user needs to ensure that the working electrode wiring does not touch the reference or counter(auxiliary) electrode wiring; otherwise, it can damage the equipment.

The circuit in Figure 1.18, with $R_1 = 10\ \Omega$, $C_2 = 10^{-5}$ F, and $R_2 = 100\ \Omega$ was used as the *dummy cell*. EIS experiments were performed in a frequency range of 100 kHz–0.1 Hz, with seven frequencies per decade. In all the examples illustrated in Section 2.4, the frequencies were logarithmically spaced. The experiments were performed in two different models, named Equip-1 and Equip-2. Two amplitudes, viz. $E_{ac0} = 0.14$ and 14 mV, which correspond to an *rms* value of 0.1 and 10 mV, respectively, were employed. The results are shown in Figure 2.2 as a complex plane plot and Bode plots.

The results show that at a perturbation amplitude of 14 mV, both instruments yield identical data. At 0.14 mV, the data are noisy, although they overlap with the 14 mV

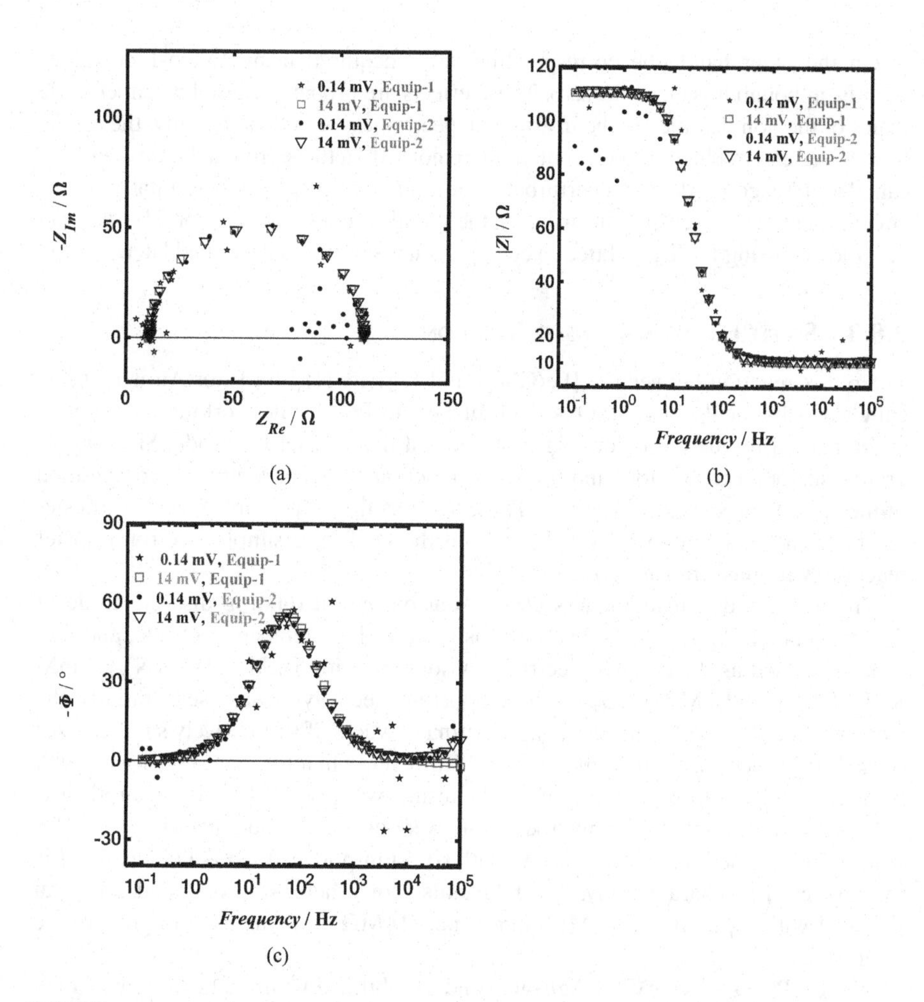

FIGURE 2.2 Impedance of a circuit measured at two amplitudes and two instruments (a). Complex plane plot and Bode plots of (b) magnitude *vs.* frequency and (c) phase *vs.* frequency. The legend shows the instrument and the perturbation amplitude employed.

data at several frequencies, indicating that 14 mV data are in the linear region. Since this circuit is made of resistors and capacitors, nonlinearities are not expected even at several volts. The scatter in the data at low amplitude perturbation shows that the signal-to-noise ratio is poor.

A careful analysis of Figure 2.2b shows that, at 0.14 mV perturbation, the low-frequency magnitude data from Equip-2 has considerable noise, while the corresponding data from Equip-1 has less noise. On the other hand, at high frequencies, the phase values from Equip-1 exhibit more noise than those from Equip-2, as seen from Figure 2.2c. At mid-frequencies, data from Equip-2 have less noise compared to those from Equip-1. We also see that even at 14 mV perturbation, the high-frequency phase data ($-\Phi$) acquired using Equip-2 does not go toward zero, but appears to increase.

On the other hand, the corresponding value acquired using Equip-1 decreases slightly, although it is closer to zero. These effects are probably caused by either cable issues or the nonidealities in the instrument electronics, or both. Generally, the results from a *real* electrochemical cell contain more noise than those from a dummy cell. We can, therefore, conclude that a perturbation amplitude of 0.14 mV is not suitable in this environment – it may ensure linearity, but the signal-to-noise ratio is poor. The data can be validated using KKT and fitted to an appropriate circuit, as described later.

2.8.3 Simple Electron Transfer Reaction

The redox reaction between $K_4[Fe(CN)_6]$ and $K_3[Fe(CN)_6]$ on Pt or Au is a classical electrochemical reaction. Oxidation of Fe^{2+} to Fe^{3+} at the working electrode is accompanied by the reduction of Fe^{3+} to Fe^{2+} at the counter electrode. Six cyanide groups surround the Fe ion, and the redox reaction does not require a reorientation of the ligands or solvent molecules. Thus, this reaction is a simple electron transfer reaction, and this example illustrates EIS at the OCP of a simple electron transfer reaction with mass transfer effects.

The above redox reaction was carried out on an Au rotating disc electrode of 5 mM diameter. Ag/AgCl (3.5 M KCl) was used as the reference electrode, and a Pt mesh was used as the counter electrode. A solution with 5 mM $K_4[Fe(CN)_6]$, 5 mM $K_3[Fe(CN)_6]$, and 1 M Na_2SO_4 (as the supporting electrolyte) was used. Initially, the working electrode was polished using alumina powders of successively smaller sizes (viz., 1, 0.3, and 0.05 μm) from Buehler®. However, when the final polish was with 0.05 μm powders, the reproducibility in the results was poor. When the electrode was polished with 1 μm powder only, the results were quite reproducible. The reason for the poor reproducibility of the results with 0.05 μm polished Au is not clear and is the subject of a separate study. All the results shown here were obtained using Au polished with 1 μm alumina. Millipore water (18 MΩ cm) was used to prepare the electrolyte.

The OCP was −241 mV *vs.* Ag/AgCl, and it stabilized within a few minutes. PDP experiments were conducted at several electrode rotational speeds. The potential was scanned from −500 mV to +500 mV *vs.* the OCP at a scan rate of 2 mV s^{-1}, and the results are shown in Figure 2.3a.

In Figure 2.3a, anodic current and potential are shown to be positive. An increase in potential results in an initial increase in the current, followed by saturation. The saturation current values increase with an increase in electrode rotational speed, indicating that in the saturation region, the reaction is diffusion-limited. The results in the cathodic region are *almost* symmetric to those in the anodic region. In the diffusion-limited region, the cathodic current magnitude at a given potential and electrode rotational speed is slightly more than the corresponding anodic current magnitude, reflecting the slightly different diffusion coefficients of the Fe^{2+} and Fe^{3+} species. The impedance was acquired at the OCP, with an amplitude of 28 mV (an *rms* value of 20 mV). The electrode rotation was maintained at 900 rpm, and the results are shown as a complex plane plot in Figure 2.3b and as Bode plots in Figure 2.3c. The high-frequency data show a capacitive loop, and the low-frequency data show a loop that is described by Bounded Warburg impedance (described in Section 6.2.1). When the

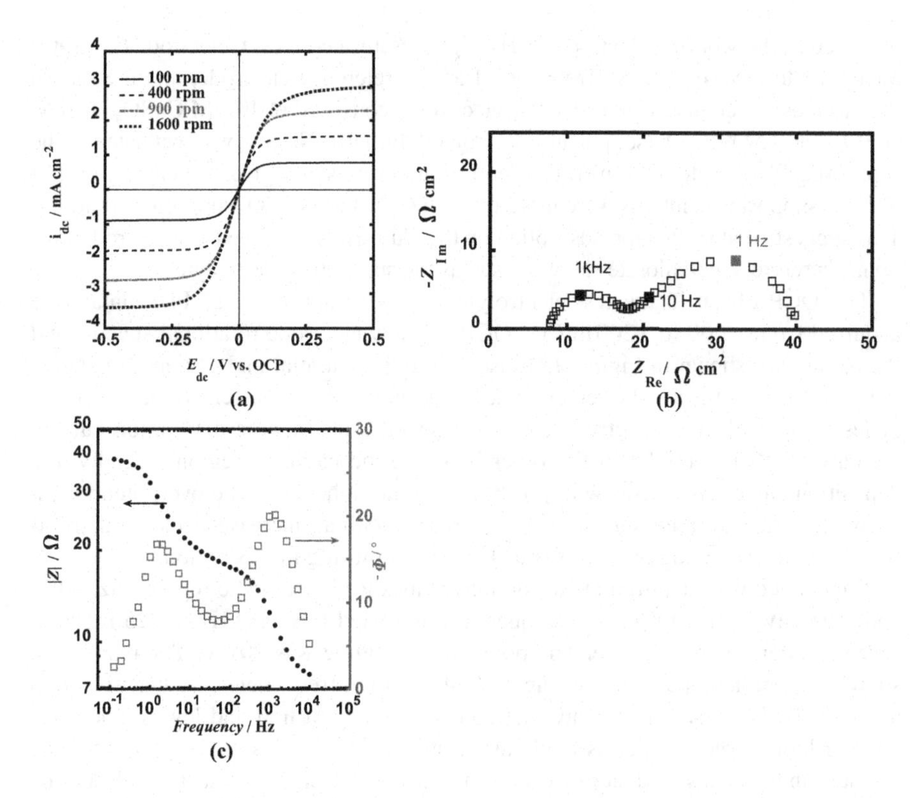

FIGURE 2.3 Results of the Fe^{2+}/Fe^{3+} redox reaction. (a) Potentiodynamic polarization, impedance at 900 rpm as (b) complex plane plot and (c) Bode plots. (Adapted from Pachimatla (2020) with permission.)

experiments were repeated at other values of electrode rotational speeds, the high-frequency loop remained the same (results not shown). In contrast, the low-frequency loop changed, clearly indicating that it is affected by mass transfer.

2.8.4 Metal Deposition in Acidic Media

In this example, we illustrate the electrodeposition of Cu from an acidic solution. Here, Cu^{2+} ions in the solution are surrounded by water dipoles. In the Cu electrodeposition process, Cu^{2+} ions from the bulk first diffuse through the boundary layer and arrive at the electrode. Next, the water dipoles attached to the ion are disengaged, and the Cu^{2+} may adsorb on the Cu surface with or without electron transfer. Since two electrons are involved, and since Cu^{+} is in a stable oxidation state, the reaction will likely occur in multiple steps. This example illustrates the EIS of a multi-step reaction with mass transfer effects. Here, a dc bias was applied, i.e., the impedance experiments were not conducted at the OCP.

The experiments were conducted with an EC301 potentiostat (SRS, USA) in conjunction with an SR780 FFT analyzer (SRS, USA). A Cu electrode of 5 mm diameter was used as the working electrode. A $Hg/HgSO_4$/saturated K_2SO_4 electrode (i.e., saturated sulfate electrode or SSE) was used as the reference electrode, and a Pt mesh was used as the counter electrode. The electrolyte consisted of 10 mM $CuSO_4$, 10 mM H_2SO_4, and 1 M $NaClO_4$ (supporting electrolyte). Initially, Na_2SO_4 was evaluated as the supporting electrolyte. However, the deposition currents at a given potential in 0.1 M Na_2SO_4 supported solutions were more than those in 1 M Na_2SO_4 supported solutions. This suggests that sulfate ion adsorption on the Cu surface inhibited Cu electrodeposition. Therefore, a perchlorate salt was used as the supporting electrolyte.

The OCP of Cu in the test electrolyte was −365 mV *vs.* SSE. PDP data were acquired from −600 to +200 mV *vs.* OCP at a few electrode rotational speeds, and the results are shown in Figure 2.4a. Here, anodic potential and current are shown to be positive. In the anodic region, where Cu dissolves, the current values are more or less independent of electrode rotational speeds. The anodic dissolution current increases with potential. On the other hand, in the cathodic region, initially, the deposition current increases with potential, and at higher cathodic overpotentials, it saturates. The saturation current value increases with an increase in electrode rotational speed, indicating that mass transfer is rate-limiting in this region.

Impedance was acquired at two dc bias values in the cathodic region, viz. −100 and −300 mV *vs.* the OCP. The frequency was varied from 10 kHz to 125 mHz, at seven frequencies per decade. The potential amplitude was 20 mV. The results are shown as complex plane plots in Figure 2.4b. At high frequencies, a capacitive loop associated with double-layer capacitance is observed. At mid and low frequencies, another loop is seen, which is likely associated with the relaxation of intermediate species, and the mass transfer of Cu^{2+} from bulk to the electrode surface. When the cathodic overpotential is changed from −100 to −300 mV *vs.* the OCP, the diameter of the high-frequency loop decreases, while that of the low-frequency loop increases.

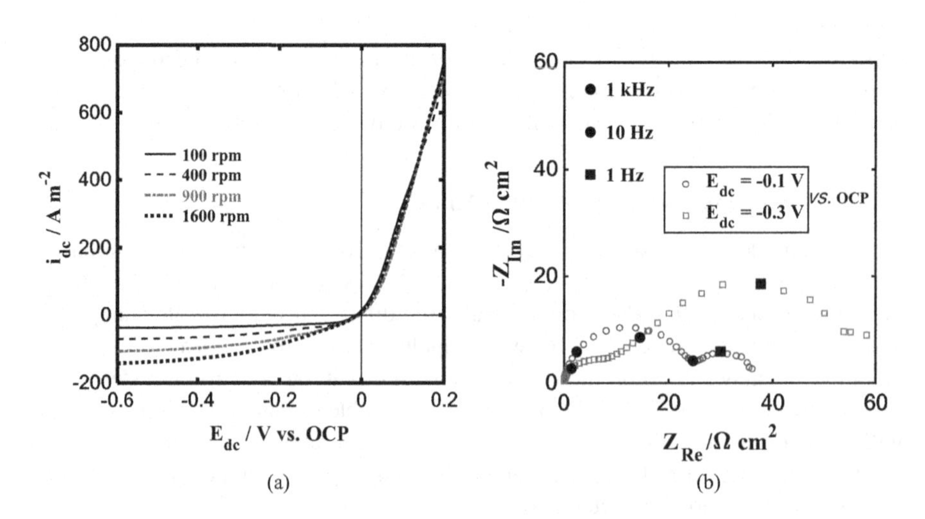

FIGURE 2.4 Cu electrodeposition from acidic media. (a) Potentiodynamic polarization and (b) impedance complex plane plot. (Adapted from Pachimatla (2020) with permission.)

In the case of simple electrical circuits, applying a dc bias will not change the impedance spectrum. In the case of electrochemical reactions, the impedance spectrum strongly depends on the dc bias.

2.8.5 Metal Dissolution in Acidic Media

The dissolution of Zr in hydrofluoric acid was investigated using EIS. A Zr electrode of 5 mm diameter was used as the working electrode. An Ag/AgCl (3.5 M KCl) was used as the reference electrode, and a Pt mesh was used as the counter electrode. The working electrode was rotated at 900 rpm, and the impedance spectrum was acquired at the OCP. The solution contained 0.1 M Na_2SO_4 as the supporting electrolyte and 5 mM HF as the active component. HF is a weak acid, but is a dangerous chemical, and must be handled with caution. Parstat 2263 (Ametek, USA) was used to acquire the data. The frequency was varied from 100 kHz to 10 mHz, and the *rms* value of the perturbation was 10 mV.

The results are shown as complex plane plots in Figure 2.5. A capacitive loop is seen at high frequencies, a second capacitive loop at mid frequencies, and an inductive loop at low frequencies. Note that if the measurement were restricted to >100 mHz, the inductive loop would not be seen, and this illustrates the importance of acquiring data at low frequencies. Below 10 mHz, the data were very noisy and not reproducible. Therefore, the experiments were restricted to 10 mHz. This example demonstrates that an inductive loop can appear in the impedance spectrum of an electrochemical reaction. The data can be fitted to an appropriate equivalent electrical circuit to extract the value of a parameter known as polarization resistance (Chapters 3 and 4). Polarization resistance can also be obtained using other methods such as PDP and LP, and EIS is used to validate those results. This experimental result and analysis form a part of a corrosion study of Zr in HF solution (Rao et al. 2020).

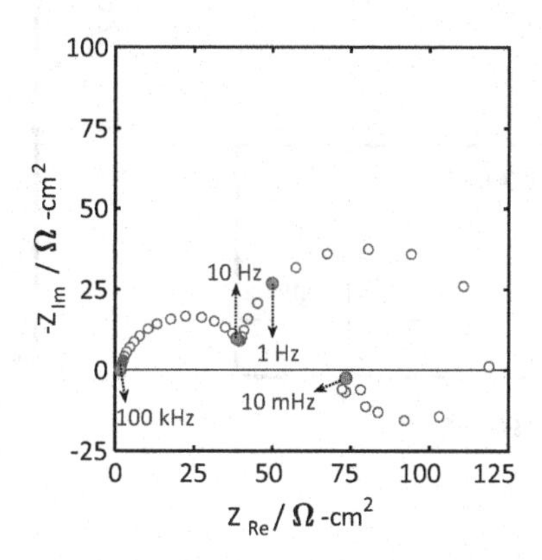

FIGURE 2.5 Complex plane plot of an impedance spectrum of Zr in 5 mM HF and 1 M Na_2SO_4 (supporting electrolyte), at the OCP. (Adapted from Rao et al. (2020), under open source license CC BY.)

2.8.6 Passivation

As a final example, we present the impedance spectra of Ti anodic dissolution in HF solutions. A Ti cylinder of 5 mm diameter, rotated at 900 rpm, was used as the working electrode. Ag/AgCl/3.5 M KCl was used as the reference electrode, and a Pt mesh was used as the counter electrode. The electrolyte consisted of 100 mM HF and 1 M Na_2SO_4 (supporting electrolyte). The OCP was −1 V *vs.* Ag/AgCl. PDP was acquired by scanning the potential from the OCP to +1.5 V *vs.* the OCP at 10 mV s^{-1}. The results are shown in Figure 2.6a and the anodic current

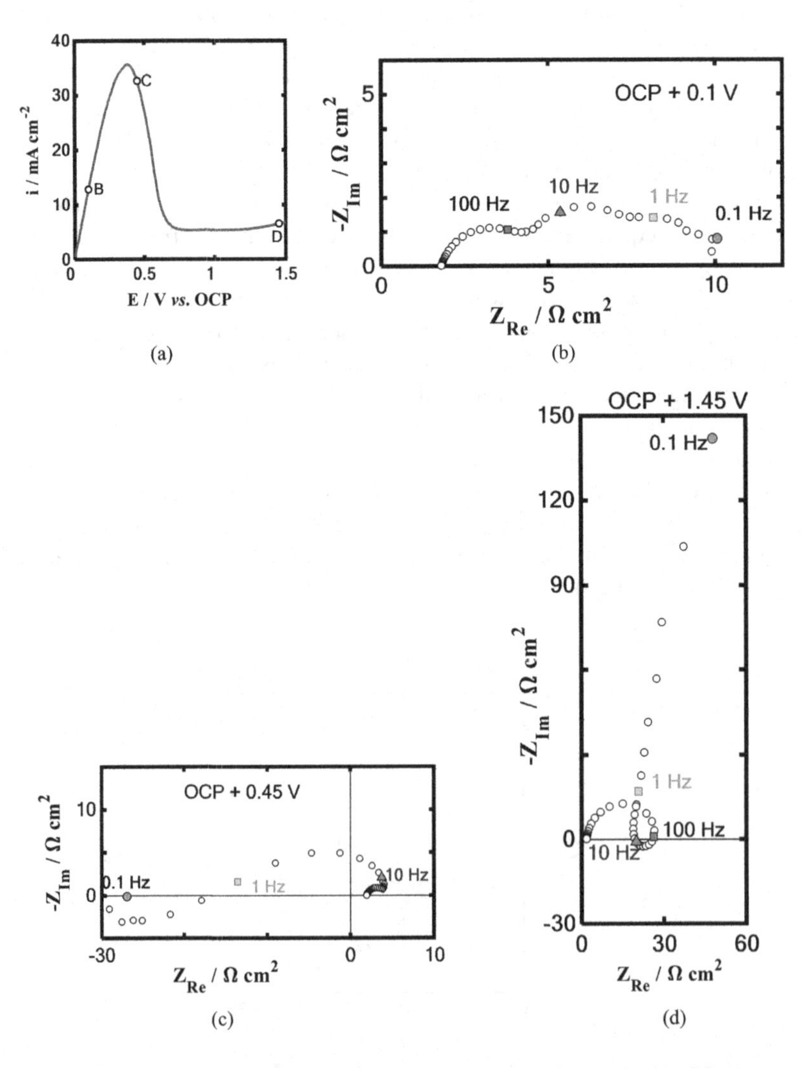

FIGURE 2.6 Results of Ti anodic dissolution in 100 mM HF and 1 M Na_2SO_4 (supporting electrolyte). (a) Potentiodynamic polarization, and complex plane plot of impedance spectra in the (b) active region, (c) passive region, and (d) transpassive region. (Adapted from Fasmin, Praveen, and Ramanathan (2015), under open source license CC BY.)

and potential are shown to be positive. When the potential is increased from 0 to 400 mV *vs.* the OCP, the current increases. This indicates that the dissolution rate of Ti increases and this is described as the *active region*. At higher potentials, from 400 to 850 mV *vs.* the OCP, the current decreases, indicating that Ti is passivating in this region. At even higher potentials, the current increases slowly in the *transpassive region*, where a 3D film is present on the surface, and dissolution continues to occur.

EIS was acquired at various dc bias values, from 30 kHz to 100 mHz, at seven frequencies per decade. The results are shown as complex plane plots in Figure 2.6b–d. In the active region, three capacitive loops are seen (Figure 2.6b). The high-frequency loop corresponds to the double-layer, while the mid and low-frequency loops correspond to the faradaic process (described in Chapter 5). The data can be modeled using an equivalent electrical circuit as well (Chapter 4). The results in the passive region show that at high frequencies, a capacitive loop manifests (Figure 2.6c). At mid and low frequencies, the real part of the impedance exhibits negative values. This example illustrates that negative resistance values can manifest in the impedance spectra. At low frequencies, $-Z_{Im}$ values also go below the real axis, i.e., an inductive behavior is seen. The major features of the spectra in the active and passive regions can be explained by mechanistic analysis (Fasmin, Praveen, and Ramanathan 2015). Figure 2.6d shows that in the transpassive region, the spectrum shows a capacitive behavior at high frequencies, an inductive loop at mid frequencies, and a capacitive behavior at low frequencies. Transpassive dissolution of valve metals in acidic fluoride media has been modeled in the literature, and they predict the main characteristics well (Kong 2010; Bojinov et al. 2003; Cattarin, Musiani, and Tribollet 2002). This experiment also serves as an example of applying a dc bias to acquire EIS data of electrochemical systems and interpreting the results.

2.9 GENERAL ISSUES WITH EIS DATA ACQUISITION

2.9.1 Reproducibility

EIS is a very sensitive technique. In many cases, a small change in the electrode–electrolyte interface causes significant changes in the resulting spectrum, and this is very useful in detecting small changes in the system. The other side of the story is that it is not easy to obtain reproducible results. Small changes in electrode preparation, or electrolyte purity, or the reference electrode position in the cell can significantly impact the results (Ziino, Marnoto, and Halpern 2020). A long electrode stabilization time may be required to get reproducible results (Xu et al. 2019). If an oxide film grows on the electrode during the experiment, then as the film thickness changes, the impedance response will vary as a function of time. Minor variations in conducting the experiments, such as different wait times being used between measurements at each frequency, can result in a difference in the results. In other words, EIS being a very sensitive technique also makes it harder to obtain reproducible data in some cases. Techniques such as potentiodynamic polarization (often referred to as Tafel experiments) or cyclic voltammetry or chronoamperometry yield data with good repeatability in most of the cases.

For a rapidly changing system, it is very difficult to get reproducible results. However, a few attempts have been made in the literature to acquire the impedance spectra of such systems (Popkirov 1996; Koster et al. 2017). The utility of such data is further impaired by the fact that any analysis that accounts for the system changes is very complex (Stoynov 1993; Victoria and Ramanathan 2011; Stoynov 1992).

2.9.2 Signal-to-Noise Ratio, Linearity Requirements, and Experiment Duration

A small perturbation amplitude will lead to a poor signal-to-noise ratio, as seen in the example in Section 2.4.2. On the other hand, a large perturbation amplitude will lead to nonlinear effects being incorporated in the response. For a given system, a series of experiments with varying perturbation amplitudes need to be performed to identify the range of amplitudes where the linearity approximation is applicable, and the signal-to-noise ratio is good, and this is time consuming.

The duration of a single EIS run depends on the frequency range, in particular, the lowest frequency and the number of frequencies per decade. It takes a particularly long time to measure the impedance at very low frequencies. While the high-frequency data contain information about the double-layer structure, the data at mid and low frequencies are essential to characterize the reaction and mass transfer effects. Thus, it becomes imperative to acquire EIS at low frequencies, and the experiment duration will be long. The noise level tends to be higher at lower frequencies. Thus, several sinusoidal cycles have to be employed to obtain a good signal-to-noise ratio, and this increases the duration of the experiment even more.

2.10 EXERCISE – EXPERIMENTAL ASPECTS

Q2.1 Typically, in which systems are potentiostatic mode EIS employed, and in which systems are galvanostatic mode EIS employed? Why?

Q2.2 Select a few publications with EIS in the last 5 years and identify the commonly used supporting electrolytes. What is the range of concentration of the supporting electrolyte employed?

Q2.3 Select a few publications with EIS in the last 5 years and summarize the amplitude and frequency range employed in EIS studies on (a) batteries or fuel cells (b) corrosion or electrodeposition and (c) sensors. What inferences can we draw from this summary?

Q2.4 (a) Create a data set consisting of two sine waves of frequencies 10 and 20 Hz, with amplitudes of 10 and 5 mV respectively, for 0–0.1 seconds, at 1 ms time intervals. (b) Add them together to synthesize a multi-sine wave. Plot the multi-sine wave as a function of time.

Q2.5 In the above example, introduce a phase difference of 30° between the waves at 10 and 20 Hz and recalculate the multi-sine wave and plot.

Lab Exercises

The following exercises require access to a potentiostat with a FRA. Some of the experiments require access to suitable wet labs. Please follow the appropriate safety standards.

Q2.6 Connect the reference and counter electrode leads of a potentiostat to one end of a resistor of 20 Ω resistance, and the working electrode lead to the other end of the resistor. Measure the impedance from 100 kHz to 1 mHz at 7 points per decade, logarithmically spaced. Use a single sine with a 10 mV amplitude. Plot the results in a complex plane and Bode format.

Q2.7 Repeat the above with a 10 μF capacitor.

Q2.8 Create a circuit shown in Figure 1.18 and measure the impedance spectrum, as described in Q2.6. Record the time for acquiring data. If the software permits selecting the 'quality' of data acquisition, e.g., fast *vs.* high quality, evaluate the results in both modes.

Q2.9 Repeat Q2.8 with an amplitude of 0.1, 1, 20, and 50 mV. What can we conclude from these results?

Q2.10 Repeat the experiment described in Q2.9 but with an applied dc bias of (1) 100 mV and (2) 200 mV. Compare the results with those of Q2.9.

Q2.11 Prepare a three-electrode system with an Au working electrode, an Ag/AgCl (3.5 M KCl) (or any other suitable) reference electrode, and a Pt mesh counter electrode. Connect the appropriate leads from the potentiostat. Prepare an electrolyte with 5 mM $K_3[Fe(CN)_6]$ + 5 mM $K_4[Fe(CN)_6]$ and 0.1 M $NaClO_4$. Polish the electrode using 2000 grit paper, followed by alumina slurries of decreasing particle sizes. Measure the OCP *vs.* time.

Q2.12 In the above setup, run CV at 5 and 50 mV s^{-1} while keeping the electrode stationary. Conduct an EIS experiment with parameters (frequency and amplitude) described in Q2.6.

Q2.13 In the above setup, conduct the experiments in Q2.12 at various electrode rotational speeds (100, 400, 900 rpm, and so on). Compare the results with those of Q2.12.

Q2.14 In the above system, at a fixed rpm, conduct an EIS experiment with a dc bias of +0.1 V and −0.1 V *vs.* the OCP, with parameters described in Q2.6.

Q2.15 Repeat the experiments described in Q2.11–2.14, after increasing the concentration of the supporting electrolyte from 0.1 to 1 M. Note: If 1 M $NaClO_4$ is used, then NaCl must be used in the reference electrode solution instead of KCl. Otherwise, $KClO_4$ will precipitate and may block the reference electrode. Alternatively, Na_2SO_4 may be used as a supporting electrolyte, and Ag/AgCl (3.5 M KCl) can be used as a reference electrode.

Q2.16 What is the duration of the experiment when the impedance of a circuit is measured *vs.* when the impedance of an electrochemical cell with a non-zero dc bias is measured for the same set of frequencies and perturbation amplitudes? If there is a significant difference, what could be the reason?

Q2.17 If the instrument has the capability for multi-sine measurement, perform the experiments described in Q2.13 and Q2.14 in multi-sine mode and compare both the measured results and experimental duration with those of single-sine measurement mode.

Q2.18 Perform the experiments described in Sections 2.4.3–2.4.5 and compare the results with those shown.

Q2.19 Repeat Q2.12, but with an electrode polished using 600 grit paper, and compare the results. What is the effect of changing surface preparation?

3 Data Validation

Accuracy of observation is the equivalent of accuracy of thinking.

—Wallace Stevens

3.1 KRAMERS–KRONIG TRANSFORMS

In 1926, **R. De L. Kronig** published the KK equation relating the real and imaginary parts of a complex quantity, while working on the dispersion of X-rays. Independently, in 1927, **H. A. Kramers** published the equation in a study of dispersion and absorption of X-rays. The equivalent relationship between the magnitude and phase was disclosed by **Hendrik W. Bode** in 1937 in a US patent application on amplifiers. In 1986, **Macdonald and Urquidi-Macdonald** illustrated the application of KKT to EIS data.

Kramers–Kronig transforms (KKT) are mathematical relationships between the real and imaginary parts of responses obtained from a system obeying certain assumptions. They are sometimes referred to as Kramers–Kronig relations. In EIS, the real and imaginary parts, or the phase and magnitude, of the impedance at a given frequency, are the relevant responses. In theory, for a given system, the impedance data would be available in the complete frequency interval of [0 ∞]. Kramers–Kronig transforms allows one to calculate the real part if the imaginary part is known over the entire frequency range. Likewise, one can calculate the imaginary part if the real part is known over the entire frequency range.

$$Z_{\mathrm{Im}}(\omega) = -\frac{2\omega}{\pi}\int_0^\infty \frac{Z_{\mathrm{Re}}(x) - Z_{\mathrm{Re}}(\omega)}{x^2 - \omega^2}\,\mathrm{d}x \tag{3.1}$$

$$Z_{\mathrm{Re}}(\omega) = Z_{\mathrm{Re}}(\infty) + \frac{2}{\pi}\int_0^\infty \frac{xZ_{\mathrm{Im}}(x) - \omega Z_{\mathrm{Im}}(\omega)}{x^2 - \omega^2}\,\mathrm{d}x \tag{3.2}$$

There are many other forms of this equation such as relating the phase and the magnitude of the impedance or the admittance (Orazem and Tribollet 2011).

The paired response will obey KKT if the system is (a) linear, (b) causal, and (c) stable. Earlier it was believed that the response at zero and infinite frequency must be finite for KKT to be obeyed, but later it was shown that this is not necessary (Agarwal, Orazem, and Garcia-Rubio 1995). Even the other three conditions are sufficient but not necessary; e.g., if a system response is linear, causal, and stable, then it *will be* KKT compliant, but if the system response is nonlinear, then it *may or may not be* KKT compliant. We will elaborate more on this in Section 3.1.4.

In practice, it is not likely that only the real (or imaginary) part of the EIS data is measured, and the imaginary (or real) part has to be calculated using KKT. Instead, both the real and imaginary parts of the impedance spectrum would be measured. Then, the KKT can be used to check if the system has obeyed the three rules. That is, KKT is used as a data validation tool. Obviously, the data are never available from 0 to ∞ frequency and at all the intermediate frequencies. Thus, the transformation integrals can be evaluated only approximately. Certain methods are proposed in the literature to extrapolate the data to zero and infinite frequencies (Kendig and Mansfeld 1983, Macdonald and Urquidi-Macdonald 1985, Haili 1987, Esteban and Orazem 1991, Orazem, Esteban, and Moghissi 1991), but none of them are universally applicable.

"An inadequately analyzed data set is not worth generating" (Macdonald 1997), but then "Is the un-validated data worth analyzing?" (Boukamp 2008). Typically, EIS data are obtained in the range of 10^{-3}–10^{5} Hz at discrete frequencies. KKT or an equivalent technique *should be* used to validate the data before *any other* analysis is done.

3.1.1 KKT Validation – Example of a Good Quality Spectrum

We begin with the remark that impedance spectra generated by electrical circuits containing only passive elements, i.e., resistance, capacitance, and inductance, are always KKT compliant. If the circuit contains an active element such as a power source, then the resulting spectrum may or may not be KKT compliant. Now let us consider a simple electrical circuit and the associated impedance spectrum, given in Figure 3.1. This contains only passive elements, and the spectrum associated with this is expected to be KKT compliant. For the spectra presented in Sections 3.1.1–3.1.4, the data were obtained in the frequency range of 10 mHz–100 kHz, at ten frequencies per decade, logarithmically spaced.

We employed a KKT software, written by Prof. DD Macdonald, to calculate the transformed values and compare them with the original values. The software has an option to extrapolate beyond the frequency range, and this option was utilized in this transformation. The transformation was done on the real and imaginary parts of the data. This corresponds to the formulas given in Eqs. 3.1 and 3.2. The variations of |Z|

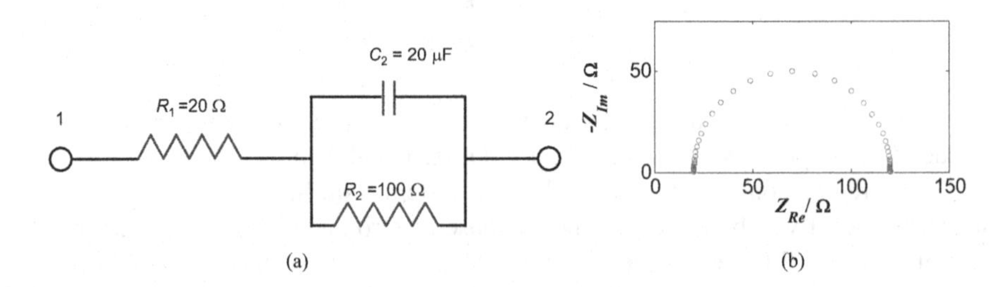

FIGURE 3.1 (a) Schematic of an electrical circuit comprising two resistors and a capacitor and (b) the complex plane plot of the corresponding impedance spectrum.

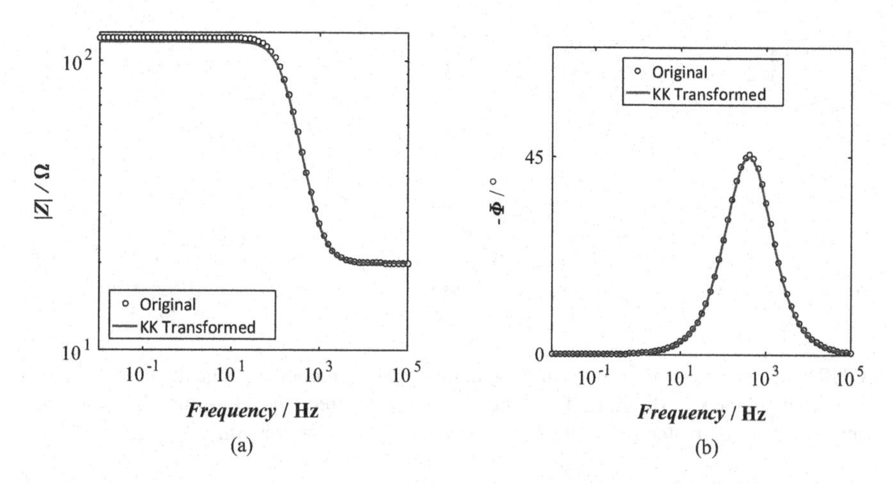

FIGURE 3.2 (a) |Z| *vs.* frequency and (b) Φ vs. frequency (original and KK transformed). The transformation was done using the real and imaginary components using the formula given in Eqs. 3.1 and 3.2 and the values are presented in Bode plots.

and angle (Φ) as a function of frequency are shown in Figure 3.2 as Bode plots. Note that the frequency and |Z| are shown in the log scale. The original and transformed data match well, as expected.

3.1.2 KKT Validation – Example of an Incomplete Spectrum

An implicit assumption in the direct integration of KKT is that data are available in a very wide frequency range, ideally from 0 to ∞, and at small frequency intervals. Many a time, data at low frequencies are not available, due to the long integration time required, or poor signal-to-noise ratio. The data analyzed in Figure 3.1 were truncated to 100 Hz–100 kHz and then subjected to KKT validation. It must be noted that the limit of 100 Hz is rather high, and EIS data are routinely obtained well below 0.1 Hz. This example is used only to illustrate the danger of attempting to validate incomplete data without suitable extrapolation. While applying KKT, the option of extrapolation beyond the frequency range of 100 Hz–100 kHz is not employed. The truncated data and the results are presented in Figure 3.3a as a complex plane plot and in Figure 3.3b and c as Bode plots. The data are flagged as not KKT compliant. While the user is encouraged to use the extrapolation option in the KKT software, there are limitations to adopting this method. The best solution is, of course, to obtain data at low frequencies and validate the data without extrapolation. Alternate methods, such as the measurement model approach and linear KKT, described in Section 3.5, will correctly flag this data as KKT compliant.

3.1.3 KKT Validation – Examples of Spectra of Unstable Systems

Experiments and simulations show that if the system has been unstable during EIS data acquisition, then KKT successfully flags the data as *noncompliant*. Urquidi-Macdonald, Real, and Macdonald (1990) measured the EIS of Fe in 1 M H_2SO_4 and

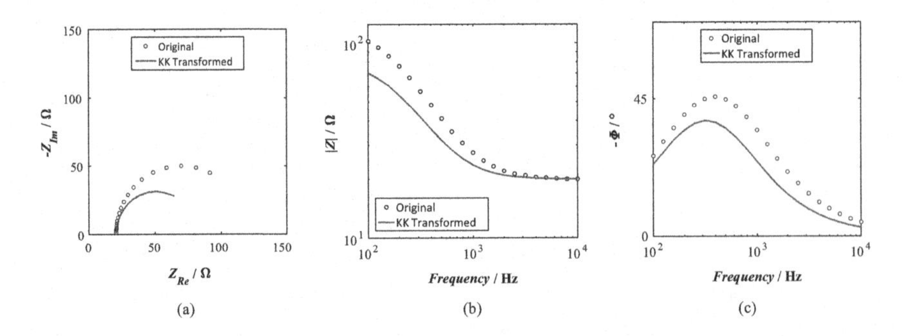

FIGURE 3.3 Issues with transforming data in a limited frequency range (truncated data) (a) the complex plane plot (b) |Z| *vs.* frequency and (c) Φ vs. frequency. Original data and transformed data do not match well, although the system is KKT compliant.

artificially introduced a drift in the system by superimposing a potential ramp of 0.133 mV s^{-1}. In the absence of a potential ramp, the complex plane plot of the data was a semicircle; however, when the potential ramp was introduced, the low-frequency data appeared more or less like a straight line, as shown in Figure 3.4a. The data were subjected to KKT, and the results, presented in Figure 3.4b and c, show that KKT flags the violation of stability. Victoria and Ramanathan (2011) simulated the effect of change in the dc potential on the EIS of a simple electron transfer reaction. Ideally, the EIS of a simple electron transfer reaction without any mass transfer limitations is expected to yield a semicircle in complex plane representation. However, when the potential drifts, the low-frequency data deviate significantly from the ideal result. Here, the instability is introduced in the form of a potential variation given by

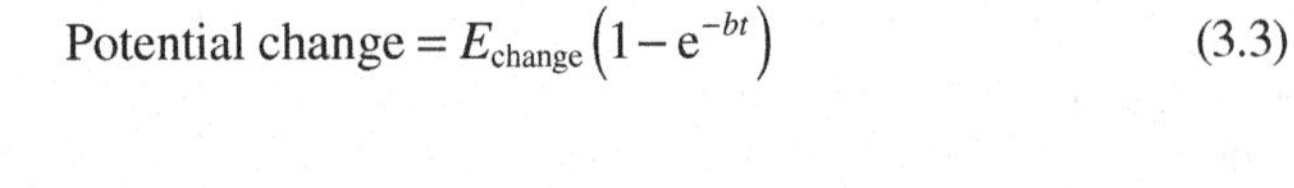

$$\text{Potential change} = E_{\text{change}}\left(1 - e^{-bt}\right) \tag{3.3}$$

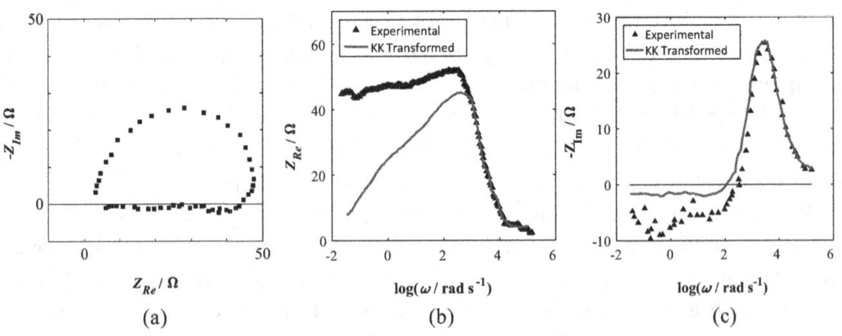

FIGURE 3.4 (a) Experimental results for Fe in 1 M H_2SO_4. EIS acquired in the presence of a ramp voltage of 0.133 mV s^{-1}. Corresponding KKT results showing (b) Z_{Re} *vs.* angular frequency and (c) $-Z_{Im}$ *vs.* angular frequency. (Adapted from Mirna Urquidi-Macdonald, Silvia Real, Digby D. Macdonald, Applications of Kramers–Kronig transforms in the analysis of electrochemical impedance data—III. Stability and Linearity. *Electrochimica Acta.* 35:1559–1566, Copyright (1990), with permission from Elsevier.)

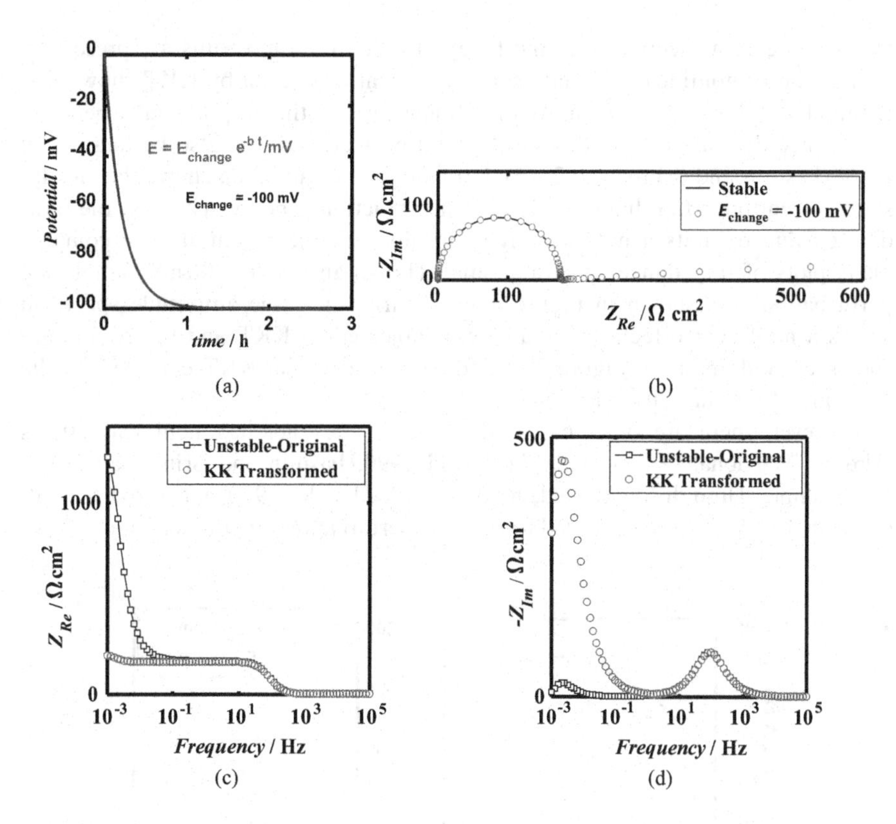

FIGURE 3.5 (a) A potential drift artificially introduced in the system. Complex plane plots of simulated impedance spectra, in the presence of a drift voltage given by "potential change" $= E_{\text{change}}\left(1-e^{-bt}\right)$, with $E_{\text{change}} = -100\,\text{mV}$ and $b = 1.3\times10^{-3}\,\text{s}^{-1}$. The spectrum corresponding to the unstable case was subjected to KKT. The original unstable results along with the corresponding KKT results are shown as (b) Z_{Re} *vs.* frequency and (c) $-Z_{\text{Im}}$ *vs.* frequency. (Adapted from Victoria, S. N., and S. Ramanathan. 2011. Effect of potential drifts and ac amplitude on the electrochemical impedance spectra. *Electrochimica Acta*. 56:2606–2615. Copyright (2011), with permission from Elsevier.)

where $b = 1.3\times10^{-3}\,\text{s}^{-1}$, i.e., the potential change is rapid in the beginning and is slow at later times, as shown in Figure 3.5a. The results are presented as complex plane plots in Figure 3.5b. The KK-transformed data and original data are shown in Figure 3.5c and d as Bode plots, and it is clear that the data are not KKT compliant. In general, KKT appears to be very successful in flagging violations of stability.

3.1.4 KKT Validation – Example of Spectra with Nonlinear Effects

Experimentally, a large amplitude perturbation, i.e., large E_{ac0} or I_{ac0}, can be used to acquire EIS data. If it differs from the EIS data acquired using small E_{ac0}, then we can conclude that the nonlinear effects manifest in the large E_{ac0} EIS data. In the case of simulations, it is possible to incorporate the nonlinear effects and calculate the EIS response of an electrochemical reaction. The details are described in a later section (Section 8.1), and here, only the relevant results are presented. Simulation of nonlinear EIS data and subsequent validation by KKT shows that if the plot of log$|i_F|$ *vs.* potential is nonlinear around the *dc* potential where the impedance data are acquired, then the nonlinear effects will also be flagged by KKT (Fasmin and Srinivasan 2015). For example, Figure 3.6a shows the steady-state current potential diagram of a faradaic reaction. The logarithm of the faradaic current exhibits a nonlinear relationship with the potential. The complex plane plots of impedance spectra at small (1 mV) and large (100 mV) amplitude perturbations are shown in Figure 3.6b. Clearly, the large amplitude spectrum shows a nonlinear effect. The data were subjected to KKT and the results are shown as Bode plots in Figure 3.6c and d. It is seen that KKT can identify the violation of the linearity criterion.

However, there are other cases where KKT does not flag nonlinear effects (Urquidi-Macdonald, Real, and Macdonald 1990, Fasmin and Srinivasan 2015). For example, Urquidi-Macdonald, Real, and Macdonald (1990) measured the EIS of Fe in 1 M H_2SO_4 at OCP and varied the perturbation amplitude (Figure 3.7a).

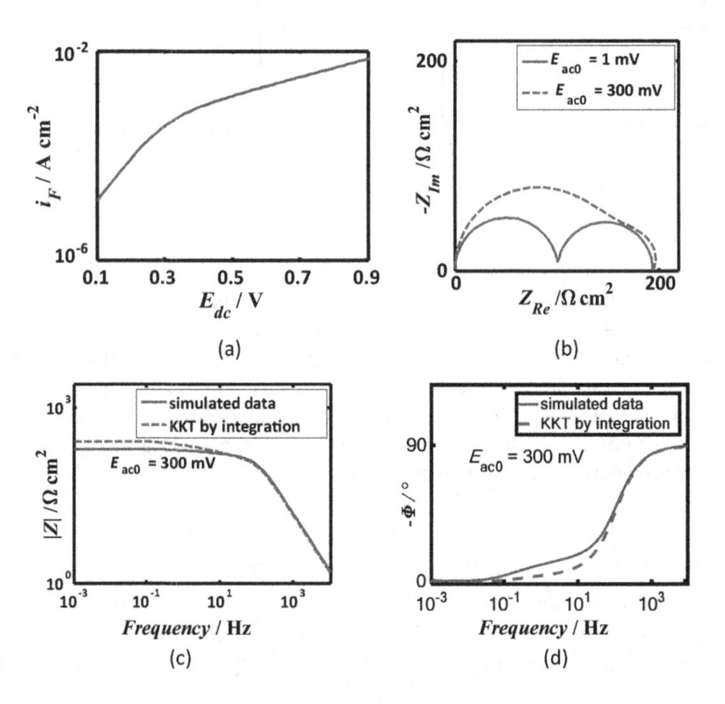

FIGURE 3.6 (a) Steady-state current potential diagram of another faradaic reaction. The logarithm of the faradaic current exhibits a nonlinear relationship with the potential. (b) Complex plane plots of impedance spectra at small and large amplitude perturbations. Bode plots of (c) |Z| vs. frequency and (d) $-\Phi$ vs. frequency of the nonlinear EIS data and the transformed data. (Adapted with permission from Springer Nature Customer Service Centre GmbH, Springer, Journal of Solid State Electrochemistry, Detection of nonlinearities in electrochemical impedance spectra by Kramers–Kronig transforms, Fathima Fasmin and Ramanathan Srinivasan, Copyright (2015).)

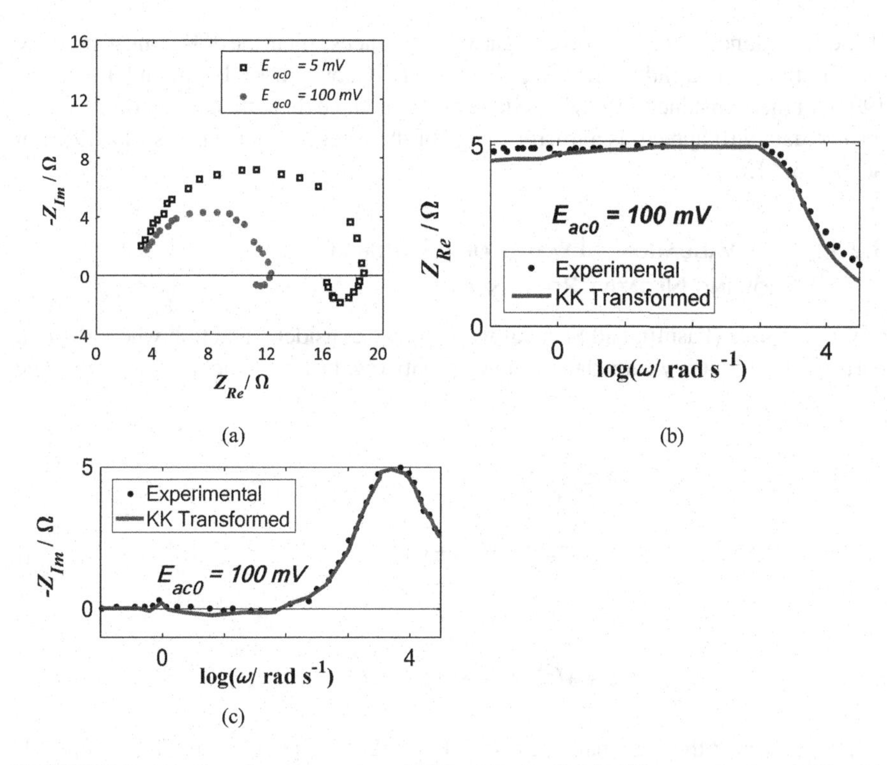

FIGURE 3.7 (a) Complex plane plot of EIS of Fe in 1 M H_2SO_4, acquired at OCP with E_{ac0} = 5 mV and E_{ac0} = 100 mV. Original and KK-transformed data showing (b) Z_{Re} vs. frequency and (c) $-Z_{Im}$ vs. frequency, corresponding to E_{ac0} = 100 mV. (Adapted from Mirna Urquidi-Macdonald, Silvia Real, Digby D. Macdonald, Applications of Kramers–Kronig transforms in the analysis of electrochemical impedance data—III. Stability and Linearity. *Electrochimica Acta*. 35:1559–1566, Copyright (1990), with permission from Elsevier.)

The complex plane plots show that the results are different at E_{ac0} = 5 and 100 mV, indicating that the nonlinear effects are present in the data acquired at E_{ac0} = 100 mV. However, a comparison of the original data at E_{ac0} = 100 mV and the corresponding KK-transformed data (Figure 3.7b and c) shows that KKT does not flag the violation of the linearity criterion. In summary, KKT, as a data validation tool, is sensitive to violations of causality and stability criteria but is not always sensitive to violations of linearity criterion.

3.2 TRANSFORMATION OF DATA IN IMPEDANCE *VS.* ADMITTANCE FORM

In 1993, **Gabrielli, Keddam, and Takenouti** showed that when the real part of the impedance has negative values, then KKT must be performed on the admittance form of the data rather than on the impedance data.

If the impedance values involve negative resistances, then the KKT must be done on admittance data rather than impedance data (Gabrielli, Keddam, and Takenouti 1993). In these instances, if KKT is done on impedance data, then a part of the transformed data will appear as a mirror image of the original data. This is illustrated in Section 3.2.1.

3.2.1 KKT Validation – Example of a Spectrum Showing Negative Resistance

In this example (Fasmin and Srinivasan 2015), we consider a reaction where the real part of low-frequency impedance shows negative values at some *dc* potentials. The reaction is

$$\begin{aligned} &M \underset{k_{-1}}{\overset{k_1}{\rightleftarrows}} M_{\text{ads}}^{+} + e^{-} \\ &M_{\text{ads}}^{+} \underset{k_{-2}}{\overset{k_2}{\rightleftarrows}} M_{\text{ads}}^{2+} + e^{-} \\ &M_{\text{ads}}^{2+} \xrightarrow{k_3} M_{\text{sol}}^{2+} \\ &M + M_{\text{ads}}^{2+} \xrightarrow{k_4} M_{\text{ads}}^{2+} + M_{\text{sol}}^{2+} + 2e^{-} \end{aligned} \tag{3.4}$$

and the details of the mass balance and charge balance equations are given in the reference. The impedance spectrum is presented as a complex plane plot in Figure 3.8a.

When the data in impedance form is subjected to KKT, the original and transformed data match well in the high and mid-frequency regime. Still, in the low-frequency regime, the transformed data appear as the mirror image of the original data (Figure 3.8b and c). In admittance form, the data will transform correctly. If the system is stable under potentiostatic mode, then KKT validation must be done on

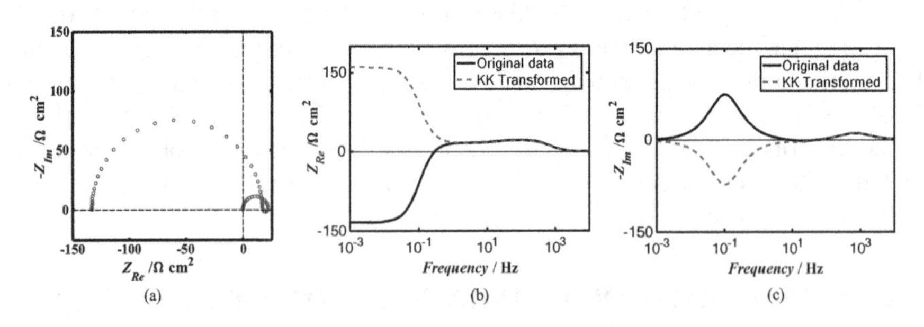

FIGURE 3.8 (a) Complex plane plot of the impedance spectrum for a catalytic mechanism. Note that the real part of low-frequency impedance shows negative resistance values. When KKT is applied on the impedance form of data, (b) Z_{Re} vs. frequency and (c) $-Z_{\text{Im}}$ vs. frequency appear as mirror images at frequencies below ~10 Hz. (Adapted with permission from Springer Nature Customer Service Centre GmbH, Springer, *Journal of Solid State Electrochemistry*, Detection of nonlinearities in electrochemical impedance spectra by Kramers–Kronig transforms, Fathima Fasmin and Ramanathan Srinivasan, Copyright (2015).)

the admittance form; if the system is stable under the galvanostatic mode, then KKT validation must be done in the impedance form. If the system is stable under both modes, then the validation can be done on either the admittance or the impedance form of the data (Gabrielli, Keddam, and Takenouti 1993).

If the impedance values are very close to zero, then they will have significant round-off errors since floating-point data are stored at finite precision. The corresponding admittance values will have large *relative errors*, and this can lead to a mismatch between the original and transformed data, even though the original data are KKT compliant. One way to alleviate this problem is to add a suitable resistance in series and then validate the data using KKT. This will reduce the impact of the numerical problems associated with inverting very small numbers. The results, presented in Figure 3.9a and b as Bode plots, show that except for a minor deviation of $|Y|$ in the low-frequency range, the original and transformed data do match well.

An alternative method, proposed by Prof. Boukamp (1995), is to add a suitably small resistance in parallel to the impedance data. This will cause the entire data set to have positive real values, as shown in Figure 3.10a. These new data can be validated directly in the impedance form, as shown in Figure 3.10b and c.

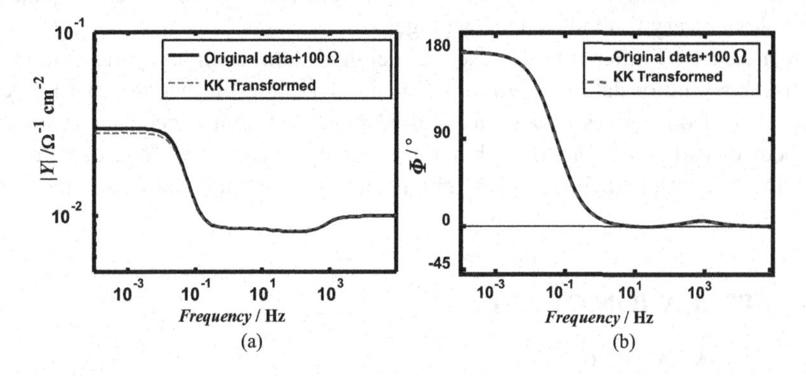

FIGURE 3.9 KKT was applied to the admittance form of data in Figure 3.6 and presented as Bode plots showing magnitude and phase (a) |Y| *vs.* frequency and (b) Φ vs. frequency. Note that the original and transformed data are practically indistinguishable.

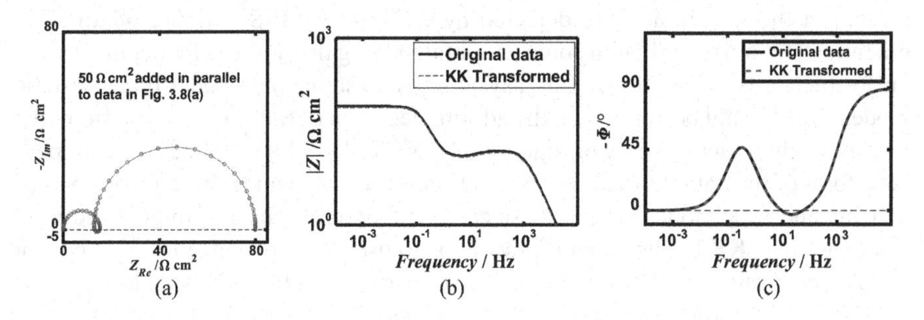

FIGURE 3.10 (a) Complex plane plots of impedance spectra obtained by adding after adding a parallel resistance of 50 Ω cm^2 to the data in Figure 3.8a. When KKT is applied to this data, the original and transformed data completely overlap and cannot be distinguished (b) |Z| *vs.* frequency and (c) Φ *vs.* frequency.

3.3 CHALLENGES IN KKT VALIDATION

In summary, the direct integration of KKT is an important data validation method, which can be used to validate the impedance data. As illustrated in the examples, the data validation must be performed with care. Sufficient care must be taken to avoid numerical errors cropping in, and the data must be validated in the appropriate (impedance or admittance) form. If the data are available only in a very limited frequency range, even if the data are obtained from a linear, causal, and stable system, KKT *may* flag the data as noncompliant. Although extrapolations can be performed, the best solution is to acquire data in a wider frequency range.

One problem in this validation methodology is that the assumptions are *sufficient conditions* and not *necessary conditions*. Thus, any violation of KKT indicates that one or more of the assumptions is violated. But compliance with KKT does not automatically imply that the system confirms all the assumptions. There are known cases where the linearity condition is violated, and yet the data were KKT compliant. Experimental and simulation results suggest that KKT is always sensitive to the presence of noise (causality) or instability (stability). Still, only in a limited number of cases, it is sensitive to nonlinear effects. Another problem is when the data are available in a very limited frequency range.

In the case of other types of electrochemical data, such as cyclic voltammograms, this kind of data validation tool is not available. In the case of EIS, KKT validation of data gives at least a certain amount of confidence that the system has been causal and stable, though not necessarily linear. Therefore, despite these challenges mentioned above, KKT remains an essential data validation technique for EIS.

3.4 APPLICATION OF KKT

3.4.1 Data Validation

If the data are not KKT compliant, then the usual culprits are noise or system instability. The third possibility is that the data are not acquired in a wide enough frequency range. The last and the least likely case is that the nonlinear effects are present in the system and are detected by KKT. Many EIS data are obtained in potentiostatic mode, although some are obtained in galvanostatic mode, in particular for fuel cells and batteries. For a system that is stable only under potentiostatic mode, KKT should be applied on the admittance form of the data. For a system that is stable only under the galvanostatic mode, KKT should be applied to the impedance form of the data. In practice, most systems are stable under both potentiostatic and galvanostatic modes, and in these cases, either admittance or impedance form can be used in KKT. The general practice is to use the impedance form. If part of the data contains impedance whose real part is negative, then one should apply the KKT on the admittance form of the data, or modify the impedance data by adding a small resistance in parallel so that the real values of the impedance are positive for the entire data set. Then KKT should be performed on the modified data in the impedance form.

3.4.2 Extrapolation

If one assumes that the system has indeed obeyed the rules of linearity, causality, and stability, but the data are acquired in a limited frequency range, perhaps due to some limitation in the experimental setup, then KKT can be used to *extrapolate* the data and estimate the impedance values at other, usually lower, frequencies (Esteban and Orazem 1991, Orazem, Esteban, and Moghissi 1991). We reiterate that extrapolation should always be performed with caution.

3.5 ALTERNATIVES TO DIRECT INTEGRATION OF KKT – MEASUREMENT MODELS

3.5.1 Introduction to the Measurement Model Approach

In 1995, **Pankaj Agarwal, Mark E. Orazem, and Luis H. Garcia-Rubio** demonstrated that the measurement model approach can be used as an alternative to KKT integration. In 1995, **Bernard Boukamp** presented linear KKT.

There are a few practical issues in applying KKT. In many instances, it may not be possible to acquire data over a wide enough frequency range. Often, noise at the low-frequency range forces us to limit the frequency range in which the results are reproducible. If the frequency range is not wide enough, then the application of KKT directly can be challenging (Figure 3.3). Ideally, we should acquire data at small frequency intervals, i.e., many frequencies in a decade, especially if the impedance varies rapidly with frequency. This is necessary so that the numerical integration can be performed with reasonable accuracy. However, it will increase the time needed to acquire the data, and the system may not be stable over a long time. In some cases, these constraints make it difficult to apply the KKT method directly to the impedance or admittance data.

An alternative approach is to fit the data to an equivalent electrical circuit (EEC), with passive elements. Specifically, one can use circuits containing resistances, capacitances, and inductances (Agarwal et al. 1995, Agarwal, Orazem, and Garcia-Rubio 1992, 1995, Shukla, Orazem, and Crisalle 2004). An electrical circuit with passive elements will generate data that are inherently KKT compliant. Therefore, if an electrical circuit with passive elements can model the EIS data 'very well,' then we can conclude that the data are KKT compliant. This method is called the *measurement model approach*.

Now, we have to define what 'very well' means. In the measurement model approach, a Voigt circuit, as shown in Figure 3.11, can be used to fit the data. A resistor in parallel with a capacitor is called a Voigt element. The number of Voigt elements can be increased, and the model fit can be improved, i.e., the difference between the model data and the experimental data (residues) can be reduced. Beyond a certain point, increasing the number of Voigt elements will not result in a significant improvement in the model fit. At this stage, the residues should be plotted as a

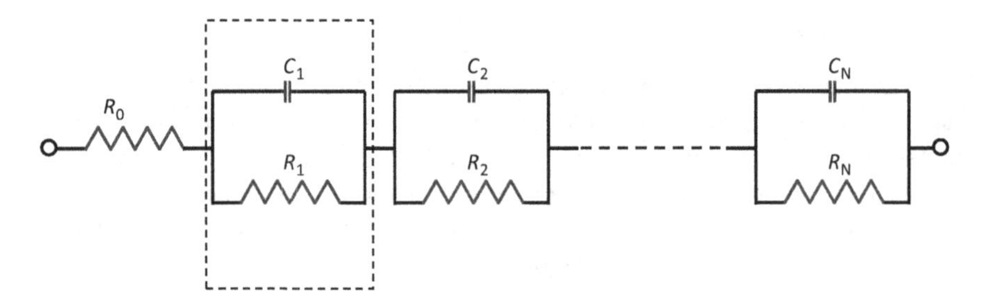

FIGURE 3.11 Example Voigt circuit. A pair of capacitance and resistance in parallel, shown in the dashed rectangle, is a Voigt element, and N such elements can be arranged in series. R_0 can represent the solution resistance.

function of frequency. If they are randomly distributed across the frequencies and are within the expected noise level, then the fit is considered good. Then the data are certainly KKT compliant.

Although the Voigt circuit consists of only resistance and capacitance, it can also be used to model data that arise from circuits containing inductance or other elements such as Warburg impedance, which corresponds to diffusion effects, or CPE (Agarwal, Orazem, and Garcia-Rubio 1992). The nature of the impedance of these elements is described in detail in Chapter 6. To model the data arising from the inductor, a Voigt element (resistance and capacitance) should be allowed to take negative values. The data arising from Warburg impedance or CPE with a positive exponent can be modeled with the Voigt circuit with only positive resistance and capacitance if a sufficient number of elements are used.

On the other hand, if the residues at the 'best fit' conditions are large and/or show a clear trend *vs.* frequency, then we can conclude that a Voigt circuit cannot successfully model the data. It does not automatically imply that the data will not be KKT compliant; perhaps, we did not choose the circuit elements correctly or determine the best-fit parameter values for the given circuit elements. However, if we ensure that the model fit has been done correctly, then we can conclude that most likely, the data are not KKT compliant.

It must be noted that the measurement model is only a data validation technique. A good fit does not automatically imply that a Voigt circuit with N elements can be correlated with the underlying physical process. Likewise, instead of Voigt circuits, other circuits such as Maxwell circuits (Figure 3.12a) or Ladder circuit (Figure 3.12b) can also be employed for data validation by the measurement model approach. For data validation, any one of the circuits shown in Figure 3.11 or 3.12 can be employed.

3.5.2 Advantages of the Measurement Model Approach

To fit the electrical circuit model to the data, it is not necessary, although it is certainly desirable that the data be available in a very wide frequency range. It is also not necessary, but again desirable that the data be acquired in a closely spaced set of frequencies. Usually, acquiring data at seven or more frequencies per decade, logarithmically spaced, is considered sufficient for analysis and KKT integration. If the

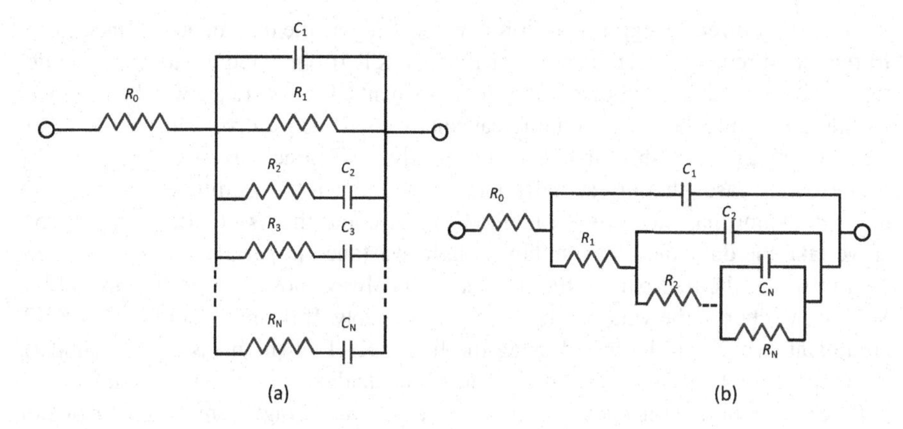

FIGURE 3.12 (a) Example Maxwell circuit. (b) Example ladder circuit.

frequencies are very widely spaced, e.g., three frequencies per decade, then the direct integration will not be accurate. In the measurement model approach, frequencies may be widely spaced. These are clear advantages over traditional direct integration of the KKT method.

3.5.3 Linear KKT

Another approach that is similar to the measurement model approach is referred to as linear KKT (Boukamp 1995, 2004). Here also, electrical circuits with Voigt elements are used. However, instead of allowing complete freedom in choosing the value of R and C in the Voigt elements, the product of R and C, called the *time constant*, is restricted. The method is described below.

The advantage of linear KKT over the measurement model analysis is that complex nonlinear least square (CNLS) fit is not required, and only linear equations need to be solved. This means that there is no need to choose any starting values. In modern computes, both CNLS fit and linear fit are quite fast, but occasionally CNLS fit may converge to a local minimum; i.e., the parameter values will not be optimal. The user may conclude, incorrectly, that the model is not good. This problem is alleviated in the linear KKT method.

Both measurement model analysis and linear KKT suffer from the following problem. If the data are KKT compliant but, for whatever reason, cannot be modeled using Voigt elements, then the residues can be large and may appear to show a systematic trend, leading the user to believe that the data are actually not KKT compliant. The direct integration method, on the other hand, does not attempt to fit the data to any particular model and is free from this deficiency. However, this is more of a theoretical issue. To the best of the authors' knowledge, any system that is flagged as noncompliant by the measurement model violates one or more of the linearity, causality, and stability criteria.

There are some occasions when the real part of the impedance shows negative values (e.g., data in Figure 3.8a). In those cases, the following approach should be used.

If we employ direct integration of KKT, we should use the data in admittance form. In the measurement model analysis or linear KKT, if one attempts to fit the model to the data, then at least in one of the Voigt elements, the resistance will be negative. At the same time, the corresponding capacitance will be positive. This would mean that the system is unstable, and hence inherently KKT noncompliant. This would be true even for cases that are actually KKT compliant in the admittance form of the data. Boukamp has suggested a method to overcome this issue (Boukamp 1995). If we take the data and add a suitably small resistance in parallel, we can get new data in which the real part of the impedance is always positive (e.g., Figure 3.10a). We have to choose the value of the resistance suitably. If the new set of data is KKT compliant (which can be tested using the linear KKT or the measurement model), then the original data are KKT compliant. While analyzing the new set of data, if we still need to employ a negative time constant (i.e., in a Voigt element, either one of the R or C is negative while the other is positive), we can conclude that the data are not KKT compliant.

3.5.4 Summary

If you have data that are acquired in a wide range of frequencies and with seven or more frequencies per decade, you should attempt direct integration of KKT. What qualifies as a 'wide range of frequencies'? At the high-frequency end, the impedance should approach a finite real value. For many aqueous electrochemical systems, it is in the range of 10^4–10^5 Hz. At the low-frequency end, the impedance should either approach a finite real value or show a constant phase element behavior (where the real and imaginary values tend to infinite as the frequency approaches zero). For many systems, it is at least 0.1 Hz or lower. Then we can consider that the frequency range is wide enough. In the latter case (where low-frequency data show the CPE behavior), the direct integration of KKT equations must extrapolate the data. In addition to the direct integration of KKT, one can also employ the *linear KKT* or measurement model approach to validate the data. In some cases, the data may not be available in a wide range of frequencies and/or may be acquired in sparse mode (i.e., <7 frequencies per decade). Then, the linear KKT or measurement model approach would be the only suitable choice of data validation.

3.6 EXPERIMENTAL VALIDATION METHODS

In addition to the above methods, it is possible to experimentally check for the linearity and certain type of stability issues. The linearity assumption can be checked by varying the perturbation amplitude (e.g., E_{ac0} of 5, 10, 15, and 20 mV) and comparing the spectra. If the spectra change significantly beyond a certain amplitude, then we can conclude that nonlinear effects manifest beyond up to that amplitude, and the spectra acquired using a smaller amplitude can be assumed to be from a linear system. For potentiostatic measurements, a guideline value of 10 mV (*rms*) is commonly used.

To check if the system has been stable, at the end of one experiment, one should reacquire the data, and if both data sets match, at least some forms of instabilities can be ruled out. (See, for example, Kaisare et al. (2011) for an illustration of instability that cannot be ruled out by this comparison.) To check for causality, often what is done is to plot the data in different forms and see if it *appears* to be noisy or not. This *chi by eye* is obviously not rigorous. The correct method is to acquire data at least twice under identical conditions and check for repeatability.

Also, the low-frequency limit of impedance must be equal to the inverse of the slope of the potentiodynamic polarization curve. The low-frequency limit of the faradaic impedance is called polarization resistance (R_p), and this is described in detail in Section 4.4.4. The low-frequency limit of the total impedance is the sum of solution resistance (R_{sol}) and polarization resistance (R_p). When the solution resistance (R_{sol}) is negligible, the low-frequency limit of the total impedance is the same as R_p. This value can also be estimated from the slope of the potentiodynamic polarization curve, i.e., $\left(R_{sol}+R_p\right)^{-1}=\left({}^{\mathrm{d}i}\!/\!_{\mathrm{d}E}\right)$. Within the experimental noise limits, both estimates of R_p should match well.

3.7 SOFTWARE

To verify KKT compliance, evaluating the integral is the direct method, but other methods such as the measurement model or linear KKT have also been proposed. Spectra from any electrical circuit containing only passive elements such as resistance, capacitance, and inductance are automatically KKT compliant. Therefore, if the data can be fit well into such a circuit, we can be sure that it is KKT compliant. We find that direct integration is good enough in most of the cases that we have analyzed. We have created a simple software, written in visual basic, to check for KKT compliance. It uses an approximate integration method (trapezoidal integration), and it is freely downloadable from the website http://www.che.iitm.ac.in/~srinivar/Software.html. Commercial impedance acquisition software such as ZSimpWin® (Princeton Applied Research) and ZView® have the facility to test impedance data for KKT compliance. The linear KKT data validation software, for Windows® operating system, is available for free at the website https://www.utwente.nl/en/tnw/ims/publications/downloads/.

It is worth noting that KKT is a purely mathematical relationship, and its development does not involve any phenomenon related to electrochemistry. It has applications in other fields, and it is frequently used in optical materials research. A dielectric constant is a complex number, which depends on frequency. Sometimes, experimentally only one part, i.e., real or imaginary of the dielectric constant, may be measured over a wide frequency range, and KKT can be used to estimate the other part.

3.8 EXERCISE – IMPEDANCE DATA VALIDATION USING KKT

Q3.1 In Eqs. 3.1 and 3.2, how can we handle the case when $x=\omega$?

Q3.2 (a) Generate impedance data for the circuit given in Figure 3.1a, in the frequency range of 1 mHz–1 MHz. (b) Apply the integration given in Eqs. 3.1 and 3.2 to calculate KK-transformed data, assuming that the lower and upper limits can be approximated by 1 mHz and 1 MHz, respectively.

Q3.3 (a) Select a few journal publications in the last 5 years, which show impedance spectra in the Bode format. Digitize the plots and generate the impedance data points and the frequencies. (b) Validate the above impedance data using any KKT software.

Q3.4 Use the data generated in Q3.2(a) but limit the frequency range to 10 Hz–100 kHz and validate using *linear KKT* software and one more software that uses direct integration of KKT equations. Describe your observations and explain the results.

Q3.5 Reanalyze the data chosen in Q3.4 using linear KKT software. Now, change the values of 'τ' and rerun the validation code. What happens to the residual when the value is (a) increased and (b) decreased?

Q3.6 Digitize the 'original' data given in Figure 3.8. Validate the data using linear KKT software in (a) impedance mode and (b) admittance mode. Plot the residuals as a function of frequency. What conclusions can be drawn from these results?

Q3.7 Consider the circuit shown in Figure 3.1a. Generate the impedance in a frequency range of 1 mHz–1 MHz, seven frequencies per decade, logarithmically spaced, for $R_1 = 10\ \Omega$, $C_2 = 10^{-5}$ F, and $R_2 = -100\ \Omega$. Plot the complex plane plot. Validate the data using a suitable KKT program.

Q3.8 Find a suitable resistance value that can be added in parallel to the above circuit such that the data fall in the first quadrant in the complex plane plot. Validate the data using a suitable KKT program.

Q3.9 Digitize the experimental data shown in Figure 3.4b and c. Validate the data using a suitable KKT program.

Q3.10 Repeat the above process using the data shown in Figures 3.5–3.7.

4 Data Analysis – Equivalent Electrical Circuits

There is nothing more deceptive than an obvious fact.

—**Arthur Conan Doyle**

First, the method of using an equivalent electrical circuit (EEC) to model the data is described. Then the issue of distinguishability of different circuits is discussed, and the other limitations of EEC analyses are presented.

4.1 EQUIVALENT ELECTRICAL CIRCUITS: WHAT CIRCUIT TO CHOOSE?

Although the emphasis of this book is on the electrochemical impedance spectroscopy (EIS) of *reactions*, at first, we introduce only electrical circuits and analyze the impedance spectra of those circuits. This is because it is easy to (a) simulate the spectra of electrical circuits comprising resistors, capacitors, and inductors, as we have seen in Chapter 1 and, (b) given an experimental spectrum and an electrical circuit model, obtain the 'best' estimate of electrical circuit parameter values so that the simulated and 'experimental' data match well. After becoming familiar with these ideas using electrical circuits model the data, when the equations are developed to obtain impedance of reactions, it will be easier to understand RMA methodology.

Recall (Section 1.8) that it is possible to analyze the EIS data using the EEC model and determine the parameter values of a circuit. There, we assumed that we know the correct circuit to use and that we only did not know the values of the resistances and capacitance. In that case, one can guess the initial parameter values and then simulate the spectrum and compare it with the *experimental* data. The values of the parameters can be adjusted by a suitable optimization program so that the simulated and experimental spectra match. In a more general scenario, we first need to identify a circuit that is suitable to model the data. Only then does it make sense to estimate the values of the parameters, i.e., the resistances and capacitances, which would minimize the error between the simulated and 'experimental' data. This leads to the following question: How do we arrive at a suitable circuit to model a given data?

4.1.1 What Circuit to Choose?

First, we have to examine the data in the complex plane plot (sometimes called a Nyquist plot) and choose a circuit that can be evaluated first. If the plot of $-Z_{im}$ *vs.* Z_{re} has one loop as in Figure 4.1a, then it needs at least one capacitor and one resistor in parallel. If the impedance at high frequency is not zero, then a resistor is added to this circuit to model the solution resistance (Figure 4.1b).

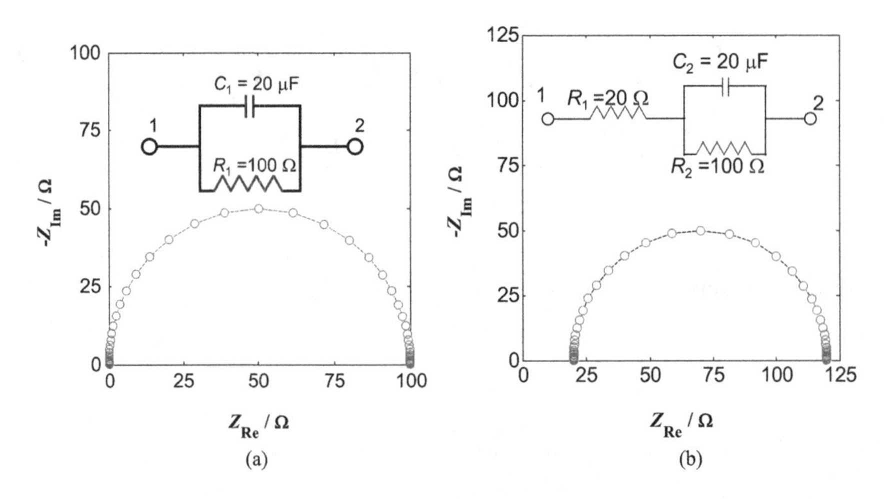

FIGURE 4.1 Sample complex plane plots of impedance spectra (a) without solution resistance (b) with solution resistance. The insets show an electrical circuit that can be used to model the data.

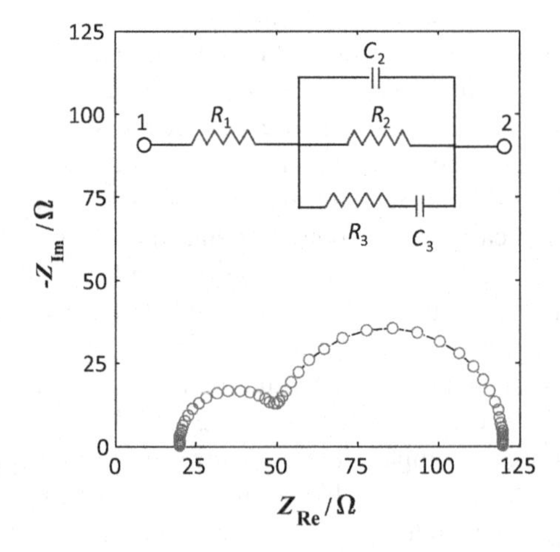

FIGURE 4.2 Sample complex plane plot on an impedance spectrum showing two loops. The inset shows a circuit that can be used to model the data.

If it has two loops as in Figure 4.2, it would need a circuit that has a capacitor in parallel with a resistor, and then a pair of resistor–capacitor that is in parallel with this, as shown in the inset of Figure 4.2. If the complex plane plot has three loops, as in Figure 4.3, it would need at least three capacitors, as shown in the inset of Figure 4.3, and so on. In summary, the first capacitor is in parallel with a resistance, but the second, third, and subsequent capacitors are always drawn in series with a resistance, and this pair (of resistor–capacitor) can be arranged in parallel with the first capacitor and resistor. The total number of resistors/capacitors needed would equal the total

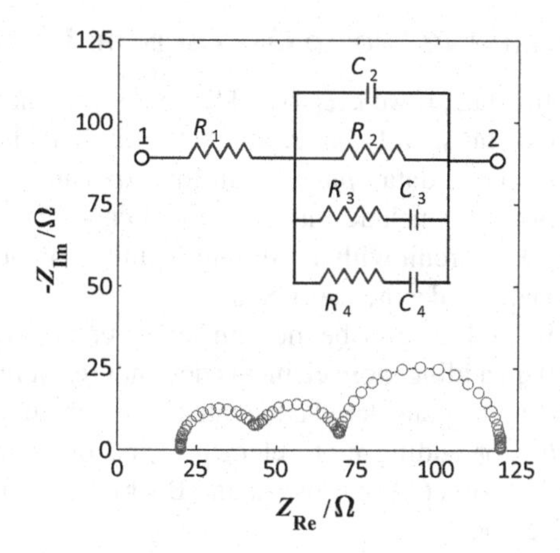

FIGURE 4.3 Sample complex plane plot of impedance spectrum, exhibiting three loops. The inset shows an electrical circuit that can model the data.

number of loops. The high-frequency limit resistance is always added in series to the entire electrical circuit model.

There are some challenges in using the above methodology solely based on visual inspection of the plots of the data. Sometimes, two loops will merge and appear as one loop. This depends on the actual values of the capacitances and resistances involved. The product of the resistance (*R*) and the corresponding capacitance (*C*) is also called as time-constant ($\tau = R \times C$); if two time-constants (τ_1 and τ_2) are close enough, the corresponding loops will tend to merge, and it will be difficult to identify the individual loops by visual inspection. For example, the two plots in Figure 4.4 were generated from the same circuit shown in Figure 4.2, but with different values of C_2. The filled circle clearly shows two loops, whereas the open squares appear to be a single deformed loop. In these cases, initially, the user may employ a simple model with one capacitor. However, if it does not fit the data well, then the user has to add another resistor–capacitor pair and verify if the fit is considerably better.

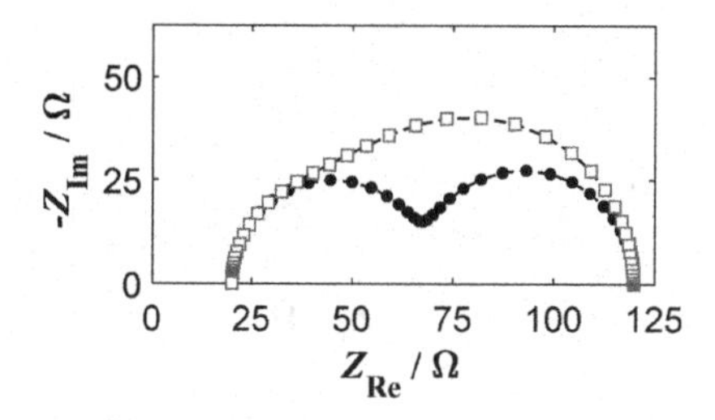

FIGURE 4.4 Example of complex plane plots of impedance spectrum arising out of the circuit shown in Figure 4.2. The closed circles arise from a circuit with two well-separated time constants while the open squares arise from a circuit with similar time constants.

4.1.2 How Many Elements Should One Use in the Electrical Circuit?

While modeling real-life data, if we keep on adding more and more resistor–capacitor pairs, one may obtain a statistically better and better fit, but when should one stop? Many times, if the simulated data *appear* to fit the experimental data to the naked eye, the practitioner would conclude that the model captures the *major processes* adequately. In that case, a circuit with a minimum number of elements that can reasonably fit the data would be deemed sufficient.

The term *residue* is used to describe the sum square error between the model and experimental data. When adding more elements does not result in a significant reduction in residue, the fit can be considered adequate to describe the data. Alternatively, one can use a *penalty* for adding more elements, and one such method uses the Akaike Information Criterion (AIC) (Posada and Buckley 2004). The expression for calculating AIC is given below.

$$\text{AIC} = 2k + n\left[\ln(\text{RSS})\right] \tag{4.1}$$

where n is the total number of observations and k is the total number of independently adjusted parameters in the given model. Here RSS denotes the residual sum squared value. Adding more elements will lead to an increase in AIC, while it will also tend to reduce the residue. The model corresponding to the minimum in AIC can be taken as the most suitable model.

When the number of observations is few ($n/k < 40$), a modified version of the AIC (called AIC with correction) is used (Hurvich and Tsai 1989), and it is given by

$$\text{AIC}_c = \text{AIC} + \frac{2k(k+1)}{n-k-1} \tag{4.2}$$

Now let us consider an example, where *experimental* data are actually numerically synthesized; it is indeed created using the electrical circuit given in Figure 4.2, with up to 5% random noise added. The data are presented in Figure 4.5. We used the commercial software Zsimpwin® (Princeton Applied Research, AMETEK, USA) to fit various circuits shown in Figures 4.1b, 4.2, and 4.3 to the data, and the results are shown in Table 4.1.

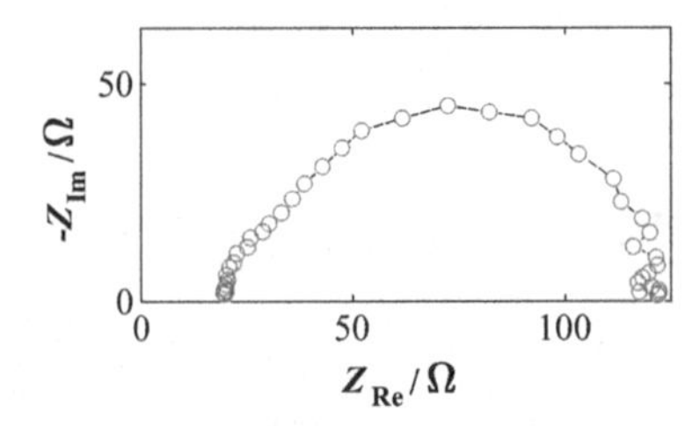

FIGURE 4.5 Complex plane plot of 'experimental' data, which is actually simulated from the circuit in Figure 4.2 and (up to 5%) random noise is added. The total number of data points is equal to 41. The circuits in Figures 4.1b, 4.2, and 4.3 are fit to the data. The results are summarized in Table 4.1.

TABLE 4.1
Best Fit Values of Elements in Various Circuits for the Data Shown in Figure 4.5

Element	Figure 4.1b	Figure 4.2	Figure 4.3
R_1 (Ω)	22.4 (<3%)	19.9 (<1%)	19.97 (<1%)
R_2 (Ω)	95.2 (<3%)	100.6 (<1%)	101 (<4%)
C_2 (μF)	16.6 (<5%)	10 (3%)	10 (<3%)
R_3 (Ω)	-	52.2 (<6%)	52.1 (<6%)
C_3 (μF)	-	9.8 (<3%)	9.8 (<3%)
R_4 (Ω)	-		23,370 (~767%)
C_4 (μF)	-		7.5 (~1704%)

The values in the parenthesis are the relative errors. The large relative errors for the two elements in Figure 4.3 indicate that they are not reliable estimates.

The relative error values in the parameters are given in the parenthesis. A large relative error indicates that one cannot be confident that the estimated value is accurate. When the circuit in Figure 4.1b was chosen, the fit was poor. The circuit in Figure 4.2 gave a good fit. While the circuit in Figure 4.3 also gave a good fit, the relative errors in two parameters (in this case, R_4 and C_4) are very large, and these values are not reliable. What has happened is this: the circuit in Figure 4.2 is sufficient to model the data, and the addition of one more *RC* pair does not enhance the fit. Any large value for resistance R_4 will essentially make it an open circuit, and it is as if the elements R_4 and C_4 were not present at all. Likewise, an extremely small capacitance value for the element C_4 will again make the impedance very large and hence equal to an open circuit.

Sometimes, depending on the software used and constraints employed, it may not be possible to get the relative error values in each of the parameters estimated. Then, AIC can be used to choose the circuits. Once a model fit is obtained, the residues, calculated as sum-squared errors, can be determined. Then using the number of parameters in the model and the residue, AIC_c can be calculated. As the residue decreases, the AIC_c value decreases. On the other hand, as the number of parameters increases, the AIC_c value increases. The best model is chosen based on the minimum of AIC_c values. The values for the model fit to the data in Figure 4.5 are shown in Table 4.2.

TABLE 4.2
Residue and AIC Value of Various Model Fits to Data in Figure 4.8

Circuit in Figures	# of Parameters (k)	Residuals	AIC	AIC_c
Figure 4.1b	3	1196	296.6	297.2
Figure 4.2	5	76	187.3	189.0
Figure 4.3	7	76	191.5	194.9

The AIC_c values in Table 4.2 show that the model corresponding to Figure 4.2 has the lowest AIC_c among the choices evaluated and is the most suitable model as per this criterion. Whenever one attempts to select the best model among many choices, one should also keep in mind the repeatability and the noise level in the data. In all the cases, ascribing a physical meaning to the circuit elements is very important, for only then can we get a better understanding of the actual system. Otherwise, the data fitting to the electrical circuit become a mere exercise with no tangible outcome.

4.2 DISTINGUISHABILITY

Even if a simple circuit fits the data accurately, this circuit choice is not exactly unique. For example, data shown in Figure 4.2 can also be modeled using three other circuits, as shown in Figure 4.6. All four circuits in Figure 4.6 are completely interchangeable. That is, if the values of the elements (R, L, and C) of one circuit is known, then we can calculate the values of elements of other circuits such that they will produce the exact same spectrum in the entire frequency range $[0\ \infty]$. The only downside is that negative values are allowed for the R, L, and C. There is a unique relationship between the values of the elements in these circuits, as shown below.

The relationship between the group R_1, L_1, and R_2 and the group R_3, C_3, and R_4 is given below.

$$R_3 = -R_1, C_3 = -{L_1}/{R_1^2} \text{ and } (R_4)^{-1} = (R_1)^{-1} + (R_2)^{-1} \tag{4.3}$$

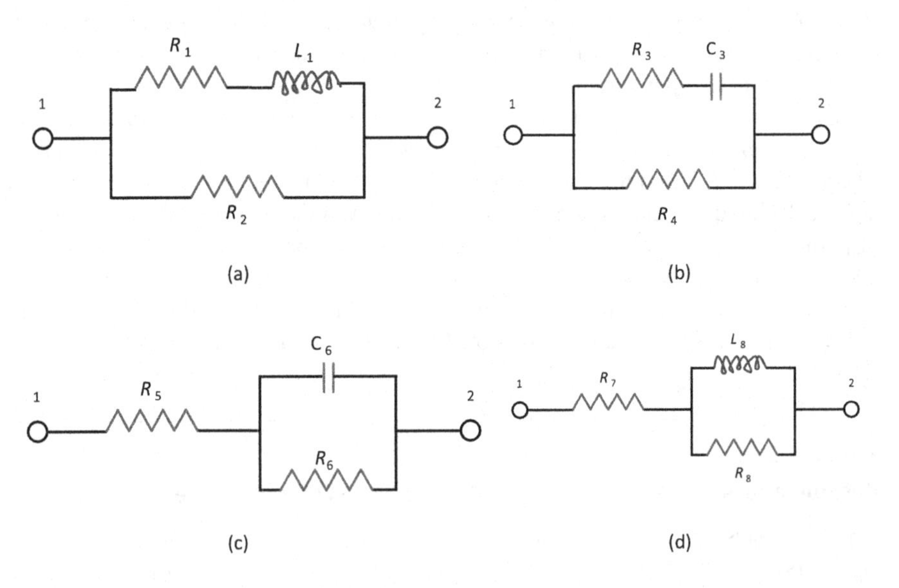

FIGURE 4.6 Four equivalent electrical circuits, which cannot be distinguished in the entire frequency domain. (a) Maxwell with an inductor. (b) Maxwell with a capacitor. (c) Ladder with a capacitor. (d) Ladder with an inductor.

The inverse relationship can be written as

$$R_1 = -R_3, L_1 = -C_3 R_3^2 \text{ and } (R_2)^{-1} = (R_3)^{-1} + (R_4)^{-1} \tag{4.4}$$

The admittance of the first circuit (Figure 4.6a) is given by

$$Y_{4.6a} = \frac{1}{R_2} + \frac{1}{R_1 + j\omega L_1} \tag{4.5}$$

The admittance of the second circuit (Figure 4.6b) is given by

$$Y_{4.6b} = \frac{1}{R_4} + \frac{1}{R_3 + \left(\frac{1}{j\omega C_3} \right)} = \left(\frac{1}{R_1} + \frac{1}{R_2} \right) + \frac{1}{-R_1 + \left(\frac{-R_1^2}{j\omega L_1} \right)} \tag{4.6}$$

$$= \frac{1}{R_1} + \frac{1}{R_2} - \frac{1}{R_1} \left(\frac{j\omega L_1}{R_1 + j\omega L_1} \right) = \frac{1}{R_2} + \frac{1}{R_1 + j\omega L_1} = Y_{4.6a} \tag{4.7}$$

Similarly, the relationship between the group R_3, C_3, and R_4 and the group R_5, R_6, and C_6 is given below.

$$(R_5)^{-1} = (R_4)^{-1} + (R_3)^{-1}, R_6 = \frac{R_4}{1 + \frac{R_3}{R_4}} \text{ and } C_6 = C_3 \left(1 + \frac{R_3}{R_4} \right)^2 \tag{4.8}$$

The inverse relationship can be written as

$$R_3 = R_5 \left(1 + \frac{R_5}{R_6} \right), C_3 = C_6 \left(1 + \frac{R_5}{R_6} \right)^{-2} \text{ and } R_4 = R_5 + R_6 \tag{4.9}$$

Likewise, the relationship between the group R_5, C_6, and R_6 and the group R_7, R_8, and L_8 is given as

$$R_7 = R_5 + R_6, R_8 = -R_6 \text{ and } L_8 = -C_6 R_6^2 \tag{4.10}$$

The inverse relationship can be written as

$$R_5 = R_7 + R_8, C_6 = \frac{-L_8}{R_8^2} \text{ and } R_6 = -R_8 \tag{4.11}$$

For example, the simulated data in Figure 4.2 can be obtained from the circuits in Figures 4.6a or b or c or d, with the values given in Table 4.3. The reader can verify that they yield the same spectrum. A simple utility, which allows one to calculate the values, can be downloaded from the web site (http://www.che.iitm.ac.in/~srinivar/Software.html).

TABLE 4.3
Electrical Element Values for the Circuits in Figure 4.6, Which Yield Identical Spectra

Figure 4.6a			Figure 4.6b		
R_1 (Ω)	L_1 (H)	R_2 (Ω)	R_3 (Ω)	C_3 (F)	R_4 (Ω)
−25	-5×10^{-3}	20	25	8×10^{-6}	100
Figure 4.6c			**Figure 4.6d**		
R_5 (Ω)	R_6 (Ω)	C_6 (F)	R_7 (Ω)	L_8 (H)	R_8 (Ω)
20	80	1.25×10^{-5}	100	−0.08	−80

As a more general example, a list of equivalent circuits (comprising up to six elements) and the relationship between those circuits, are given by Fletcher (Fletcher 1994).

Now, in the case of a simple electrochemical reaction, we would choose Figure 4.6c to model the system because we can assign the solution resistance to R_5, double-layer capacitance to C_6, and the faradaic impedance to R_6. This is physically intuitive and conforms to our understanding of the system. We can model the data equally well with the circuit in Figure 4.6a or b, but assigning physical meaning to the elements is not easy. Thus, among many equivalent circuits that model the data equally well, which have the same number of elements *and which are not distinguishable at all even in theory*, we should choose the circuit that gives a physically meaningful picture and enhances our understanding of the system.

4.2.1 Equivalent Circuits

It must be noted that the equivalency of the circuits given in Figure 4.6 is applicable even when we consider it as a part of a larger circuit. For example, consider the circuit given in Figure 4.7a. It is the same circuit shown in Figure 4.2.

Now the part enclosed in the dashed rectangle is the same as the one given in Figure 4.7b, i.e., it has a resistor in series with a capacitor, and this segment is connected in parallel with a resistor. This enclosed part can be replaced by the circuit in

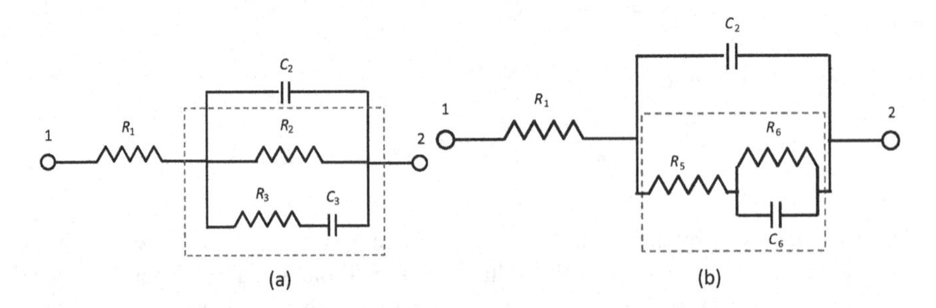

FIGURE 4.7 Illustration of two equivalent circuits, with faradaic impedance given in (a) Maxwell representation and (b) Ladder representation. The parts in the dashed rectangle can be interchanged using the equivalence relationship.

Figure 4.6c, i.e., a resistor in parallel with a capacitor, and this segment is connected in series with a resistor. The entire segment is given in Figure 4.7b.

Note that the other elements can be left as such, and the formula to calculate the equivalent element values needs to be applied only on the segment that is replaced. Other examples can be found in http://www.che.iitm.ac.in/~srinivar/EIS-Book.html.

4.3 ZEROS AND POLES REPRESENTATION

It is possible to represent a circuit uniquely using the zeros and poles of the impedance written as a function of angular frequency (Sadkowski 1999, 2004). For example, the impedance equation of the data shown in Figure 4.6c is written as

$$Z(\omega) = R_5 + \frac{1}{1/R_6 + (j\omega C_6)} = R_5 + \frac{R_6}{1 + j\omega C_6 R_6} = \frac{R_5 + R_6 + j\omega C_6 R_6 R_5}{1 + j\omega C_6 R_6} \quad (4.12)$$

This equation has a zero and pole respectively at

$$\omega = -\frac{R_5 + R_6}{jC_6 R_5 R_6} \text{ and } \omega = \frac{-1}{jR_6 C_6} \quad (4.13)$$

This is unique and does not depend on the type of EEC employed. That is, if we employ the values given in Table 4.3 for R_5, R_6, and C_6, the zero occurs at $\omega = 5000j$, and the pole occurs at $\omega = 1000j$. Even if we use the other circuits in Figure 4.9, as long as we employ the corresponding element values, the zero and pole values will remain $5000j$ and $1000j$, respectively. The disadvantage of the zero/pole representation is that it is not easy to relate the values of the zeros and poles to physical quantities.

The reader would have noticed that we have employed negative values for resistances and inductances in Table 4.3. We can also generate plots with negative capacitance values. Now, what do the negative resistances and capacitances (or inductances) mean, physically? In Figure 4.6, we can avoid that question by claiming that only the circuits shown in Figure 4.6b or c are the appropriate circuits to represent the data, and the other two circuits (shown in Figure 4.6a or d) do not have physical meaning. However, this issue cannot be avoided in some other cases, as described in the example in Section 4.5.

4.4 MODEL FITTING

In 1970, **R. J. Sheppard, B. P. Jordan, and E. H. Grant** published on complex nonlinear least square (CNLS) analysis to fit models to dielectric permittivity measurements, which are complex quantities. In 1977, **J. Ross Macdonald and J. A. Garber** applied the CNLS technique to fit the equivalent circuit to impedance data.

4.4.1 Software Choices

To the best of the author's knowledge, all the EIS equipment manufacturers provide EEC modeling software along with their equipment. A few open-source codes and software, as well as a few commercial versions, are also available. A partial list, as of 2020 is given here. (a) LEVMW, (b) ZView, (c) MEISP, (d) Zsimpwin, (e) EIS Spectrum Analyzer, and (f) ZFIT (MATLAB® based). A few software employ the CNLS algorithm. A few others employ the simplex algorithm. In some cases, the simplex algorithm is used initially, and when the solution is near the final value, the CNLS algorithm is employed.

4.4.2 Parameter Values – Initial Guess

In all the cases, optimization algorithms use initial guess values of the parameters. In some cases, the program itself may estimate the initial guess and perform the optimization. This works well when the data appears to be a semicircle in the complex plane plots. If the data exhibit many complex features, then the optimization programs do not always yield a good fit, and this may be because the initial parameter estimates are incorrect. Then the initial parameter values should be given manually. Literature shows that when electrochemical reactions are involved, the capacitance or inductance values that should be used to model the data can be very large (e.g., many milli Farads, or many Henries). One should also note that at very low frequencies, the capacitance will behave as an open circuit and that the inductance will behave as a short circuit. The circuit can be simplified in that case, and the initial guesses of the resistance values should be such that the impedance predicted by the circuit should be close to the low-frequency impedance of the measured data. A few examples illustrating the importance of an appropriate initial guess can be found in http://www.che.iitm.ac.in/~srinivar/EIS-Book.html.

4.4.3 Circuit Choices

In the context of electrochemical reactions, with no significant mass transfer resistance and without any film on the electrode surface, the impedance can be modeled by the circuit shown in Figure 4.8.

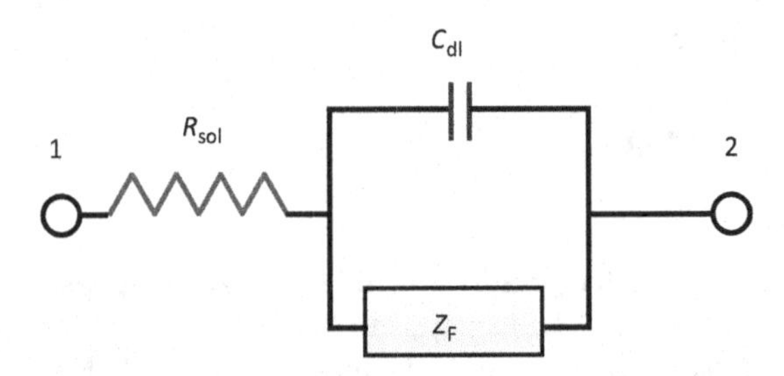

FIGURE 4.8 Equivalent electrical circuit for an electrode–electrolyte interface with a faradaic process.

Here, Z_F represents the faradaic impedance and can be as simple as a resistor, but many times, it is more complex. Z_F can be represented by a Voigt circuit (Figure 3.11) or Maxwell circuit (Figure 3.12a) or a ladder circuit (Figure 3.12b), with capacitors. It is also possible to construct a Voigt, Maxwell, or ladder circuit with inductors for modeling purposes, although it is less common. As explained in Section 4.2, these representations are equivalent, i.e., as long as the number of electrical elements is the same (number resistances should be equal among the circuits, and the number of capacitances or inductances should be equal among the circuits) in two representations, and negative values are allowed, the data fit should also be equally good.

4.4.4 High and Low-Frequency Limits of Impedance

Depending on the level of complexity of the data, a circuit with several electrical elements may be needed to model the results. Based on the knowledge of the physical system under investigation, we have to choose the appropriate circuit representation.

Two important parameters that are related to the faradaic impedance, and often extracted from impedance spectra, are charge-transfer resistance (R_t) and polarization resistance (R_p). They are defined (Diard, Le Gorrec, and Montella 1997a), respectively, as

$$R_t = \lim_{E_{ac0}\to 0} \lim_{\omega\to\infty} Z_F \quad (\text{for potentiostatic EIS}) \tag{4.14}$$

$$= \lim_{i_{ac0}\to 0} \lim_{\omega\to\infty} Z_F \quad (\text{for galvanostatic EIS}) \tag{4.15}$$

and

$$R_p = \lim_{E_{ac0}\to 0} \lim_{\omega\to 0} Z_F \quad (\text{for potentiostatic EIS}) \tag{4.16}$$

$$= \lim_{i_{ac0}\to 0} \lim_{\omega\to 0} Z_F \quad (\text{for galvanostatic EIS}) \tag{4.17}$$

However, this is not the only definition employed in the literature. A more rigorous definition of charge-transfer resistance is (Harrington 2015)

$$R_t = \left(\frac{\partial E}{\partial i}\right)_{C_i,\theta_i} \tag{4.18}$$

which reduces, in most cases, to the definition given above. Here, C_i is the concentration of species i, and θ_i is the fractional surface coverage of adsorbed species i, described in Chapter 5. Some authors consider R_t and R_p to be the same and use them interchangeably (He and Mansfeld 2009). Therefore, when reading the literature, it is important to note the definition employed. In this book, we employ the definitions given in Eqs. 4.14–4.17.

In each of the circuit representation (Figures 3.11 and 3.12) of faradaic impedance, it is important to know the relationship between the circuit element values and the high and low-frequency limits of impedance. If Z_F is modeled by Voigt circuit with capacitors (Figure 3.11), then

$$R_t = R_0 \text{ and} \tag{4.19}$$

$$R_p = R_0 + R_1 + R_2 + R_3 + \cdots + R_N. \tag{4.20}$$

If instead, a Voigt circuit with inductors is employed to represent Z_F, then

$$R_t = R_0 + R_1 + R_2 + R_3 + \cdots + R_N \text{ and} \tag{4.21}$$

$$R_p = R_0 \tag{4.22}$$

Likewise, if a Maxwell element with capacitors (Figure 3.12a) is employed to model Z_F, then

$$(R_t)^{-1} = (R_0)^{-1} + (R_1)^{-1} + (R_2)^{-1} + \cdots + (R_N)^{-1} \text{ and} \tag{4.23}$$

$$R_p = R_0 \tag{4.24}$$

If a ladder circuit with capacitors (Figure 3.12b) is used to model Z_F, then

$$R_t = R_0, \text{ and} \tag{4.25}$$

$$R_p = R_0 + R_1 + R_2 + R_3 + \cdots + R_N. \tag{4.26}$$

These relationships can help choose the initial guesses of at least some of the circuit elements during the model fit. Note that if Z_F is modeled using a circuit containing both inductors and capacitors, then these simplifications are not possible. There are alternate methods to calculate R_p using linear polarization or potentiodynamic polarization experiments, as described in http://www.che.iitm.ac.in/~srinivar/EIS-Book.html.

4.5 LIMITATIONS

4.5.1 Inductance and Negative (Differential) Resistance

In 1960, **Israel Epelboin and G. Loric** studied the anodization of aluminum in perchlorate solution and published an impedance spectrum showing an inductive loop in the complex plane plot.

If some data fall in the fourth quadrant (Figure 4.9a), then an inductance is needed to model the data. Alternatively, a resistor or a capacitor with negative parameter values can be used to describe the data. Either way, the situation is unpleasant (Franceschetti and Macdonald 1977). We can easily understand the idea behind using a resistor to model conduction in the electrolyte and a capacitor to model the double-layer.

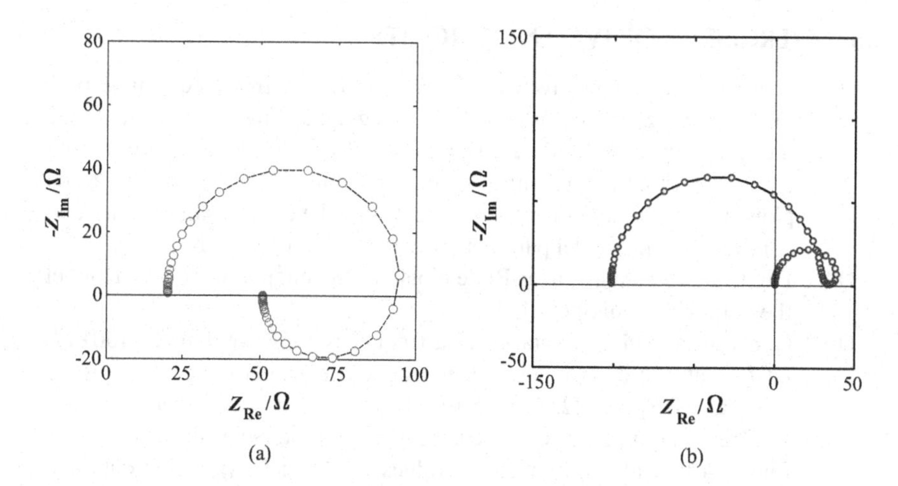

FIGURE 4.9 Complex plane plots of impedance data exhibiting the presence of (a) an inductance or (b) a negative (differential) resistance.

After all, a simple ionic conductor can be visualized as a resistor. Likewise, two parallel closely spaced and oppositely charged surfaces do look like a capacitor.

Where does an inductance come in? The physical meaning of this inductance remains to be examined. There is no coil or magnetic field in these electrochemical systems. If we choose to use negative resistance or capacitance, what does that mean? To add to this discomfort, sometimes EIS data fall in the second quadrant, as shown in Figure 4.9b. There are also reports of EIS data falling in the third quadrant (Bojinov 1997b). In these cases, it becomes necessary to use a negative resistance in the circuit to model the data, regardless of whether one chooses to use an inductor or a capacitor to complete the circuit.

4.5.2 Challenges in EEC Analysis

The optimal element values in an EEC model fit depend on the choice of initial values. No algorithm available at present ensures global optimization. If the initial guesses are poor, in some cases, the optimization of errors (i.e., the difference between the model and experimental values) may result in a local minimum and lead to an incorrect conclusion about the suitability of a model to describe the data. Recall that several circuits can model a given data equally well, and choosing the correct circuit to interpret the results can be challenging.

Finally, even if the appropriate circuit is identified and the best-fit values are obtained, assigning physical meaning to the large values of capacitance or inductances observed or to the negative resistance values encountered is next to impossible, and this is the major limitation of EEC modeling. Thus, the circuit analogy should not be pushed too far. However, it is possible to explain these results in the framework of reaction mechanism analysis. A few reaction mechanisms and some of their variants are discussed in detail in Chapter 5.

4.6 EXERCISE – EQUIVALENT CIRCUITS

Q4.1 Consider the circuit given in Figure 4.2. In the frequency range of 1 mHz–100 kHz, at 7 points per decade, log-spaced frequencies, calculate the impedance if $R_1 = 25\ \Omega$, $C_2 = 20\ \mu F$, $R_2 = 100\ \Omega$, $C_3 = 100\ \mu F$, and $R_3 = 40\ \Omega$. Plot the impedance spectra in a complex plane plot and Bode plots. Using a model fitting software, fit the data to the same circuit and confirm that the model parameters are recovered correctly.

Q4.2 Digitize EIS data given (as Bode plots) in a recent publication and model the data with a suitable EEC.

Q4.3 Generate impedance data using a frequency range of 1 mHz–100 kHz (at 7 points per decade, log-spaced frequencies) and the circuit in Figure 4.10a using $R_1 = 10\ \Omega$, $C_2 = 15\ \mu F$, $R_2 = 150\ \Omega$, $L_3 = 5$ H, and $R_3 = 100\ \Omega$. Draw the impedance as a complex plane plot. Next, fit the circuit in Figure 4.10b (allowing negative values for R_3 and C_3) to that data and extract the corresponding element values.

Q4.4 Repeat the above problem, but this time with $R_3 = 1000\ \Omega$. What inference can we draw from these results?

Q4.5 Generate impedance data using a frequency range of 1 mHz–100 kHz (at 7 points per decade, log-spaced frequencies) and the circuit in Figure 4.11a using $R_0 = 10\ \Omega$, $C_1 = 2\ \mu F$, $R_1 = 500\ \Omega$, $C_2 = 50\ \mu F$, $R_2 = 120\ \Omega$, $C_3 = 1$ mF, and $R_3 = 800\ \Omega$. Draw the impedance as a complex plane plot. Next, fit the circuit in Figure 4.11b and c to that data.

Q4.6 Generate impedance data using a frequency range of 1 mHz–100 kHz (at 7 points per decade, log-spaced frequencies) and the circuit in Figure 4.11a using $R_0 = 10\ \Omega$, $C_1 = 15\ \mu F$, $R_1 = -1500\ \Omega$, $C_2 = 2$ mF, $R_2 = 300\ \Omega$, $C_3 = 500\ \mu F$, and $R_3 = 200\ \Omega$. Explore the effect of varying the value of each element.

Q4.7 Generate impedance data using a frequency range of 1 mHz–100 kHz (at 7 points per decade, log-spaced frequencies) and the circuit in Figure 4.11a using $R_0 = 15\ \Omega$, $C_1 = 20\ \mu F$, $R_1 = 10\ \Omega$, $C_2 = 0.2$ mF, $R_2 = 20\ \Omega$, $C_3 = -20$ mF, and $R_3 = -8\ \Omega$. Explore the effect of varying the value of each element. (Note: This generates a spectrum that exhibits a pattern known as hidden negative impedance.)

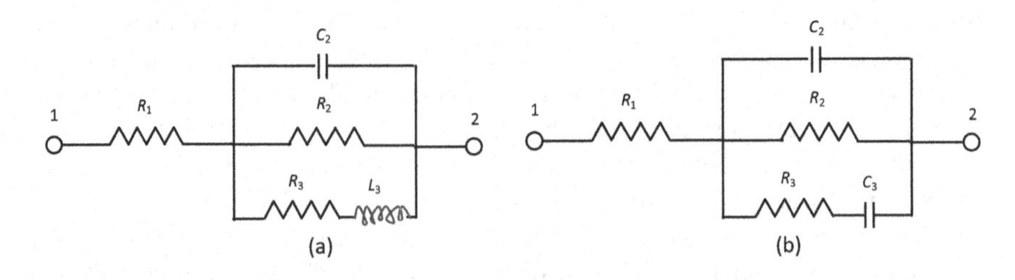

FIGURE 4.10 (a) Circuit with an inductor in the Maxwell pair. (b) Circuit with a capacitor in the Maxwell pair.

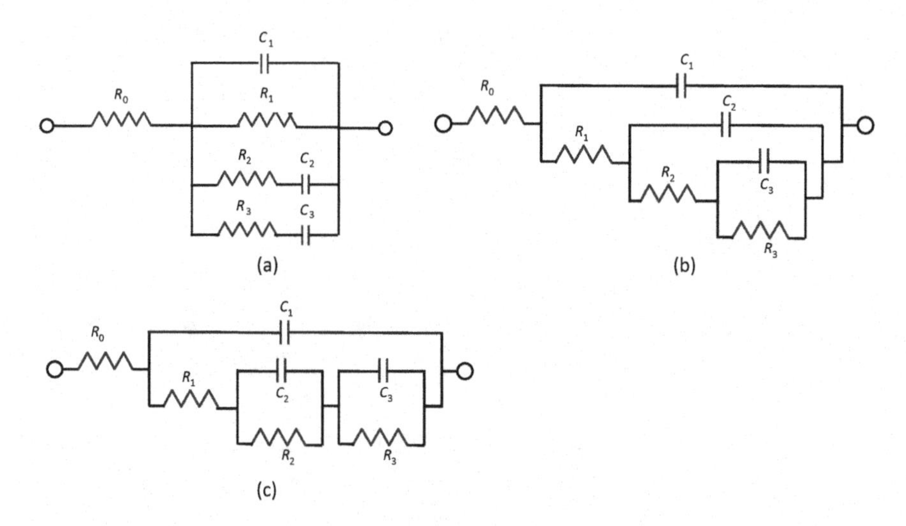

FIGURE 4.11 (a) Maxwell circuit. (b) Ladder circuit. (c) Voigt circuit.

Q4.8 Generate impedance data using a frequency range of 1 mHz–100 kHz (at 7 points per decade, log-spaced frequencies) and the circuit in Figure 4.10b using $R_1 = 10\ \Omega$, $C_2 = 10\ \mu F$, $R_2 = -20\ \Omega$, $C_3 = -1$ mF, and $R_3 = 15\ \Omega$. Explore the effect of varying the value of each element.

Q4.9 Repeat Q4.3–Q4.5, but this time, add a few % of random noise to the real and imaginary parts of the impedance generated and then fit the model. What are the relative standard errors in each element? What are the χ^2 values in each case?

Q4.10 Repeat the question Q4.1, after adding a few % random noise, and fit the circuit given in Figure 4.10b. What are the relative standard errors in each element? What are the χ^2 values in each case?

Q4.11 Use the data generated in Q4.10 and fit the circuit given in Figure 4.11a. What are the relative standard errors in each element? What are the χ^2 values in each case?

Q4.12 Compare the results of Q4.10 and Q4.11. In addition to comparing residual sum square values, calculate the AIC value, as given in Eq. 4.2. What inferences can we draw from this?

5 Mechanistic Analysis

Nothing in life is to be feared; it is only to be understood. Now is the time to understand more, so that we may fear less.

—Marie Curie

Reaction mechanism analysis (RMA) is introduced in this chapter, and the linearization of the relevant equations is presented in detail. RMA is the method of identifying the reaction mechanism from the electrochemical impedance spectroscopic (EIS) data. We consider a few reactions and illustrate the method in detail. The first example is a simple reaction, serving as an illustration of the linearization of the rate constants. The second example is a reaction involving an adsorbed intermediate. Here, we learn about the linearization of variation in the surface coverage of the adsorbed intermediate. We also learn how, depending on the parameter values, an inductive loop or a capacitive loop can arise from this mechanism. The third case is a reaction with two adsorbed intermediates. Once we learn how to determine the expression for impedance for this case, we can handle any reaction mechanism with any number of intermediates. The fourth example is a second-order reaction, involving what is known as a catalytic mechanism. It shows how negative resistance (negative differential resistance, to be precise) can naturally arise in some cases. The fifth case is an electron-transfer and electroadsorption reaction mechanism, where again, negative resistance can arise. A few other reactions are also touched upon. Examples in published literature where EIS data are analyzed using RMA are given. The issues in choosing a suitable reaction mechanism to model a data set and the methods employed to estimate the kinetic parameters from the experimental data are described. Finally, we compare the pros and cons of EEC and RMA methods of analysis.

While analyzing EIS data with EEC, we saw that first, we need to identify a circuit that can adequately model the data. For example, to model the data in Figure 4.2, we cannot use the circuit in Figure 4.1b but need to use the circuit in Figure 4.2 or its equivalent. Only then should we seek the values of resistances and capacitances. Likewise, while analyzing the EIS data of electrochemical reactions, we first have to propose a suitable reaction mechanism and confirm if the data can be adequately fit using that mechanism (model), at least in theory. Only if the answer is *yes*, we should seek the values of the kinetic parameters that would result in the best fit. If the answer is *no*, then we have to look for alternate reaction mechanisms that can model the data. To possess that knowledge, one has to be familiar with the type of spectra that each mechanism can generate, and one of the goals of this chapter is to help in this regard.

5.1 REACTION MECHANISM ANALYSIS – LINEARIZATION OF EQUATIONS

The following sections describe the development of equations relating the faradaic impedance to the electrochemical reactions.

5.1.1 Simple Electron Transfer Reaction

Consider the redox reaction between Fe^{2+} and Fe^{3+}. This is one of the simplest electrochemical reactions, and as part of introductory EIS studies, it is carried out using 5 mM of $K_3[Fe(CN)_6]$ and 5 mM of $K_4[Fe(CN)_6]$ in a supporting electrolyte of 100 mM of $KClO_4$. The reaction is often written as

$$Fe^{2+} \rightleftarrows Fe^{3+} + e^- \tag{5.1}$$

although the actual reaction is

$$\left[Fe(CN)_6\right]^{4-} \rightleftarrows \left[Fe(CN)_6\right]^{3-} + e^- \tag{5.2}$$

The rate constant corresponding to the forward reaction is written as k_f (in mol cm^{-2} s^{-1}), and that for the reverse reaction is written as k_r. Although these are referred to as *rate constants* in the literature, they actually change with potential and temperature. They also depend on what material is used as the electrode. For a given electrode, under isothermal conditions, at a given dc potential, they do not change. Their dependence on potential is exponential; i.e. $k_i = k_{i0}e^{b_i V}$ (where i stands for 'f' or 'r'), and in this case, b_f is positive while b_r is negative. The net oxidation rate per unit area is given by

$$\text{rate} = k_f\left[Fe^{2+}\right] - k_r\left[Fe^{3+}\right] \tag{5.3}$$

Here the square brackets represent the concentrations of the respective species. We should ideally use activity, but we assume that the activity coefficient is one and use concentrations instead. We further assume that the solution is always well mixed and that the concentrations of the species Fe^{2+} and Fe^{3+} everywhere, including the electrode surface, do not change significantly with time, even though the reaction continues to occur at the electrodes.

Note that E is measured with respect to the *open-circuit potential* (OCP, the potential where the forward and reverse reactions are equal in magnitude). At the OCP, the net current will be zero. If E is positive, then the forward (anodic) reaction dominates, and if E is negative, then the reverse (cathodic) reaction dominates. The electrode is rotated fast so that the mass transfer is very fast. This implies that the concentrations of species on the surface do not change significantly during the experiment, i.e., we are in the kinetically limited regime. We further assume that the solution resistance is negligible. As an example, the OCP of an Au electrode in

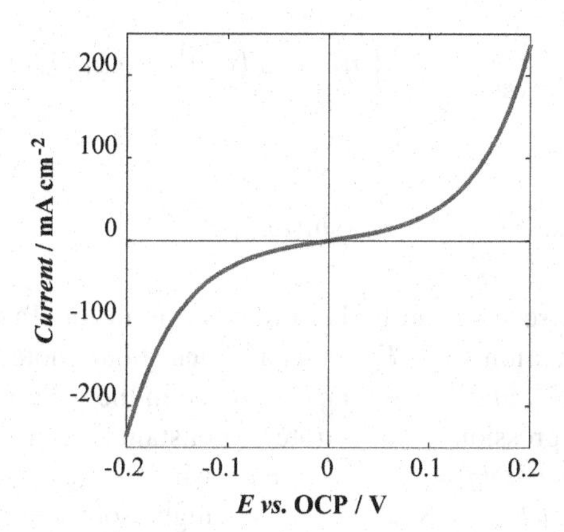

FIGURE 5.1 Simulated potentiodynamic polarization plot for a simple electron transfer reaction, with $k_{f0} = k_{r0} = 10^{-4}$ m s^{-1}, $[Fe^{2+}] = [Fe^{3+}] = 5$ mM, $b_f = 19.47$ V^{-1}, and $b_r = -19.47$ V^{-1}. It is assumed that solution resistance and mass transfer resistance are negligible.

5 mM of an Fe^{2+}/Fe^{3+} redox couple is 0.220 V *vs.* Ag/AgCl (3.5 M KCl) reference electrode. If the working electrode is held at 0.300 V *vs.* Ag/AgCl reference electrode, then $E_{dc} = 0.300 - 0.220 = 0.080$ V. Here, if the mass transfer is very fast, then the current as a function of potential will appear as shown in Figure 5.1. Note that in actual experiments, solution resistance and mass transfer resistance cannot always be neglected.

Let the electrode be held at some potential (E_{dc} *vs.* OCP). Then the net rate can be written as

$$\text{rate} = k_{f0} e^{b_f E_{dc}} \left[Fe^{2+} \right] - k_{r0} e^{b_r E_{dc}} \left[Fe^{3+} \right] \tag{5.4}$$

The current density due to the reaction can easily be calculated as

$$i_F = F \times \text{rate} \tag{5.5}$$

where F is the Faraday constant. The term i_F denotes the faradaic current density, i.e., current density due to the electrochemical reaction.

If a small amplitude sinusoidal wave of a certain frequency (f) is applied along with the dc potential on the electrode, then along with the dc current, a small ac current will also ensue. Let E_{ac0} be the amplitude and $\omega = 2\pi f$ be the angular frequency. Then the potential applied is $E = E_{dc} + E_{ac} = E_{dc} + E_{ac0} \sin(\omega t)$. As the potential changes, the rate constants and hence the reaction rate also change.

5.1.1.1 Linearization

If we confine ourselves to small-amplitude perturbations, the following approximation can be made.

$$k_f = k_{f0} e^{b_f E} = k_{f0} e^{b_f (E_{dc} + E_{ac})} = k_{f0} e^{b_f E_{dc}} e^{b_f E_{ac}} = k_{f0} e^{b_f E_{dc}} e^{b_f E_{ac0} \sin(\omega t)} \tag{5.6}$$

$$= k_{f0}e^{b_f E_{dc}}\left(1 + b_f E_{ac0}\sin(\omega t) + \frac{\{b_f E_{ac0}\sin(\omega t)\}^2}{2!} + \frac{\{b_f E_{ac0}\sin(\omega t)\}^3}{3!} + \cdots\right) \tag{5.7}$$

$$\simeq k_{f0}e^{b_f E_{dc}}\left(1 + b_f E_{ac0}\sin(\omega t)\right) = k_{f0}e^{b_f E_{dc}}\left(1 + b_f E_{ac}\right) \tag{5.8}$$

Here we neglect the second and higher-order terms in the Taylor series expansion of an exponential function since E_{ac0} is small. One should note the difference in the subscripts between E_{ac0} and E_{ac} and the difference in the meaning.

Thus the expression for rate constant can be written as $k_f \simeq k_{f\text{-dc}}\left(1 + b_f E_{ac0}\sin(\omega t)\right) = k_{f\text{-dc}}\left(1 + b_f E_{ac}\right)$, where $k_{f\text{-dc}} = k_{f0}e^{b_f E_{dc}}$. Similarly, we can write $k_r \simeq k_{r\text{-dc}}\left(1 + b_r E_{ac}\right)$. Since we neglect higher-order terms and keep only the linear term, this is referred to as *linearization*. The faradaic current density can then be expressed as

$$i_F = F \times \text{rate} \simeq F\left(k_{f\text{-dc}}\left(1 + b_f E_{ac}\right)\left[\text{Fe}^{2+}\right] - k_{r\text{-dc}}\left(1 + b_r E_{ac}\right)\left[\text{Fe}^{3+}\right]\right) \tag{5.9}$$

In the absence of an ac potential, the faradaic dc current density is given by

$$i_{F\text{-dc}} = F\left(k_{f\text{-dc}}\left[\text{Fe}^{2+}\right] - k_{r\text{-dc}}\left[\text{Fe}^{3+}\right]\right) \tag{5.10}$$

Hence, in the presence of an ac potential, the faradaic current density can be rewritten as

$$i_F \simeq i_{F\text{-dc}} + F\left(b_f k_{f\text{-dc}}\left[\text{Fe}^{2+}\right] - b_r k_{r\text{-dc}}\left[\text{Fe}^{3+}\right]\right)E_{ac} \tag{5.11}$$

The second part on the right side can be considered as $i_{F\text{-ac}}$, the ac component of the faradaic current density

$$i_{F\text{-ac}} = F\left(b_f k_{f\text{-dc}}\left[\text{Fe}^{2+}\right] - b_r k_{r\text{-dc}}\left[\text{Fe}^{3+}\right]\right)E_{ac} \tag{5.12}$$

Then the ratio

$$\frac{E_{ac}}{i_{F\text{-ac}}} = \frac{1}{F\left(b_f k_{f\text{-dc}}\left[\text{Fe}^{2+}\right] - b_r k_{r\text{-dc}}\left[\text{Fe}^{3+}\right]\right)} \tag{5.13}$$

can be easily calculated.

The ratio of E_{ac} to $I_{F\text{-ac}}$ at the angular frequency ω is the faradaic impedance of the system at the frequency ω. Since we use the current density instead of the current, the impedance calculated has the unit Ω cm^2 instead of Ω. Here we are using only the faradaic current density, and hence this is the faradaic impedance, denoted by Z_F.

In the literature, various notations have been employed to denote the faradaic impedance, e.g.,

$$Z_F = \frac{dE}{di_F} \text{ or } \frac{E_{ac}}{i_{F\text{-ac}}} \text{ or } \Delta E/\Delta i_F \text{ or } \delta E/\delta i_F \text{ or } \frac{E(s)}{i(s)} \tag{5.14}$$

where s indicates the frequency plane or the Laplace parameter domain. Note that although dE/di_F or similar expressions have been used to calculate the faradaic impedance, the faradaic impedance is not the derivative obtained from the steady-state potential-(faradaic) current diagram; rather, it is calculated by applying a potential wave of some frequency (frequencies) and by measuring the phase and magnitude of the current at the frequency (frequencies). That is, Z_F is a function of ω.

Thus, at a given dc potential (and at a constant temperature, with negligible solution resistance, and without any mass transfer limitations), the faradaic impedance for this particular reaction is a constant and is independent of frequency. This is the characteristic of a simple resistance.

To calculate the total impedance (Z_T), the double-layer capacitance is also taken into consideration.

$$Z_T = \frac{1}{1/Z_c + 1/Z_F} \tag{5.15}$$

where

$$Z_c = 1/j\omega C_{dl} \tag{5.16}$$

is the impedance of the double-layer capacitance. Thus, for the simple electron transfer reaction,

$$Z_T = \frac{1}{j\omega C_{dl} + F\left(b_f k_{f\text{-dc}}\left[Fe^{2+}\right] - b_r k_{r\text{-dc}}\left[Fe^{3+}\right]\right)} \tag{5.17}$$

Here, we assume that the reaction does not affect the double-layer capacitance, which may or may not be true. When the dc potential is changed, the value of the resistance in Eq. 5.13 will also change. An example plot from simulated data is given in Figure 5.2. Recall that we have assumed that mass transfer is very fast and that the concentrations of Fe^{2+} and Fe^{3+} are maintained throughout the experiment. Under these conditions, the simple electron transfer reaction will result only in semicircles in the complex plane plots of EIS data. When the dc potential is made more positive with respect to the OCP, we can see that the diameter of the corresponding semicircle decreases. When the dc potential is made more negative with respect to the OCP also, the diameter of the semicircle will decrease. This happens because we have employed symmetrical values for the parameters and the variables. The reader is encouraged to change the values of the kinetic parameters, concentrations, and the dc potential and plot the resulting complex plane plots.

Note that the impedance spectra, shown in Figure 5.2, can be modeled using simple Randles circuit with zero solution resistance (circuit shown in the inset of

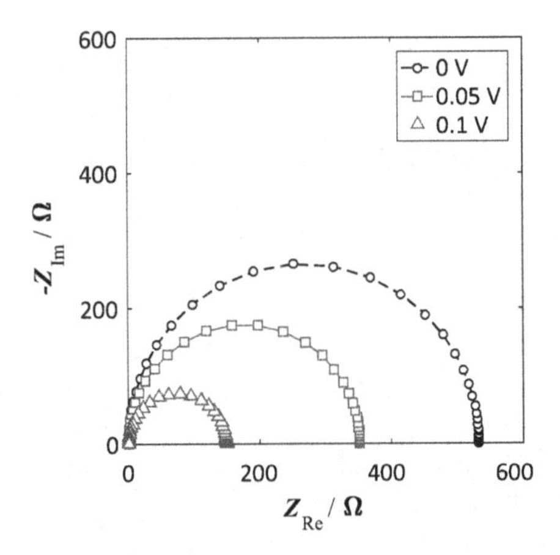

FIGURE 5.2 Complex plane plots of simulated impedance of a simple electron transfer reaction, at different dc potentials. The values employed are $k_{f0} = k_{r0} = 10^{-4}$cm s^{-1}, $b_f = 19.47$ V^{-1}, $b_r = -19.47$ V^{-1}, $[Fe^{2+}] = [Fe^{3+}] = 5$ mM, and $C_{dl} = 10$ μF cm^{-2}. The values in the legend are *dc* potentials with respect to OCP.

Figure 4.1). In this particular case of a simple electron transfer reaction without any mass transfer limitations, R_t is equal to R_p.

5.1.2 Reaction with an Adsorbed Intermediate

In 1960, **Von H. Gerischer** and **W. Mehl** studied hydrogen evolution reaction and derived an expression for the faradaic admittance of a reaction with an adsorbed intermediate.

Let us consider another reaction. Here, the overall reaction is the dissolution of metal, but it occurs through an intermediate species.

$$M \xrightarrow{k_1} M^{+}_{ads} + e^{-} \tag{5.18}$$

$$M^{+}_{ads} \xrightarrow{k_2} M^{+}_{sol} \tag{5.19}$$

Both steps are only in the forward direction, and the reverse reactions are neglected. This approximation is valid only when the potentials are very anodic (above the OCP). Near the OCP, or at potentials below the OCP, the reverse steps cannot be neglected. A pictorial representation is given in Figure 5.3. The surface of the electrode contains metal atoms. When the potential is anodic (positive with respect to the OCP), some of the atoms lose an electron. They are shown as shaded atoms. These are still present on the surface of the electrode. These are called M^{+}_{ads} species and are

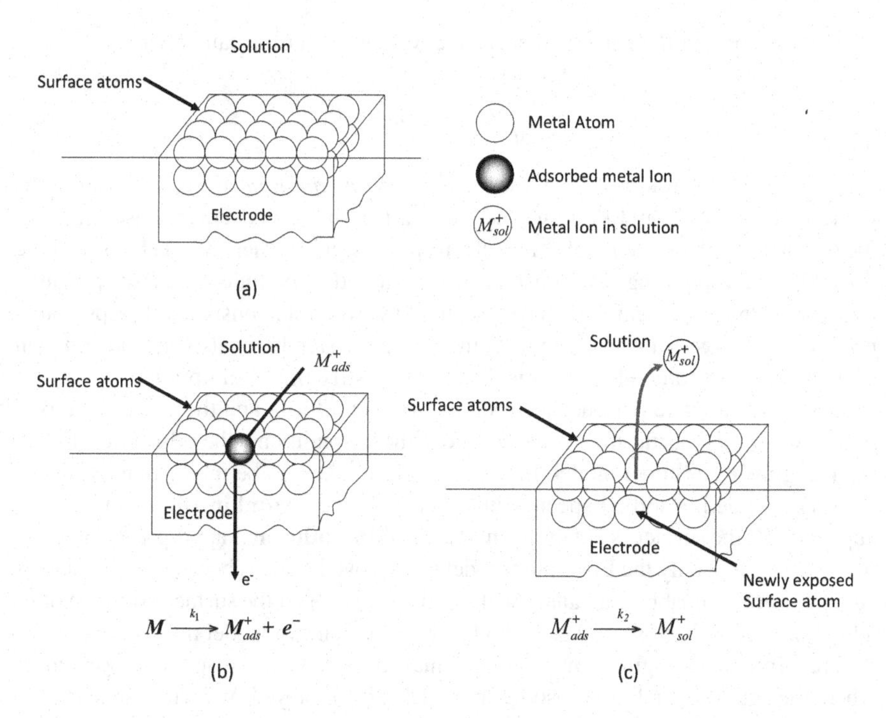

FIGURE 5.3 A pictorial representation illustrating the two-step mechanism Eqs. 5.18 and 5.19. (a) A metal surface is exposed to a solution. (b) The first step where a surface atom loses an electron and becomes an 'adsorbed' intermediate ion is shown. (c) The second step, where the adsorbed intermediate ion goes into the solution, leaving a new surface atom (which was below this ion) exposed to the solution is shown.

considered as adsorbed species. Normally, the term *adsorbed species* is used to represent an entity that comes from the liquid phase onto the solid phase. However, here it is used to represent the metal ions, which were originally metal atoms present on the electrode and have now lost an electron. These *adsorbed* species will go out into the solution in the next step. Now, the atoms below these shaded adsorbed species are not likely to lose electrons. On the other hand, the other *neutral* atoms or bare metal atoms on the surface, exposed to the solution, may lose electrons at the anodic potential. If the fraction of the surface covered by the adsorbed species is called θ, then the bare metal surface covers $(1 - \theta)$ of the surface. The bare metal sites are also called as *vacant* metal sites. θ does not have any units. Thus, the rate of the first reaction can be written as $k_1 (1 - \theta)$. In addition, we assume that the solution resistance is negligible. The effect of solution resistance will be dealt with subsequently in the final chapter.

The adsorbed species can go into the solution as an ion by the second reaction. This reaction rate is equal to $k_2\theta$. Note, first, that when the adsorbed species dissolves and goes into the solution, the metal atom below is exposed to the solution again, i.e., a vacant metal site is gained. We shall denote by Γ the total number of atoms (or surface sites), in moles per unit area (e.g., mol cm^{-2}).

The variation of the fractional surface coverage with time can be written as

$$\Gamma \frac{d\theta}{dt} = k_1(1-\theta) - k_2\theta \tag{5.20}$$

This is sometimes referred to as the *mass balance equation*. Here, we assume that the rate constants k_1 and k_2 do not depend on the surface coverage. Also, note that the second step does not involve any electron transfer, and hence is a chemical reaction. The corresponding rate constant k_2 is assumed to be independent of potential. In some instances, researchers have assumed that the rate constants of steps that do not involve charge-transfer (such as k_2 in the second step here) also depend exponentially on the potential, although it is difficult to justify this assumption.

Now let us assume that the electrode surface is uniform, and the rate constants do not vary with the surface coverage θ. Under these conditions, the adsorption follows the **Langmuir isotherm** model. In reality, there is likely to be interaction (repulsion) between charged adsorbed species, in which case the **Frumkin isotherm** model is applicable. If the surface is not uniform, the **Temkin isotherm** is appropriate. However, most of the time, only the Langmuir model is employed. This is because, only then, the equations are amenable to an analytical solution. Besides, if the surface coverage of the adsorbed species is close to 0 or 1, the Frumkin or Temkin model results are not very different from the Langmuir model. In the final chapter, we will discuss a few examples where the Frumkin or Temkin isotherm models are employed. We will also learn how these complex equations can be solved numerically to calculate the EIS response.

Here, if only a dc potential is applied, then the rate constants $k_{1\text{-dc}}$ and k_2 will not vary with time, and hence the surface coverage also will remain steady. This steady-state surface coverage is calculated by setting Eq. 5.20 to zero. Let $k_{1\text{-dc}} = k_{10}e^{b_1 E_{dc}}$. Then,

$$k_{1\text{-dc}}(1-\theta_{SS}) - k_2\theta_{SS} = 0 \tag{5.21}$$

$$\theta_{SS} = \frac{k_{1\text{-dc}}}{k_{1\text{-dc}} + k_2} \tag{5.22}$$

The faradaic current is given by

$$i_F = Fk_{1\text{-dc}}(1-\theta_{SS}) \tag{5.23}$$

The potentiodynamic polarization curve and the fractional surface coverage as a function of potential, for a sample case, are shown in Figure 5.4. Note that at the OCP, the current should be zero, but since the reverse reaction is not taken into account, the model predicts nonzero current values. The readers are reminded that this model is valid only at potentials well above the OCP. In Figure 5.4, at relatively low potentials, the fractional surface coverage of the intermediate is well below unity, and the current shows a clear increase with potential. At large potentials, the surface is completely covered with M^+_{ads} species, and an increase in current cannot cause a significant increase in θ_{SS}. The current levels off, and although this trend appears to be similar to diffusion-limited current (to be discussed in Section 6.2), the actual cause is the saturation of θ_{SS}.

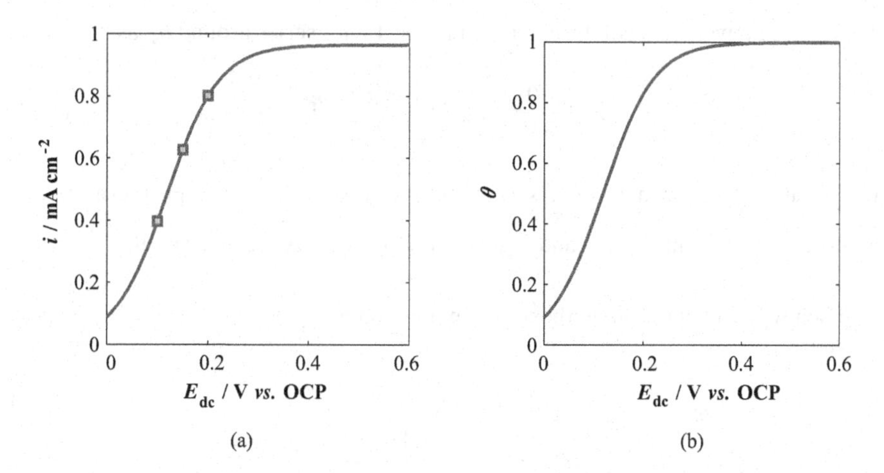

FIGURE 5.4 (a) Simulated potentiodynamic polarization plot of the reaction given in Eqs. 5.18 and 5.19. (b) Fractional surface coverage of the intermediate as a function of potential. The parameter values employed are $k_{10} = 10^{-9}$ mol cm^{-2}s^{-1}, $k_2 = 10^{-8}$ mol cm^{-2}s^{-1}, and $b_1 = 19.47$ V^{-1}. Solution resistance and mass transfer resistance are assumed to be negligible. The squares mark the locations where EIS data (Figure 5.5) were generated.

To calculate the impedance spectrum of this system, at a given *dc* potential, the charge and mass balance equations should be linearized, as shown below.

5.1.2.1 Linearization of Charge Balance Equations

To calculate the current due to these reactions, we note that only the first step involves electron transfer, and hence the second step does not directly contribute to the current. The *charge balance* equation, giving the faradaic current density, is therefore

$$i_F = Fk_1(1-\theta) \tag{5.24}$$

Remember that the rate of the first step is not just proportional to the rate constant, but also depends on θ. Now, θ depends on the potential in a complex manner, and it can vary from 0 to 1. This is one of the reaction mechanisms, where the rate does not change exponentially with the potential, although the rate constant changes exponentially with the potential.

Now let us consider the case where an ac potential ($E_{ac} = E_{ac0}\sin(\omega t)$), superimposed on a dc potential (E_{dc}), is applied. If the amplitude of the ac potential is small, then we can write

$$k_1 \simeq k_{1\text{-dc}}(1+b_1E_{ac}) \tag{5.25}$$

The current can be written as

$$i_F = Fk_1(1-\theta) \simeq Fk_{1\text{-dc}}(1+b_1E_{ac})(1-\theta) \tag{5.26}$$

The surface coverage θ can be expanded in the Taylor series around θ_{ss} as

$$\theta = \theta_{ss} + \left.\frac{\partial\theta}{\partial E}\right|_{E_{dc}} E_{ac} + \frac{1}{2!}\left.\frac{\partial^2\theta}{\partial E^2}\right|_{E_{dc}} E_{ac}^2 + \cdots \tag{5.27}$$

For small E_{ac}, we can neglect second and higher-order terms. Furthermore, we assume that the term $\left.\frac{\partial^2\theta}{\partial E^2}\right|_{E_{dc}}$ and higher-order derivatives are also small.

Then we can approximate the above expression as

$$\theta \simeq \theta_{ss} + \left.\frac{\partial\theta}{\partial E}\right|_{E_{dc}} E_{ac} \tag{5.28}$$

Hence, Eq. 5.26 can be written as

$$i_F \simeq Fk_{1\text{-dc}}(1 + b_1 E_{ac})\left(1 - \theta_{ss} - \left.\frac{\partial\theta}{\partial E}\right|_{E_{dc}} E_{ac}\right) \tag{5.29}$$

$$= Fk_{1\text{-dc}}(1 - \theta_{ss}) + b_1 Fk_{1\text{-dc}}(1 - \theta_{ss})E_{ac} \tag{5.30}$$

$$+Fk_{1\text{-dc}}\left(-\left.\frac{\partial\theta}{\partial E}\right|_{E_{dc}}\right)E_{ac} - b_1 Fk_{1\text{-dc}}\left(-\left.\frac{\partial\theta}{\partial E}\right|_{E_{dc}}\right)E_{ac}^2 \tag{5.31}$$

Since we neglected second and higher-order terms of the Taylor series to obtain this expression, here also we should neglect E_{ac}^2. Then the faradaic current can be written as

$$i_F \simeq Fk_{1\text{-dc}}(1 - \theta_{ss}) + b_1 Fk_{1\text{-dc}}(1 - \theta_{ss})E_{ac} + Fk_{1\text{-dc}}\left(-\left.\frac{\partial\theta}{\partial E}\right|_{E_{dc}}\right)E_{ac} \tag{5.32}$$

Splitting the faradaic current as the dc and ac components, we get $i_F = i_{F\text{-dc}} + i_{F\text{-ac}}$, where $i_{F\text{-dc}} \simeq Fk_{1\text{-dc}}(1 - \theta_{ss})$ and $i_{F\text{-ac}} \simeq \left\{b_1 Fk_{1\text{-dc}}(1 - \theta_{ss}) + Fk_{1\text{-dc}}\left(-\left.\frac{\partial\theta}{\partial E}\right|_{E_{dc}}\right)\right\}E_{ac}$

(5.33)

The problem is now reduced to the evaluation of $\left.\frac{\partial\theta}{\partial E}\right|_{E_{dc}}$. To do this, let us now return to the mass balance equation.

5.1.2.2 Linearization of the Mass Balance Equation

The left side of the mass balance Eq. 5.20 can be written as

$$\Gamma\frac{d\theta}{dt} = \Gamma\left(\left.\frac{\partial\theta}{\partial E}\right|_{E_{dc}}\right)\frac{\partial E}{\partial t} \tag{5.34}$$

Here

$$\frac{dE}{dt} = \frac{d\left(E_{dc} + E_{ac0}\sin(\omega t)\right)}{dt} = \omega E_{ac0}\cos(\omega t) = \omega E_{ac0}\sin\left(\omega t + \pi/2\right) \tag{5.35}$$

Now, in the complex plane, for any point Z, a counter-clockwise rotation by 90° (i.e., $\pi/2$ rad) is the same as multiplying Z by $e^{j\pi/2} = 0 + 1j$. Therefore,

$$\frac{dE}{dt} = \omega E_{ac0}\sin\left(\omega t + \pi/2\right) = j\omega E_{ac0}\sin(\omega t) = j\omega E_{ac} \tag{5.36}$$

The same result can also be obtained by writing the potential in complex form as

$$E = \left(E_{dc} + E_{ac0}e^{j\omega t}\right) \tag{5.37}$$

which leads to

$$\frac{dE}{dt} = \frac{d\left(E_{dc} + E_{ac0}e^{j\omega t}\right)}{dt} = j\omega E_{ac0}e^{j\omega t} = j\omega E_{ac} \tag{5.38}$$

Thus, the LHS of Eq. 5.20 can be written as

$$\Gamma\frac{d\theta}{dt} = j\omega\Gamma\left(\left.\frac{\partial\theta}{\partial E}\right|_{E_{dc}}\right)E_{ac} \tag{5.39}$$

Expanding the RHS of Eq. 5.20 in the Taylor series around $E = E_{dc}$ and $\theta = \theta_{ss}$ and truncating each term after the first derivative, we get

$$k_1(1-\theta) - k_2\theta \simeq \left(\begin{array}{c} \left\{k_{1\text{-dc}}\left(1 + b_1 E_{ac}\right)\right\}\left\{1 - \theta_{ss} - \left.\frac{\partial\theta}{\partial E}\right|_{E_{dc}} E_{ac}\right\} \\ -\left\{k_2\right\}\left\{\theta_{ss} + \left.\frac{\partial\theta}{\partial E}\right|_{E_{dc}} E_{ac}\right\} \end{array}\right) \tag{5.40}$$

After completing the multiplication, we get

$$\simeq k_1(1-\theta) - k_2\theta \simeq \left[\begin{array}{c} \left\{k_{1\text{-dc}}\left(1-\theta_{ss}\right) - k_2\theta_{ss}\right\} - \left(k_{1\text{-dc}} + k_2\right)\left(\left.\frac{\partial\theta}{\partial E}\right|_{E_{dc}}\right)E_{ac} \\ +\left\{b_1 k_{1\text{-dc}}\left(1-\theta_{ss}\right)\right\}E_{ac} \\ +\left\{-b_1 k_{1\text{-dc}}\left.\frac{\partial\theta}{\partial E}\right|_{E_{dc}}\right\}E_{ac}^2 \end{array}\right] \tag{5.41}$$

Since we have neglected the second and higher-order terms to arrive at this approximate relationship, here too, we should ignore the second-order term (E_{ac}^2). Hence,

$$\simeq k_1(1-\theta)-k_2\theta \simeq \left[\begin{array}{c}\left\{k_{1\text{-dc}}(1-\theta_{\text{ss}})-k_2\theta_{\text{ss}}\right\}-(k_{1\text{-dc}}+k_2)\left(\left.\dfrac{\partial\theta}{\partial E}\right|_{E_{\text{dc}}}\right)E_{\text{ac}} \\ +\left\{b_1k_{1\text{-dc}}(1-\theta_{\text{ss}})\right\}E_{\text{ac}}\end{array}\right] \tag{5.42}$$

Moreover, we know that $k_{1\text{-dc}}(1-\theta_{\text{ss}})-k_2\theta_{\text{ss}}=0$, and thus the first term on the right side can be removed.

$$k_1(1-\theta)-k_2\theta \simeq -(k_{1\text{-dc}}+k_2)\left(\left.\frac{\partial\theta}{\partial E}\right|_{E_{\text{dc}}}\right)E_{\text{ac}}+\left\{b_1k_{1\text{-dc}}(1-\theta_{\text{ss}})\right\}E_{\text{ac}} \tag{5.43}$$

Equating the LHS and RHS of the mass balance equation (Eq. 5.20), with the approximate relationships, we get

$$j\omega\Gamma\left(\left.\frac{\partial\theta}{\partial E}\right|_{E_{\text{dc}}}\right)E_{\text{ac}} \simeq -(k_{1\text{-dc}}+k_2)\left(\left.\frac{\partial\theta}{\partial E}\right|_{E_{\text{dc}}}\right)E_{\text{ac}}+\left\{b_1k_{1\text{-dc}}(1-\theta_{\text{ss}})\right\}E_{\text{ac}} \tag{5.44}$$

After dividing the above expression by E_{ac} and rearranging, we get

$$\left.\frac{\partial\theta}{\partial E}\right|_{E_{\text{dc}}} \simeq \frac{b_1k_{1\text{-dc}}(1-\theta_{\text{ss}})}{(k_{1\text{-dc}}+k_2+j\omega\Gamma)} \tag{5.45}$$

Since $k_{1\text{-dc}}(1-\theta_{\text{ss}})-k_2\theta_{\text{ss}}=0$, we can also write it as

$$\left.\frac{\partial\theta}{\partial E}\right|_{E_{\text{dc}}} \simeq \frac{(b_1)k_2\theta_{\text{ss}}}{(k_{1\text{-dc}}+k_2+j\omega\Gamma)} \tag{5.46}$$

The derivation is not difficult, but it is lengthy, and hence is prone to mistakes. It does not involve anything more than expanding the variables in the Taylor series and neglecting the higher-order terms.

After substituting for $\left.\dfrac{\partial\theta}{\partial E}\right|_{E_{\text{dc}}}$ given in Eq. 5.46, in the equation for $i_{F\text{-ac}}$, i.e., Eq. 5.33, we can write the faradaic admittance (Y_F) as

$$Y_F=\frac{i_{F\text{-ac}}}{E_{\text{ac}}}=F\left\{b_1k_{1\text{-dc}}(1-\theta_{\text{ss}})-k_{1\text{-dc}}\frac{(b_1)k_2\theta_{\text{ss}}}{(k_{1\text{-dc}}+k_2+j\omega\Gamma)}\right\} \tag{5.47}$$

The inverse of the faradaic admittance (Y_F) is the faradaic impedance (Z_F). From this and Z_C (the impedance corresponding to the double-layer capacitance), the total impedance Z_T can be calculated as

$$Z_T = \frac{1}{1/Z_c + 1/Z_F} \tag{5.48}$$

5.1.2.3 Types of Complex Plane Plots We Can Expect for This Mechanism

In Eq. 5.47, the first term within the parenthesis is independent of the frequency and is referred to as inverse of charge-transfer resistance. (For the definition and detailed discussion on charge-transfer resistance, please see Section 4.4.4). The second term on the right is, in general, dependent on the frequency ($\omega = 2\pi f$). It gives rise to the following possibility. The second term on the RHS is negative. Note that all the parameters ($k_{1\text{-dc}}$, k_2, t, and θ_{ss}) in the second term are positive quantities. An increase in ω will cause the denominator value to increase and the magnitude of the second term to decrease. Since the second term is negative, this is effectively increasing the value of the second term. Thus, faradaic admittance will increase with ω. This means the faradaic impedance will decrease with ω. This is similar to the behavior of a capacitor. An example of the impedance data arising from this reaction is shown as a complex plane plot in Figure 5.5.

This can be modeled using an equivalent electrical circuit given in Figure 4.2. The reader can verify that the spectra can be simulated with element values presented in Table 5.1.

If we were to use an actual electrical circuit to mimic these results, a large capacitor device is needed to obtain a capacitance value of the order of hundreds of μF. Here we remark that the equivalent circuit is only an *analogy*, i.e., one should not conclude from this analysis that the electrochemical system contains a huge capacitor. Besides, note that the polarization resistance is equal to R_2 while the charge-transfer resistance is given by $R_t = \left[(R_2)^{-1} + (R_3)^{-1}\right]^{-1}$. The charge-transfer resistance decreases with overpotential, while the polarization resistance increases with overpotential.

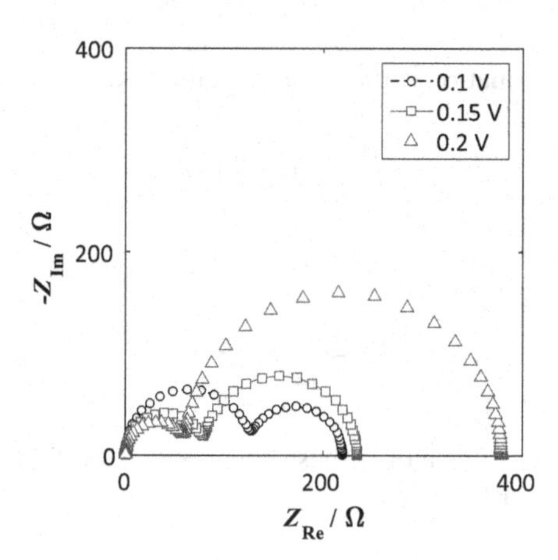

FIGURE 5.5 Complex plane plots of the impedance of reaction given in Eqs. 5.18 and 5.19, at a few dc potentials. The values employed are $k_{10} = 10^{-9}$mol cm^{-2}s^{-1}, $k_2 = 10^{-8}$mol cm^{-2}s^{-1}, $b_1 = 19.47$ V^{-1}, $\Gamma = 10^{-9}$mol cm^{-2}, and $C_{dl} = 10$ μF cm^{-2}. The values in the legends are *dc* potentials with respect to the OCP.

TABLE 5.1
Electrical Circuit Element Values That Should Be Used with the Circuit in Figure 4.2 to Model the Data in Figure 5.5

E_{dc} (V *vs.* OCP)	R_1 (Ω)	C_2 (μF)	R_2 (Ω)	C_3 (μF)	R_3 (Ω)	R_t (Ω)
0.1	0	10	219.7	187.5	313.6	129.2
0.15	0	10	233.9	277.8	126.1	81.9
0.2	0	10	378.7	219.4	77.1	64.1

One could also model the data using an equivalent ladder circuit or a Voigt circuit. While the individual electrical element values will vary with the choice of the circuit, R_p and R_t values would only depend on the actual data and not on the choice of the circuit.

5.1.3 Reaction with an Adsorbed Intermediate – Two Electrochemical Steps

Consider a slightly modified mechanism, where the second step is also electrochemical.

$$\begin{aligned} &M \xrightarrow{k_1} M^{+}_{ads} + e^{-} \\ &M^{+}_{ads} \xrightarrow{k_2} M^{2+}_{sol} + e^{-} \end{aligned} \quad (5.49)$$

Here, the second rate-constant is also dependent on the potential as $k_2 = k_{20}e^{b_2 E}$. Then the charge balance equation is given by

$$i_F = F\left[k_1(1-\theta) + k_2\theta\right] \quad (5.50)$$

and the mass balance equation is the same as before, i.e.,

$$\Gamma\frac{d\theta}{dt} = k_1(1-\theta) - k_2\theta \quad (5.51)$$

The steady-state surface coverage is given by

$$\theta_{ss} = \frac{k_{1\text{-dc}}}{k_{1\text{-dc}} + k_{2\text{-dc}}} \quad (5.52)$$

It can be shown that after expanding these equations in the Taylor series and neglecting second and higher-order terms,

$$\frac{d\theta}{dE} = \frac{b_1 k_{1\text{-dc}}(1-\theta_{ss}) - b_2 k_{2\text{-dc}}\theta_{ss}}{k_{1\text{-dc}} + k_{2\text{-dc}} + j\omega\Gamma} = \frac{(b_1 - b_2)k_{2\text{-dc}}\theta_{ss}}{k_{1\text{-dc}} + k_{2\text{-dc}} + j\omega\Gamma} \quad (5.53)$$

$$i_{F\text{-ac}} \simeq F\left\{b_1 k_{1\text{-dc}}\left(1-\theta_{ss}\right)+b_2 k_{2\text{-dc}}\theta_{ss}+\left(k_{2\text{-dc}}-k_{1\text{-dc}}\right)\left(\left.\frac{\partial\theta}{\partial E}\right|_{E_{dc}}\right)\right\}E_{ac} \tag{5.54}$$

$$= F\left\{\left(b_1+b_2\right)k_{2\text{-dc}}\theta_{ss}+\left(k_{2\text{-dc}}-k_{1\text{-dc}}\right)\left(\left.\frac{\partial\theta}{\partial E}\right|_{E_{dc}}\right)\right\}E_{ac} \tag{5.55}$$

and

$$Y_F = \frac{i_{F\text{-ac}}}{E_{ac}} = F\left\{\left(b_1+b_2\right)k_{2\text{-dc}}\theta_{ss}+\left(k_{2\text{-dc}}-k_{1\text{-dc}}\right)\left(\frac{\left(b_1-b_2\right)k_{2\text{-dc}}\theta_{ss}}{k_{1\text{-dc}}+k_{2\text{-dc}}+j\omega\Gamma}\right)\right\} \tag{5.56}$$

i. if $b_1 = b_2$, then the second term on the RHS of the above equation is zero, and hence the faradaic impedance is independent of frequency. A simple resistor can represent the faradaic impedance. An example of impedance data arising from this reaction is shown as a complex plane plot in Figure 5.6a.
ii. If the second term on the RHS is negative, then an increase in ω will cause the denominator value to increase and the magnitude of the second term to decrease. Since the second term is negative, this is effectively increasing the value of the second term. Thus, faradaic admittance will increase with ω. This means the faradaic impedance will decrease with ω. This is similar to the behavior of a capacitor. An example of impedance data arising from this reaction is shown as a complex plane plot in Figure 5.6b.
iii. If the second term on the RHS is positive, then by similar arguments, we can show that faradaic impedance will increase with ω. This is similar to the behavior of an inductor. An example of impedance data arising from this reaction is shown as a complex plane plot in Figure 5.6c.

The reader can verify that the circuit shown in the inset of Figure 4.1b can model the spectrum in Figure 5.6a. The spectra in Figure 5.6b and c can be modeled using the circuit in Figure 4.2, using element values presented in Table 5.2. The inductive loop was modeled by permitting the Maxwell pair (R_3, C_3) to hold negative values. Another choice is to model the data in Figure 5.6c using a Maxwell pair of resistor and inductor (R_3, L_3), and then only positive element values can be employed to simulate the data.

5.1.4 Electron Transfer – Electroadsorption Reaction (E-EAR) – Negative Impedance

In 1972, **I. Epelboin** and **M. Keddam** studied Ni dissolution in sulfuric acid and reported impedance with a negative real part.

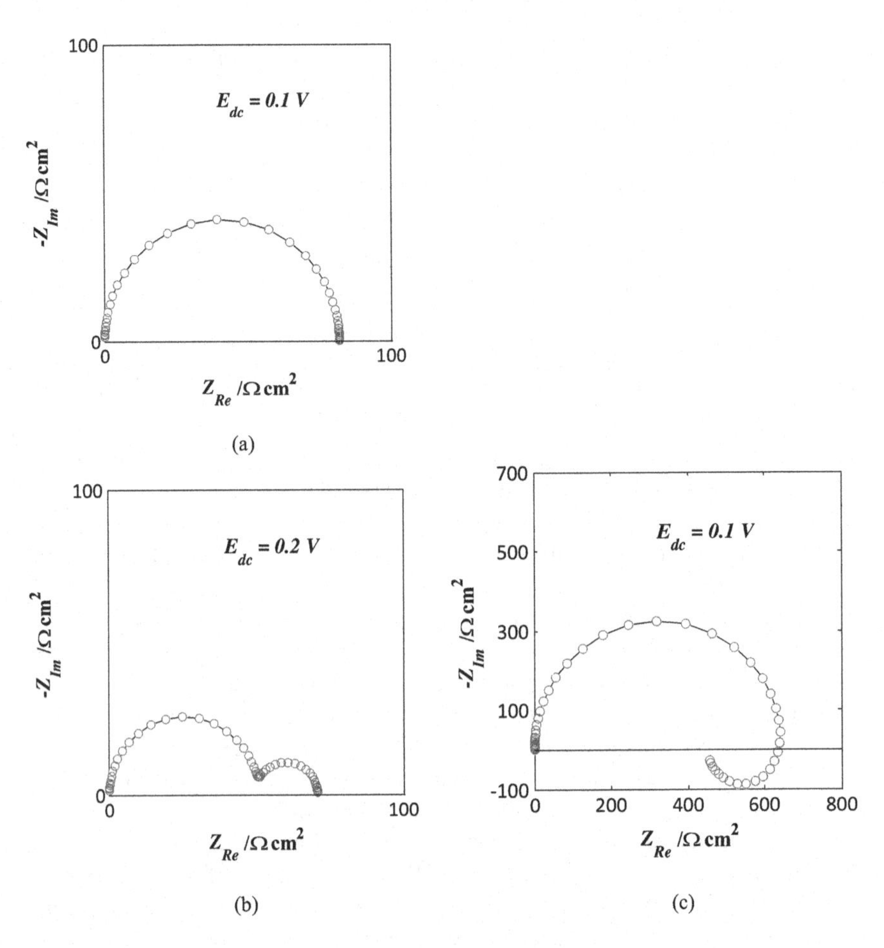

FIGURE 5.6 Sample complex plane plots of impedance of a reaction with an adsorbed intermediate. The parameter sets and *dc* potential employed are (a) $k_{10} = 10^{-7}$ mol cm^{-2}s^{-1}, $b_1 = 19\,V^{-1}$, $k_{20} = 5 \times 10^{-10}$ mol cm^{-2}s^{-1}, $b_2 = 19\,V^{-1}$, $\Gamma = 10^{-9}$ mol cm^{-2}, and $E_{dc} = 0.1$ V. (b) $k_{10} = 10^{-9}$ mol cm^{-2}s^{-1}, $b_1 = 9\,V^{-1}$, $k_{20} = 10^{-11}$ mol cm^{-2}s^{-1}, $b_2 = 37\,V^{-1}$, $\Gamma = 10^{-9}$ mol cm^{-2}, and $E_{dc} = 0.2$ V. (c) $k_{10} = 10^{-9}$ mol cm^{-2}s^{-1}, $b_1 = 9\,V^{-1}$, $k_{20} = 10^{-11}$ mol cm^{-2}s^{-1}, $b_2 = 37\,V^{-1}$, $\Gamma = 10^{-9}$ mol cm^{-2}, and $E_{dc} = 0.1$ V. In all the plots, $C_{dl} = 10\,\mu F\,cm^{-2}$. Note that the parameters are identical in the case of (b) and (c), and only E_{dc} was changed.

TABLE 5.2
Electrical Circuit Element Values That Should Be Used with the Circuit in Figures 4.1b or 4.2 to Model the Data in Figure 5.6

Figure No.	R_1 (Ω)	C_2 (μF)	R_2 (Ω)	C_3 (μF)	R_3 (Ω)	R_t (Ω)
5.6a	0	10	81.3	-	-	81.3
5.6b	0	10	70.9	245	182.2	51.0
5.6c	0	10	451.5	−235.1	−1485	648.7

The following reaction mechanism is sometimes referred to as E-EAR (Molina Concha et al. 2013). The first step is a simple electron transfer reaction. The second is an electrochemical adsorption (or electrosorption) reaction.

$$P_{sol} \xrightarrow{k_1} Q_{sol} + e^- \text{ and} \tag{5.57}$$

$$A^-_{sol} \underset{k_{-2}}{\overset{k_2}{\rightleftharpoons}} A_{ads} + e^- \tag{5.58}$$

Here P and Q are soluble species, and Q has one electron less than P, e.g., P can be Fe^{2+}, and then Q will be Fe^{3+}. The first reaction can occur only on the free metal sites of the electrode. A^- is present in the solution, and it can adsorb onto the electrode surface, effectively blocking it. In those locations, the first reaction cannot occur. The mass balance equation for this system is

$$\Gamma\frac{d\theta}{dt} = k_2 C_{A^-_{sol}}(1-\theta) - k_{-2}\theta, \tag{5.59}$$

where θ is the fractional surface coverage of A_{ads}.

The charge balance equation is

$$i_F = F\left[k_1 C_{P_{sol}}(1-\theta) + k_2 C_{A^-_{sol}}(1-\theta) - k_{-2}\theta\right] \tag{5.60}$$

From this, it can be shown that the steady-state surface coverage of the adsorbed species is

$$\theta_{ss} = \frac{k_{2\text{-dc}} C_{A^-_{sol}}}{k_{2\text{-dc}} C_{A^-_{sol}} + k_{-2\text{-dc}}} \tag{5.61}$$

$$\frac{d\theta}{dE} = \frac{(b_2 - b_{-2}) k_{-2\text{-dc}} \theta_{ss}}{k_{2\text{-dc}} C_{A^-_{sol}} + k_{-2\text{-dc}} + j\omega\Gamma} \tag{5.62}$$

$$i_{F\text{-ac}} \simeq F\left\{\begin{array}{l} b_1 k_{1\text{-dc}} C_{P_{sol}}(1-\theta_{ss}) + b_2 k_{2\text{-dc}} C_{A^-_{sol}}(1-\theta_{ss}) - b_{-2} k_{-2\text{-dc}} \theta_{ss} \\ -\left(k_{1\text{-dc}} C_{P_{sol}} + k_{2\text{-dc}} C_{A^-_{sol}} + k_{-2\text{-dc}}\right)\left(\left.\frac{\partial\theta}{\partial E}\right|_{E_{dc}}\right) \end{array}\right\} E_{ac} \tag{5.63}$$

The faradaic impedance can be written as

$$Y_F = \frac{i_{F\text{-ac}}}{E_{ac}} = F\left\{\begin{array}{l} b_1 k_{1\text{-dc}} C_{P_{sol}}(1-\theta_{ss}) + b_2 k_{-2\text{-dc}} C_{A^-_{sol}}(1-\theta_{ss}) - b_{-2} k_{2\text{-dc}} \theta_{ss} \\ -\left(k_{1\text{-dc}} C_{P_{sol}} + k_{2\text{-dc}} C_{A^-_{sol}} + k_{-2\text{-dc}}\right)\left(\frac{(b_2 - b_{-2}) k_{-2\text{-dc}} \theta_{ss}}{k_{2\text{-dc}} C_{A^-_{sol}} + k_{-2\text{-dc}} + j\omega\Gamma}\right) \end{array}\right\} \tag{5.64}$$

For a set of kinetic parameters, the fractional surface coverage of A_{ads} *vs.* E_{dc} is shown in Figure 5.7a, and the i_{dc} *vs.* E_{dc} curve is shown in Figure 5.7b. The kinetic parameter values are given in the figure caption. The complex plane plots of the impedance spectra at two different *dc* potentials are shown in Figure 5.7c and d. The low-frequency impedance of the data presented in Figure 5.7d shows negative values, and the physical meaning of this needs to be explained. We remind the reader that the impedance discussed here is the *differential impedance*, which means that if the *slope* of the i_{dc} *vs.* E_{dc} plot is negative, then at low frequencies, a negative differential impedance, sometimes called *negative impedance* or *negative resistance*, would arise. Note that here *negative resistance* does NOT mean that if we apply a *dc*

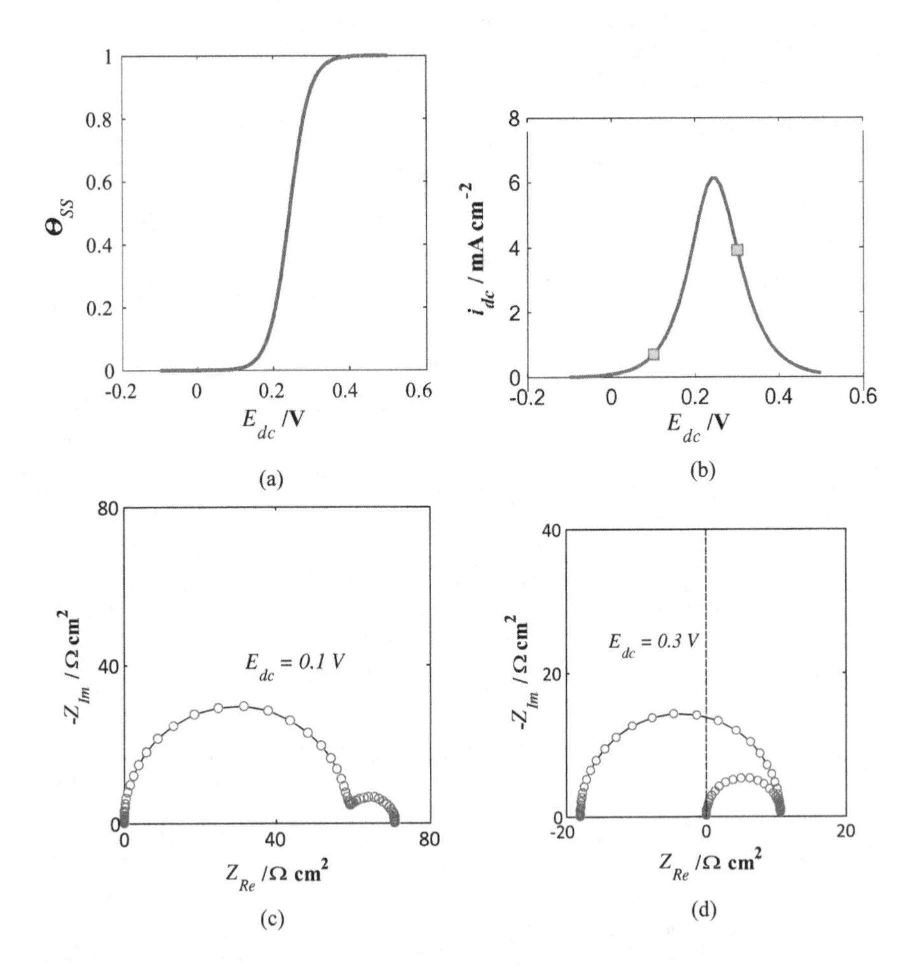

FIGURE 5.7 (a) Steady-state surface coverage of intermediate and (b) current, as a function of *dc* potential for an electrochemical reaction given in Eqs. 5.57 and 5.58 (the E-EAR mechanism). The parameter values are $k_{10} = 10^{-5}$cm s^{-1}, $b_1 = 20\,V^{-1}$, $k_{20} = 10^{-6}$cm s^{-1}, $b_2 = 20\,V^{-1}$, $k_{-20} = 10^{-6}$mol cm^{-2}s^{-1}, $b_{-2} = -18\,V^{-1}$, $\Gamma = 10^{-8}$mol cm^{-2}, $C_{A^-_{sol}} = C_{P_{sol}} = 10^{-4}$mol cm^{-3}, (i.e., 0.1 mol L^{-1}), and $C_{dl} = 10$ μF cm^{-2}. Corresponding complex plane plots of impedance spectra at (c) $E_{dc} = 0.1$ V and (d) $E_{dc} = 0.3$ V.

TABLE 5.3
Electrical Circuit Element Values That Should Be Used with the Circuit in Figure 4.2 to Simulate the Data in Figure 5.7

E_{dc} (V *vs.* OCP)	R_1 (Ω)	C_2 (μF)	R_2 (Ω)	C_3 (μF)	R_3 (Ω)	R_t (Ω)
0.1	0	10	71.1	169.7	355	59.2
0.2	0	10	−18	33,180	6.7	10.7

potential, then the dc current will flow in the opposite direction. The meaning of the negative resistance is that if we *increase* the potential, then the current will *decrease*, but it will continue to flow in the same direction.

The reader can verify that the spectra in Figure 5.7c and d can be modeled using the circuit in Figure 4.2, using element values presented in Table 5.3. The low-frequency negative resistance was modeled by permitting R_2 to hold a negative value.

5.1.5 More Reactions with One Adsorbed Intermediate

The following are some examples of reactions with one adsorbed intermediate species.

$$\begin{gathered} M \underset{k_{-1}}{\overset{k_1}{\rightleftarrows}} M^+_{ads} + e^- \\ M^+_{ads} \xrightarrow{k_2} M^+_{sol} \end{gathered} \tag{5.65}$$

In the above reaction, the first step is reversible. This can give rise to the EIS data with either a low-frequency capacitance loop or a low-frequency inductance loop. Also, it is possible that at low dc potential, this can exhibit a capacitive loop while at higher dc potentials, it can exhibit an inductive loop.

Another variation is when both steps are reversible, as shown below.

$$\begin{gathered} M \underset{k_{-1}}{\overset{k_1}{\rightleftarrows}} M^+_{ads} + e^- \\ M^+_{ads} \underset{k_{-2}}{\overset{k_2}{\rightleftarrows}} M^+_{sol} \end{gathered} \tag{5.66}$$

In this case, depending on the parameter values, it is possible to obtain a negative resistance at the low-frequency limit (Lasia 2014).

A third variation is when direct dissolution occurs in parallel to the dissolution via an adsorbed intermediate.

$$\begin{gathered} M \xrightarrow{k_1} M^+_{ads} + e^- \\ M^+_{ads} \xrightarrow{k_2} M^+_{sol} \\ M \xrightarrow{k_3} M^+_{sol} + e^- \end{gathered} \tag{5.67}$$

As the reader can imagine, many more variations are possible, depending on whether each step is reversible or irreversible, whether ions in the solution interact with the electrode and if direct dissolution can occur in parallel.

5.1.6 Reaction with Two Adsorbed Intermediates

Now let us consider the following reaction with three steps and two intermediate species:

$$M \xrightarrow{k_1} M^{+}_{ads} + e^{-} \tag{5.68}$$

$$M^{+}_{ads} \xrightarrow{k_2} M^{2+}_{ads} + e^{-} \tag{5.69}$$

$$M^{2+}_{ads} \xrightarrow{k_3} M^{3+}_{sol} + e^{-} \tag{5.70}$$

Since there are two adsorbed intermediate species M^{+}_{ads} and M^{2+}_{ads}, we can denote their fractional surface coverage values by θ_1 and θ_2, respectively. The fractional surface coverage of the vacant metal sites is then $(1 - \theta_1 - \theta_2)$, and this quantity is sometimes represented by θ_v.

The mass balance equation consists of

$$\Gamma \frac{d\theta_1}{dt} = k_1 (1 - \theta_1 - \theta_2) - k_2 \theta_1 \tag{5.71}$$

$$\Gamma \frac{d\theta_2}{dt} = k_2 \theta_1 - k_3 \theta_2 \tag{5.72}$$

Under steady-state conditions (i.e., only a dc potential E_{dc} is applied), the LHS of the mass balance equations vanishes and the steady-state surface coverage values can be written as

$$\theta_{1\text{-ss}} = \frac{k_1 k_3}{k_1 k_2 + k_2 k_3 + k_3 k_1} = \frac{1/k_2}{\sum_{m=1}^{3} 1/k_m} \tag{5.73}$$

$$\theta_{2\text{-ss}} = \frac{k_1 k_2}{k_1 k_2 + k_2 k_3 + k_3 k_1} = \frac{1/k_3}{\sum_{m=1}^{3} 1/k_m} \tag{5.74}$$

The charge balance equation gives the faradaic current density

$$i_F = F\big(k_1 (1 - \theta_1 - \theta_2) + k_2 \theta_1 + k_3 \theta_2\big) \tag{5.75}$$

5.1.6.1 Linearization of Mass Balance Equations

When a small amplitude potential perturbation (E_{ac}) is applied along with a dc potential (E_{dc}), the rate constants and the surface coverage can be expanded in the Taylor series and approximated by neglecting second and higher-order terms. In addition, the steady-state expressions for θ_1 and θ_2 can be used to simplify the terms, as we have seen in the previous sections.

$$\left(k_{1\text{-dc}} + k_{2\text{-dc}} + j\omega\Gamma\right)\left.\frac{\partial\theta_1}{\partial E}\right|_{E_{dc}} + k_{1\text{-dc}}\left.\frac{\partial\theta_2}{\partial E}\right|_{E_{dc}} \simeq (b_1 - b_2)k_{2\text{-dc}}\theta_{1\text{-ss}} \text{ and} \tag{5.76}$$

$$-k_{2\text{-dc}}\left.\frac{\partial\theta_1}{\partial E}\right|_{E_{dc}} + \left(k_{3\text{-dc}} + j\omega\Gamma\right)\left.\frac{\partial\theta_2}{\partial E}\right|_{E_{dc}} \simeq (b_2 - b_3)k_{3\text{-dc}}\theta_{2\text{-ss}} \tag{5.77}$$

The above two equations, which are linear algebraic equations with $\left.\frac{\partial\theta_1}{\partial E}\right|_{E_{dc}}$ and $\left.\frac{\partial\theta_2}{\partial E}\right|_{E_{dc}}$ as unknown values, can be solved simultaneously to obtain

$$\left.\frac{\partial\theta_1}{\partial E}\right|_{E_{dc}} \simeq \frac{\left\{(b_1 - b_2)k_{2\text{-dc}}\theta_{1\text{-ss}}\right\}\left\{k_{3\text{-dc}} + j\omega\Gamma\right\} - \left\{k_{1\text{-dc}}\right\}\left\{(b_2 - b_3)k_{3\text{-dc}}\theta_{2\text{-ss}}\right\}}{\left\{k_{1\text{-dc}} + k_{2\text{-dc}} + j\omega\Gamma\right\}\left\{k_{3\text{-dc}} + j\omega\Gamma\right\} + k_{1\text{-dc}}k_{2\text{-dc}}} \tag{5.78}$$

$$\left.\frac{\partial\theta_2}{\partial E}\right|_{E_{dc}} \simeq \frac{\left\{k_{1\text{-dc}} + k_{2\text{-dc}} + j\omega\Gamma\right\}\left\{(b_2 - b_3)k_{3\text{-dc}}\theta_{2\text{-ss}}\right\} + \left\{(b_1 - b_2)k_{2\text{-dc}}\theta_{1\text{-ss}}\right\}\left\{k_{2\text{-dc}}\right\}}{\left\{k_{1\text{-dc}} + k_{2\text{-dc}} + j\omega\Gamma\right\}\left\{k_{3\text{-dc}} + j\omega\Gamma\right\} + k_{1\text{-dc}}k_{2\text{-dc}}} \tag{5.79}$$

Obviously, both quantities are, in general, dependent on the frequency of the applied ac potential.

5.1.6.2 Linearization of Charge Balance Equations

The faradaic current density can also be linearized in the same way we did for the previous mechanism, and the equation can be written as

$$i_F = i_{F\text{-dc}} + i_{F\text{-ac}} \tag{5.80}$$

Here, $i_{F\text{-dc}} = F\left(k_{1\text{-dc}}\left(1 - \theta_{1\text{-ss}} - \theta_{2\text{-ss}}\right) + k_{2\text{-dc}}\theta_{1\text{-ss}} + k_{3\text{-dc}}\theta_{2\text{-ss}}\right)$ (5.81)

$$i_{F\text{-ac}} \simeq F\left\{\begin{bmatrix}\left(b_1k_{1\text{-dc}}\left(1-\theta_{1\text{-ss}}-\theta_{2\text{-ss}}\right)+b_2k_{2\text{-dc}}\theta_{1\text{-ss}}+b_3k_{3\text{-dc}}\theta_{2\text{-ss}}\right)\\ +\left(k_{2\text{-dc}}-k_{1\text{-dc}}\right)\left.\dfrac{\partial\theta_1}{\partial E}\right|_{E_{dc}}+\left(k_{3\text{-dc}}-k_{1\text{-dc}}\right)\left.\dfrac{\partial\theta_2}{\partial E}\right|_{E_{dc}}\end{bmatrix}\right\}E_{ac} \tag{5.82}$$

The faradaic admittance (Y_F) is then

$$Y_F = \frac{i_{F\text{-ac}}}{E_{ac}} \simeq F\left\{\begin{bmatrix}\left(b_1k_{1\text{-dc}}\left(1-\theta_{1\text{-ss}}-\theta_{2\text{-ss}}\right)+b_2k_{2\text{-dc}}\theta_{1\text{-ss}}+b_3k_{3\text{-dc}}\theta_{2\text{-ss}}\right)\\ +\left(k_{2\text{-dc}}-k_{1\text{-dc}}\right)\left.\dfrac{\partial\theta_1}{\partial E}\right|_{E_{dc}}+\left(k_{3\text{-dc}}-k_{1\text{-dc}}\right)\left.\dfrac{\partial\theta_2}{\partial E}\right|_{E_{dc}}\end{bmatrix}\right\} \tag{5.83}$$

The inverse of the faradaic admittance yields the faradaic impedance. $Z_F = 1/Y_F$.

The derivation is tedious, and the expressions are lengthy. However, the impedance can be calculated easily at any frequency, in programming environments such as MATLAB®, Mathematica®, Maple®, or Mathcad®. The impedance spectrum calculations, evaluated at many frequencies, hardly take a few seconds in modern desktop computers. With some effort, these calculations can also be performed in low-level programming languages such as C++, so that they would be even faster.

A few sample spectra evaluated for a few sets of parameters are given in Figure 5.8a–d. The reader may note that other than the capacitive loop at high frequencies, two more loops can be seen. Depending on the parameter values, and the frequency range employed, we can observe 0, 1, or 2 more loops, which may be capacitive or inductive. These data can be modeled using a circuit with two Maxwell pairs (Figure 4.3), and the modeling is left as an exercise.

5.1.7 More Reactions with Two Adsorbed Intermediate

The following is another example of reactions with two adsorbed intermediate species.

$$\mathrm{M} \underset{k_{-1}}{\overset{k_1}{\rightleftarrows}} \mathrm{M}^{+}_{\mathrm{ads}} + \mathrm{e}^{-} \tag{5.84}$$

$$\mathrm{M}^{+}_{\mathrm{ads}} \underset{k_{-2}}{\overset{k_2}{\rightleftarrows}} \mathrm{M}^{2+}_{\mathrm{ads}} + \mathrm{e}^{-} \tag{5.85}$$

$$\mathrm{M}^{2+}_{\mathrm{ads}} \xrightarrow{k_3} \mathrm{M}^{2+}_{\mathrm{sol}} \tag{5.86}$$

Here the first and second steps are reversible. Besides, the third step is a chemical step and not an electrochemical step. One can certainly come up with more variations. A parallel direct dissolution is a possibility, and a parallel pathway producing M^{2+}_{ads} directly from M is a possibility. Many more mechanisms involving two adsorbed intermediate species can be proposed.

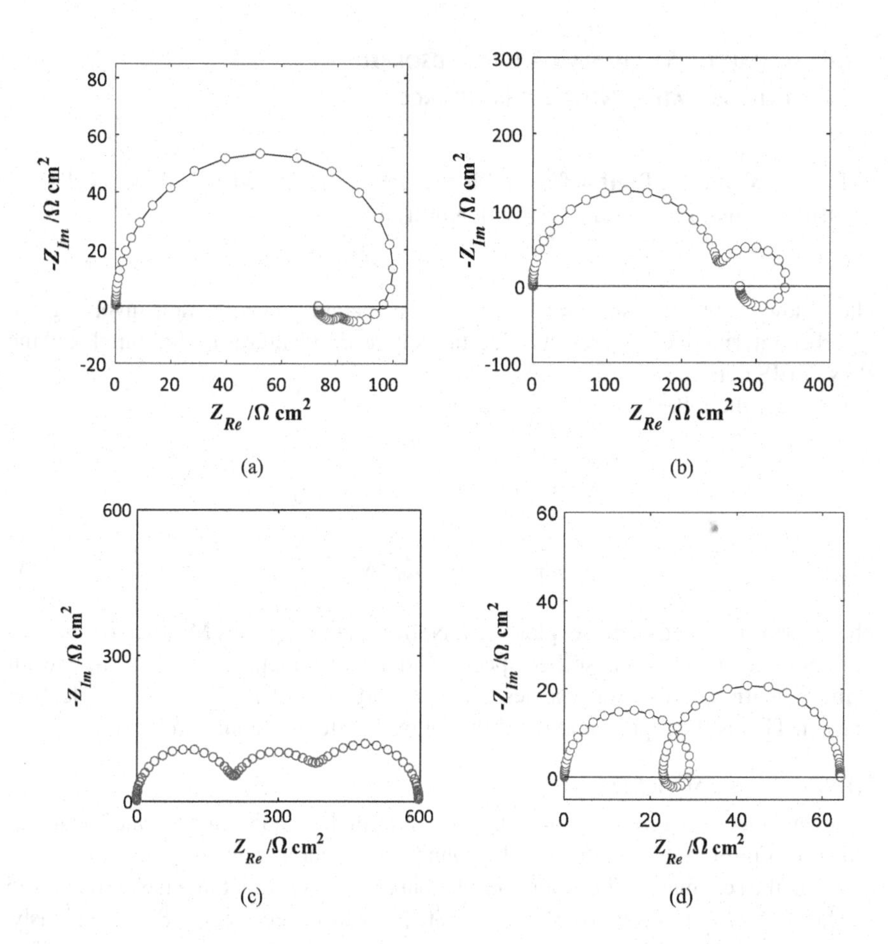

FIGURE 5.8 (a–d) Sample complex plane plots of the impedance of a reaction with two adsorbed intermediates (Eqs. 5.68–5.70). The parameter sets and *dc* potential employed are given in Table 5.4. $C_{dl} = 10\ \mu F\ cm^{-2}$.

TABLE 5.4
Kinetic Parameters and E_{dc} Values Employed in Simulating the Spectra Presented in Figure 5.8

	k_{10}	k_{20}	k_{30}	b_1	b_2	b_3	Γ	E_{dc}
Figure No.		**(mol cm^{-2}s^{-1})**			**(V^{-1})**		**(mol cm^{-2})**	**(V)**
5.8a	10^{-13}	5×10^{-8}	5×10^{-11}	28	9	19	10^{-8}	0.35
5.8b	10^{-11}	10^{-11}	10^{-9}	17	10	5	10^{-8}	0.50
5.8c	2×10^{-9}	2×10^{-10}	10^{-9}	17	13	2	2×10^{-8}	0.30
5.8d	10^{-14}	10^{-9}	10^{-8}	30	15	0	10^{-8}	0.50

5.1.8 Catalytic Mechanism – One Adsorbed Intermediate – Negative Impedance

In 1957, **Von K. F. Bonhoeffer** and **K. E. Heusler** proposed a catalytic mechanism to explain the anodic dissolution of iron.

The following mechanism is an example of a second-order reaction involving the adsorbed intermediate and results in certain interesting features in the complex plane plots of EIS data.

Consider the following two-step reaction

$$M \xrightarrow{k_1} M^+_{ads} + e^- \text{ followed by} \tag{5.87}$$

$$M^+_{ads} + M \xrightarrow{k_2} M^+_{sol} + M^+_{ads} + e^- \tag{5.88}$$

The second step needs some explanation. Notice that the species M^+_{ads} occurs on both sides of the reaction. An adsorbed species and a vacant metal site interact and result in the formation of a soluble species and an adsorbed species along with an electron transfer. This is the representation of the following steps in equation form.

5.1.8.1 Physical Picture

Imagine a metal surface with one adsorbed intermediate adjacent to a vacant site, as shown in Figure 5.9. An anion in the solution may come near the intermediate, as shown in the schematic. The anion would repel electrons present in the bare metal site and facilitate the conversion of the vacant site to an adsorbed species. Simultaneously, the anion would *pull* the adsorbed intermediate out into the solution phase as M^+_{sol}. Thus, the bare metal is converted to the adsorbed intermediate along with an electron transfer, and simultaneously, the adsorbed intermediate is converted into dissolved species, exposing a vacant metal site below. The net production/consumption of the species M^+_{ads} is effectively zero for the overall reaction, but it certainly participates in the reaction. Hence it is called a *catalytic mechanism*.

5.1.8.2 An Issue with the Steady-State Solution

If we write the charge balance equation for faradaic current density, we get

$$i_F = F\left\{k_1(1-\theta) + k_2\theta(1-\theta)\right\} \tag{5.89}$$

The mass balance equation, however, is not affected by the second step and hence it is written as

$$\Gamma\frac{d\theta}{dt} = k_1(1-\theta) \tag{5.90}$$

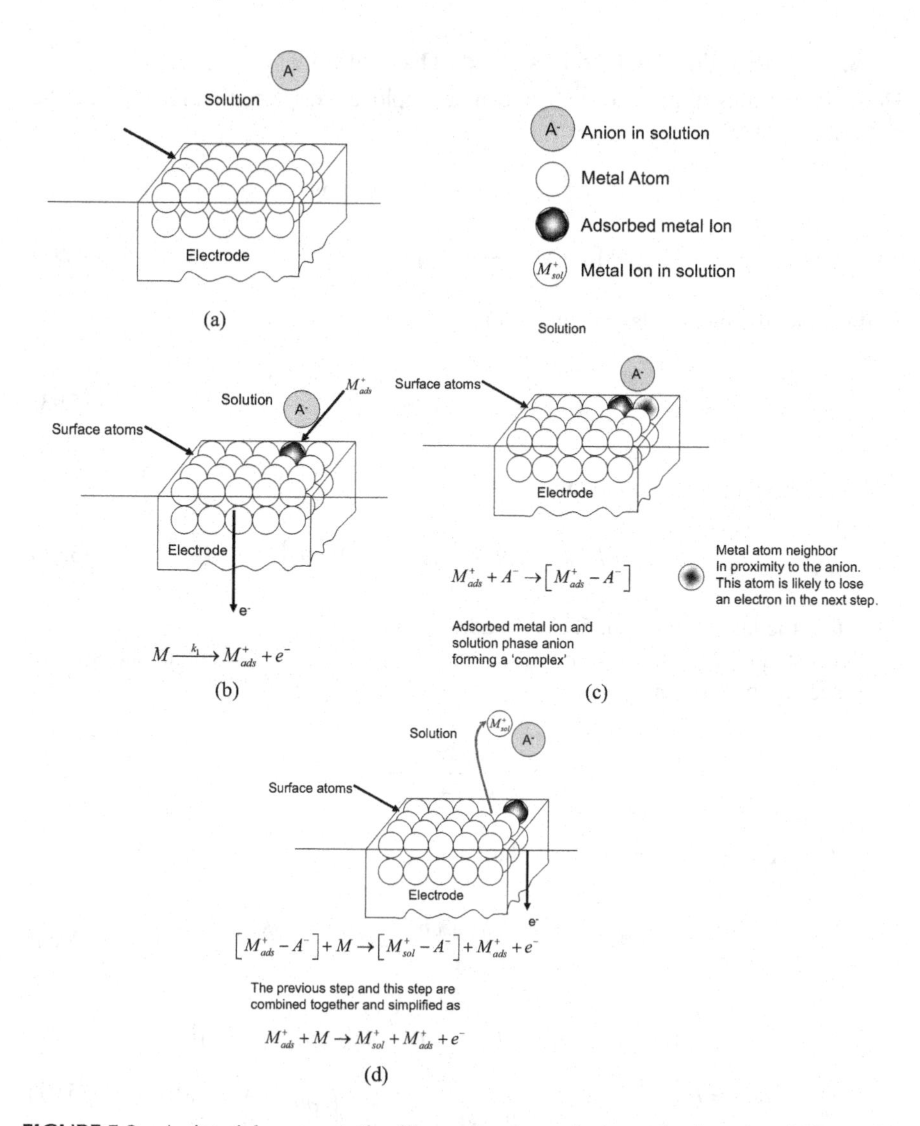

FIGURE 5.9 A pictorial representation illustrating the catalytic mechanism given in Eqs. 5.87 and 5.88. (a) A metal surface is exposed to the solution. (b) The first step where a surface atom loses an electron and becomes an 'adsorbed' intermediate ion is shown. The second step can be visualized in two parts. (c) The first part, where the anion comes close to the adsorbed ion and tends to be associated with a vacant neighbor. (d) The second part, where the adsorbed intermediate goes into the solution leaving a new surface atom (which was below this ion) exposed to the solution, is shown. Simultaneously, the 'vacant neighbor' loses an electron and becomes an adsorbed intermediate.

The only steady-state solution is that the surface coverage θ is unity, and the reaction will not proceed any further. The first step produces adsorbed intermediate, and the second step does not consume it at all. This reaction mechanism can never represent a steady dissolution.

5.1.8.3 A Variation That Admits Steady Dissolution

However, variants of this mechanism can be employed, e.g., the first reaction can be reversible.

$$\mathrm{M} \underset{k_{-1}}{\overset{k_1}{\rightleftharpoons}} \mathrm{M}^+_{\mathrm{ads}} + \mathrm{e}^- \text{ followed by} \tag{5.91}$$

$$\mathrm{M}^+_{\mathrm{ads}} + \mathrm{M} \xrightarrow{k_2} \mathrm{M}^+_{\mathrm{sol}} + \mathrm{M}^+_{\mathrm{ads}} + \mathrm{e}^- \tag{5.92}$$

In this case, the mass balance equation is

$$\Gamma \frac{\mathrm{d}\theta}{\mathrm{d}t} = k_1(1-\theta) - k_{-1}\theta \tag{5.93}$$

and the charge balance equation is

$$i_F = F\left\{k_1(1-\theta) - k_{-1}\theta + k_2\theta(1-\theta)\right\} \tag{5.94}$$

Note that the last term in Eq. 5.94 is quadratic in θ.

Here, steady dissolution is possible, and at any given dc potential, the surface coverage can be written as

$$\theta_{\mathrm{ss}} = \frac{k_{1\text{-dc}}}{k_{1\text{-dc}} + k_{-1\text{-dc}}} \tag{5.95}$$

From these equations, we can show that

$$\frac{\mathrm{d}\theta}{\mathrm{d}E} = \frac{b_1 k_{1\text{-dc}}(1-\theta_{\mathrm{ss}}) - b_{-1}k_{-1\text{-dc}}\theta_{\mathrm{ss}}}{k_{1\text{-dc}} + k_{-1\text{-dc}} + j\omega\Gamma} = \frac{(b_1 - b_{-1})k_{-1\text{-dc}}\theta_{\mathrm{ss}}}{k_{1\text{-dc}} + k_{2\text{-dc}} + j\omega\Gamma} \tag{5.96}$$

$$i_{F\text{-ac}} \simeq F\left\{\begin{bmatrix} b_1 k_{1\text{-dc}}(1-\theta_{\mathrm{ss}}) + b_{-1}k_{-1\text{-dc}}\theta_{\mathrm{ss}} + b_2 k_{2\text{-dc}}\theta_{\mathrm{ss}}(1-\theta_{\mathrm{ss}}) \\ +(k_{2\text{-dc}} - 2k_{2\text{-dc}}\theta_{\mathrm{ss}} - k_{1\text{-dc}} - k_{-1\text{-dc}})\left(\left.\dfrac{\partial\theta}{\partial E}\right|_{E_{\mathrm{dc}}}\right) \end{bmatrix}\right\} E_{\mathrm{ac}} \tag{5.97}$$

and

$$Y_F = \frac{i_{F\text{-ac}}}{E_{\mathrm{ac}}} = F\left\{\begin{bmatrix} b_1 k_{1\text{-dc}}(1-\theta_{\mathrm{ss}}) + b_{-1}k_{-1\text{-dc}}\theta_{\mathrm{ss}} + b_2 k_{2\text{-dc}}\theta_{\mathrm{ss}}(1-\theta_{\mathrm{ss}}) \\ +(k_{2\text{-dc}} - 2k_{2\text{-dc}}\theta_{\mathrm{ss}} - k_{1\text{-dc}} - k_{-1\text{-dc}})\left(\dfrac{(b_1 - b_{-1})k_{-1\text{-dc}}\theta_{\mathrm{ss}}}{k_{1\text{-dc}} + k_{2\text{-dc}} + j\omega\Gamma}\right) \end{bmatrix}\right\} \tag{5.98}$$

The catalytic step leads to some interesting behavior. When the potential is increased (i.e., made more anodic), the rate constant values will also increase for k_1 and k_2 and will decrease for k_{-1}. When the potential is just above the OCP, an increase in potential will lead to an increase in current. However, at large anodic potentials, when the surface coverage of the adsorbed intermediate is very high, there will not be many vacant sites, and the dissolution by the catalytic mechanism will decrease. Thus, the current will decrease with an increase in potential.

A plot of *dc* faradaic current *vs.* *dc* potential for one set of kinetic parameters is given in Figure 5.10a. At small dc potentials, the slope of the line representing i_{dc} *vs.* E_{dc} is positive, and at large dc potentials, the slope is negative. Correspondingly, at large dc potentials and low frequencies, the complex plane plots show a negative (differential) resistance. Sample EIS data, in complex plane plot representation, are also shown in Figure 5.10b–d. It is seen that the low-frequency loop may be capacitive (Figure 5.10b) or inductive (Figure 5.10c) or maybe capacitive with a negative

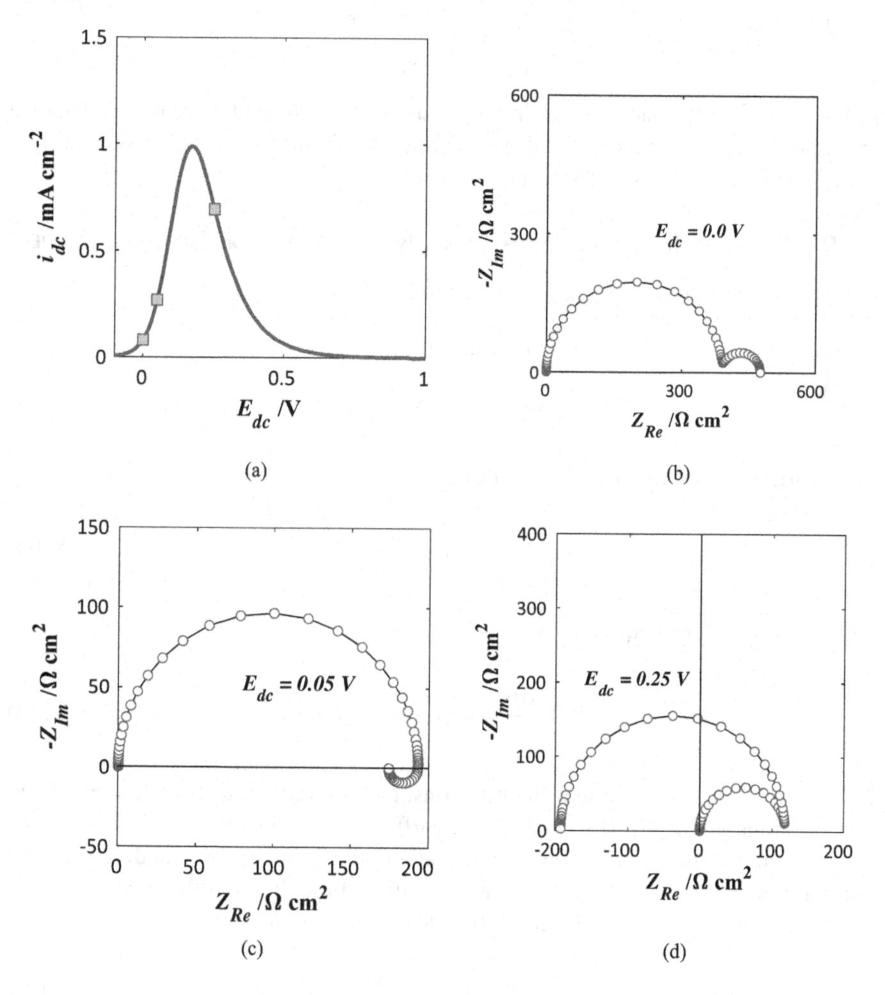

FIGURE 5.10 (a) Steady-state potential current diagram of a electrochemical reaction given in Eqs. 5.87 and 5.88 (catalytic mechanism). The parameter values are $k_{10} = 10^{-9}$mol cm^{-2}s^{-1}, $b_1 = 10\,V^{-1}$, $k_{-10} = 10^{-8}$mol cm^{-2}s^{-1}, $b_{-1} = -10\,V^{-1}$, $k_{20} = 10^{-8}$mol cm^{-2}s^{-1}, $b_2 = 10\,V^{-1}$, $\Gamma = 10^{-8}$mol cm^{-2}, and $C_{dl} = 10$ μF cm^{-2}. Corresponding complex plane plots of impedance spectra at (b) $E_{dc} = 0\,V$, (c) $E_{dc} = 0.05\,V$, and (d) $E_{dc} = 0.25\,V$ *vs.* the OCP.

resistance in parallel (Figure 5.10d). The spectra in Figure 5.10b–d can be modeled using a circuit with one Maxwell pair (Figure 4.2), as long as some of the elements are allowed to hold negative values, and this is left as an exercise to the reader.

In general, for a given system, R_p does not show a monotonic trend with overpotential. Although in the examples shown here, R_t shows a decrease with an increase in overpotential, in general, a monotonic trend cannot be expected, and the literature shows that, depending on the actual system studied, R_t can increase or decrease with overpotential.

Instead of making the first step reversible, a third step can be added to bring about a steady dissolution, i.e.,

$$M \xrightarrow{k_1} M^{+}_{ads} + e^{-} \tag{5.99}$$

$$M^{+}_{ads} + M \xrightarrow{k_2} M^{+}_{sol} + M^{+}_{ads} + e^{-} \tag{5.100}$$

$$M^{+}_{ads} \xrightarrow{k_3} M^{+}_{sol} \tag{5.101}$$

and we can show that steady dissolution is possible. The formulation of mass balance and charge balance equations and linearization to obtain the expression for impedance for this mechanism is left as an exercise.

5.1.9 Two-Step Reaction, With the Frumkin Adsorption Isotherm Model

Let us consider a two-step reaction, given in Eq. 5.49

$$\begin{aligned} &M \xrightarrow{k_1} M^{+}_{ads} + e^{-} \\ &M^{+}_{ads} \xrightarrow{k_2} M^{2+}_{sol} + e^{-} \end{aligned} \tag{5.102}$$

The charge balance equation is given by

$$i_F = F\left[k_1(1-\theta) + k_2\theta\right] \tag{5.103}$$

and the mass balance equation is

$$\Gamma\frac{d\theta}{dt} = k_1(1-\theta) - k_2\theta \tag{5.104}$$

Now, instead of assuming that the rate constants are independent of fractional surface coverage of the adsorbed intermediate (θ), let us consider the possibility that the rate constants depend on θ. The Frumkin adsorption isotherm model describes that the equilibrium constant depends exponentially on θ. Although the model does set any constraint on the dependence of the rate constants on θ, setting

$$k_i = k_{i0}e^{\beta_i g\theta}e^{b_i E} \tag{5.105}$$

yields the correct form of the equation for the equilibrium constant. If we assume that the charged intermediate species M^{+}_{ads} repel each other, then the interaction parameter

g can be set to positive values. Since the first step in Eq. 5.102 leads to the formation of M^+_{ads}, then the rate constant k_1 should decrease with an increase in θ. Therefore, the parameter β_1 is negative, and by a similar argument, the parameter β_2 is positive.

The steady-state value of θ can be found by setting Eq. 5.106 to zero, i.e.,

$$k_{10}e^{\beta_1 g\theta_{ss}}e^{b_1E_{dc}}(1-\theta_{ss})-k_{20}e^{\beta_2 g\theta}e^{b_2E_{dc}}\theta_{ss}=0 \tag{5.106}$$

This equation is nonlinear in θ_{ss}, and at a given E_{dc}, the θ_{ss} value can be determined by solving Eq. 5.106 using numerical methods.

The mass balance Eq. 5.104 can be linearized as follows. At first, we expand the rate constant in the Taylor series and linearize it.

$$k_1 \simeq k_{10}e^{\beta_1 g\left(\theta_{ss}+\frac{d\theta}{dE}E_{ac}\right)}e^{b_1(E_{dc}+E_{ac})} \tag{5.107}$$

$$= k_{10}e^{\beta_1 g\theta_{ss}}e^{\beta_1 g\left(\frac{d\theta}{dE}E_{ac}\right)}e^{b_1E_{dc}}e^{b_1E_{ac}} \tag{5.108}$$

$$= k_{1\text{-dc}}\left[1+\beta_1 g\left(\frac{d\theta}{dE}E_{ac}\right)+\cdots\right]\left[1+b_1E_{ac}+\cdots\right] \tag{5.109}$$

where

$$k_{1\text{-dc}} = k_{10}e^{\beta_1 g\theta_{ss}}e^{b_1\theta_{ss}} \tag{5.110}$$

$$k_1 \simeq k_{1\text{-dc}}\left[1+\beta_1 g\left(\frac{d\theta}{dE}E_{ac}\right)+b_2E_{ac}\right] \tag{5.111}$$

Similarly, the rate constant k_2 can also be expanded in the Taylor series, and when E_{ac} is small, the second and higher-order terms can be neglected. Then, Eq. 5.104 can be written as

$$j\omega\Gamma E_{ac}\frac{d\theta}{dE} = \left\{\begin{aligned} & k_{1\text{-dc}}\left[1+\beta_1 g\left(\frac{d\theta}{dE}E_{ac}\right)+b_1E_{ac}\right]\left(1-\theta_{ss}-\frac{d\theta}{dE}E_{ac}\right) \\ & -k_{2\text{-dc}}\left[1+\beta_2 g\left(\frac{d\theta}{dE}E_{ac}\right)+b_2E_{ac}\right]\left(\theta_{ss}+\frac{d\theta}{dE}E_{ac}\right) \end{aligned}\right\} \tag{5.112}$$

The steady-state equation shows that $k_{1\text{-dc}}(1-\theta_{ss})-k_{2\text{-dc}}\theta_{ss}=0$. Using this, we can obtain

$$j\omega\Gamma E_{ac}\frac{d\theta}{dE} = \left\{\begin{aligned} & \left(b_1k_{1\text{-dc}}(1-\theta_{ss})-b_2k_{2\text{-dc}}\theta_{ss}\right)E_{ac} \\ & +\left[k_{1\text{-dc}}(1-\beta_1 g(1-\theta_{ss}))+k_{2\text{-dc}}(1+\beta_2 g)\right]\left(\frac{d\theta}{dE}E_{ac}\right) \end{aligned}\right\} \tag{5.113}$$

The above equation can be rewritten as

$$\frac{\mathrm{d}\theta}{\mathrm{d}E} = \frac{\left\{k_{1\mathrm{dc}} b_1 (1-\theta_{\mathrm{ss}}) - k_{2\mathrm{dc}} b_2 \theta_{\mathrm{ss}}\right\}}{\left\{k_{1\mathrm{dc}}\left(1-\beta_1 g(1-\theta_{\mathrm{ss}})\right) + k_{2\mathrm{dc}}\left(1+\beta_2 g \theta_{\mathrm{ss}}\right) + j\omega\tau\right\}} \tag{5.114}$$

Next, we linearize the equation for faradaic current, Eq. 5.103.

$$i_F = i_{F\text{-dc}} + i_{F\text{-ac}} \simeq F\left\{\begin{array}{l} k_{1\text{-dc}}\left[1+\beta_1 g\left(\frac{\mathrm{d}\theta}{\mathrm{d}E}E_{\mathrm{ac}}\right)+b_1 E_{\mathrm{ac}}\right]\left[1-\theta_{\mathrm{ss}}-\frac{\mathrm{d}\theta}{\mathrm{d}E}E_{\mathrm{ac}}\right] \\ +k_{2\text{-dc}}\left[1+\beta_2 g\left(\frac{\mathrm{d}\theta}{\mathrm{d}E}E_{\mathrm{ac}}\right)+b_2 E_{\mathrm{ac}}\right]\left[\theta_{\mathrm{ss}}-\frac{\mathrm{d}\theta}{\mathrm{d}E}E_{\mathrm{ac}}\right] \end{array}\right\} \tag{5.115}$$

After subtracting the dc current terms, ($i_{F\text{-dc}}$) from the above equation, we can show that

$$i_{F\text{-ac}} \simeq F\left\{\begin{array}{l} k_{1\mathrm{dc}}\left[b_1(1-\theta_{\mathrm{ss}})+\beta_1 g(1-\theta_{\mathrm{ss}})\frac{\mathrm{d}\theta}{\mathrm{d}E}-\frac{\mathrm{d}\theta}{\mathrm{d}E}\right] \\ +k_{2\mathrm{dc}}\left[b_2\theta_{\mathrm{ss}}+\beta_2 g\theta_{\mathrm{ss}}\frac{\mathrm{d}\theta}{\mathrm{d}E}+\frac{\mathrm{d}\theta}{\mathrm{d}E}\right] \end{array}\right\} E_{\mathrm{ac}} \tag{5.116}$$

The faradaic admittance can be written as

$$Y_F = (Z_F)^{-1} = \frac{i_{\mathrm{ac}}}{E_{\mathrm{ac}}} = F\left\{\begin{array}{l} k_{1\mathrm{dc}}\left[b_1(1-\theta_{\mathrm{ss}})+\left(\beta_1 g(1-\theta_{\mathrm{ss}})-1\right)\frac{\mathrm{d}\theta}{\mathrm{d}E}\right] \\ +k_{2\mathrm{dc}}\left[b_2\theta_{\mathrm{ss}}+\left(\beta_2 g\theta_{\mathrm{ss}}+1\right)\frac{\mathrm{d}\theta}{\mathrm{d}E}\right] \end{array}\right\} \tag{5.117}$$

The charge-transfer resistance (R_t) is given by

$$R_t^{-1} = F\left\{k_{1\mathrm{dc}} b_1(1-\theta_{\mathrm{ss}}) + k_{2\mathrm{dc}} b_2 \theta_{\mathrm{ss}}\right\} \tag{5.118}$$

An example of a polarization plot and EIS data as a complex plane plot are given in Figure 5.11. Since only one adsorbed intermediate is present in this mechanism, we can expect, at the most, two loops in the complex plane plots. The high-frequency loop arising from the double-layer capacitance in parallel with the charge-transfer resistance would form the first loop. The faradaic processes can yield a loop at mid and low frequencies, and this may be capacitive or inductive, depending on the values of kinetic parameters. This example shows that, although it is more complicated than the Langmuir isotherm model, it is possible to linearize the equations arising from Frumkin (or other) isotherms. In the literature, the hydrogen oxidation in a PEM fuel cell, in the presence of CO poisoning, has been described using such isotherms (Springer et al. 2001), and expression for faradaic impedance has been reported (Fasmin and Ramanathan 2015).

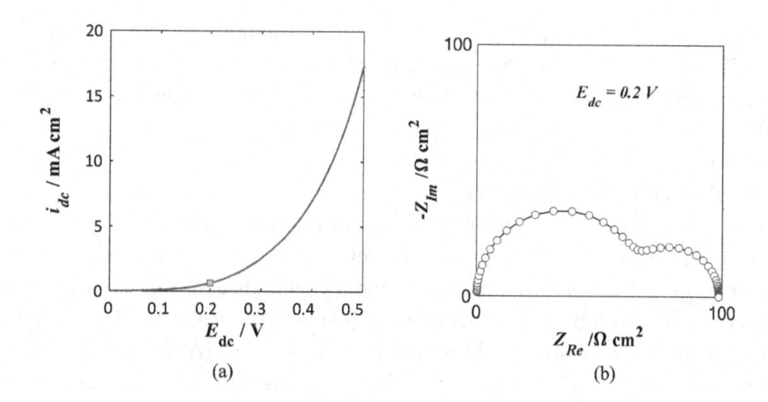

FIGURE 5.11 (a) Steady-state potential current diagram of a electrochemical reaction given in Eq. 5.102 with the Frumkin isotherm model. The parameter values are $k_{10} = 10^{-9}$mol cm^{-2}s^{-1}, $b_1 = 9\,V^{-1}$, $k_{20} = 10^{-11}$ moles cm^{-2}s^{-1}, $b_2 = 37\,V^{-1}$, $\beta_1 = -0.3$, $\beta_2 = 0.5$, $g = 20$, $\Gamma = 10^{-9}$mol cm^{-2}, and $C_{dl} = 10$ μF cm^{-2}. Corresponding complex plane plots of impedance spectra at (b) $E_{dc} = 0.2\,V$ *vs*. OCP.

5.1.10 Identification of a Reaction Mechanism – EIS Data as Complex Plane Plots

By comparing this with the results of the previous mechanism, one may propose that if the complex plane diagram of a spectrum consists of *N* loops, then the capacitive loop at high frequency can be assigned to the double-layer. We need a mechanism with at least (*N* – 1) adsorbed intermediates to simulate the other loops (either capacitive or inductive). This statement is not exactly correct (Harrington 1996), but in many cases, it is a good rule of thumb. Remember that we assume that (a) there is no film formation on the surface, for the film can give rise to additional loops, and (b) the net rate is limited by kinetics and not by diffusion. For a proper method to determine the number of loops (time constants or critical frequencies), please refer to Harrington (1996).

For example, mechanisms given in Eqs. 5.18, 5.19, 5.49, 5.57, 5.58, 5.65–67, 5.87, and 5.88 have only one adsorbed intermediate. This can give rise to an inductive or capacitive loop at the low frequencies. At high frequencies, a capacitive loop, corresponding to the double-layer, would be present. Thus, these mechanisms can give rise to, *at the most*, two loops. If experimental data show that there are three loops, and if there is evidence to suggest that there is no film formation on the surface, then we can eliminate these mechanisms from consideration. We should consider only the mechanisms that have two or more adsorbed intermediates to model the data.

Likewise, if the data show negative resistance (Figure 5.7d or 5.10d), then we can eliminate the mechanisms given in Eqs. 5.18, 5.19 5.49, and 5.65–5.67, because they cannot give rise to negative resistance. It does not mean that we can conclude that mechanism in Eqs. 5.57 and 5.58 as the right choice. This is because, even the mechanism given in Eqs. 5.87 or 5.88 can give rise to negative resistance. Based on the system analyzed, we have to consider what the likely candidate mechanisms are and evaluate the candidates that can give rise to the observed patterns. If more than one candidate mechanism can give rise to the observed patterns, all of them should be evaluated, and the one which fits the data best should be selected.

> **Examples from the Literature**
> A few examples of EIS patterns, either simulated or observed experimentally, are presented in Figure 5.12 for reference. Some of them include the presence of a film on the electrode surface, which also contributes to the total impedance. The list is given here just to give a flavor of the type of complex plane plots observed and simulated using RMA. In the figures, the arrow tip points to the direction of higher frequency, and the base of the arrow is at the lower frequency

5.1.10.1 Challenges in Identifying a Reaction Mechanism

EIS RMA modeling can show that a particular mechanism is a possibility; i.e., it can explain the observed results. It can also show, in many cases, that a few other mechanisms will not be able to explain the observed results regardless of the kinetic parameters one chooses. However, it cannot unequivocally show that a particular mechanism is the only mechanism (with a particular kinetic parameter set) that can explain the data.

More complicated mechanisms (i.e., a superset of the mechanism under consideration) can always provide an equally good or better match. In some cases, other mechanisms with the same number of parameters may give an equally good match. In all the cases, in addition to "matching" (i.e., modeling) the impedance data, the "dc current *vs.* potential" simulated data must also be "matched" with the experimentally measured values. Although one cannot guarantee that the model identified is unique, fitting the impedance data at multiple dc potentials, along with fitting the dc current *vs.* potential, will enhance the confidence in the model identification.

5.2 PARAMETER ESTIMATION

5.2.1 Error Calculation

In the previous section, we did not elaborate on the method used to estimate the parameters. Given a mechanism and a set of parameters, it is easy to simulate the spectrum at a potential. Since impedance is expressed as a complex quantity and the equations' relative impedance to the potential (E_{dc}) and frequency are nonlinear, the regression method used is CNLS. Let $Z_{\text{Re-exp}}$ and $Z_{\text{im-exp}}$ represent the real and imaginary parts of the experimental data, respectively. Let $Z_{\text{Re-sim}}$ and $Z_{\text{im-sim}}$ represent the real and imaginary parts of the simulated data, respectively. Then the total residual sum squared error (RSS) is given by

$$\text{RSS} = \sum_{\text{all } f, \text{all } V_{dc}} w_{\text{Re}} \left(Z_{\text{Re-exp}} - Z_{\text{Re-sim}} \right)^2 + \sum_{\text{all } f, \text{all } V_{dc}} w_{\text{Im}} \left(Z_{\text{Im-exp}} - Z_{\text{Im-sim}} \right)^2 \quad (5.119)$$

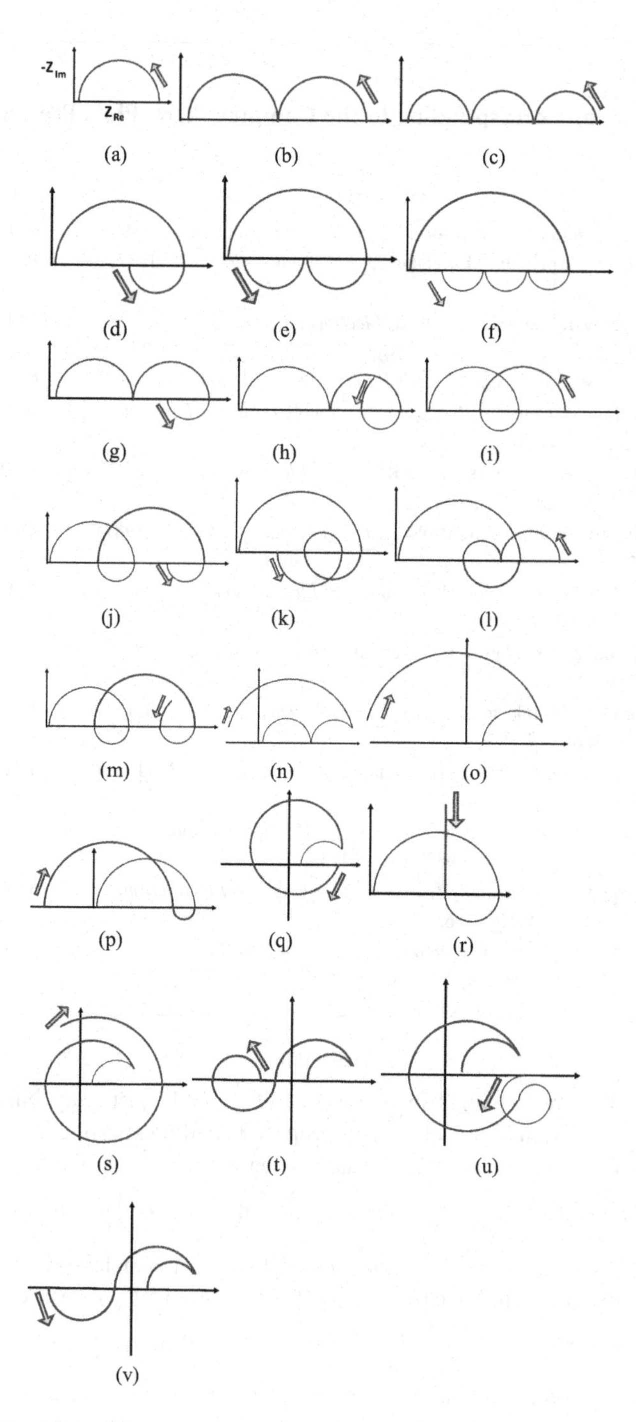

FIGURE 5.12 Approximate representation (not drawn to scale) of various patterns of complex plane plots reported in the literature. The references are listed in Table 5.5.

TABLE 5.5
List of References Corresponding to the Complex Plane Plots Presented in Figure 5.12

No.	Reference	Patterns
1	R.D. Armstrong, K. Edmonson, *Electrochim. Acta* 18 (1973) 937–943	A, B, D, O
2	B. Bechet, I. Epelboin, M. Keddam, *J. Electroanal. Chem.* 76 (1977) 129–134	B, E, I
3	M. Bojinov, I. Betova, R. Raicheff, *Electrochim. Acta* 41 (1996) 1173–1179	D, I
4	M. Bojinov, *J. Solid State Electrochem.* 1 (1997) 161–171	G, O, P, R, V,
5	M. Bojinov, S. Cattarin, M. Musiani, B. Tribollet, *Electrochim. Acta* 48 (2003) 4107–4117	B, I, P
6	G.G.O. Cordeiro, O.E. Barcia, O.R. Mattos, *Electrochim. Acta* 38 (1993) 319–324	A, B, I,
7	I. Epelboin, M. Joussellin, R. Wiart, *J. Electroanal. Chem.* 119 (1981) 61–71	D, E, I, K,
8	M. Keddam, O.R. Mattos, H. Takenouti, *J. Electrochem. Soc.* 128 (1981) 257–266	E, F, G, I, J, L, N, O, P
9	M. Keddam, O.R. Mattos, H. Takenouti, *J. Electrochem. Soc.* 128 (1981) 266–274	C
10	M. Keddam, H. Takenouti, N. Yu, *J. Electrochem. Soc.* 132 (1985) 2561–2566	G, P, Q
11	M. Keddam, O.R. Mattos, H. Takenouti, *Electrochim. Acta.* 31 (1986) 1147–1158	B, C, H, I, L, M, N, O, S, T
12	D.D. Macdonald, S. Real, S.I. Smedley, M. Urquidi-Macdonald, *J. Electrochem. Soc.* 135 (1988) 2410–2414	B, I
13	M. Urquidi-Macdonald, S. Real, D. D. Macdonald, *J. Electrochem. Soc.* 35 (1990) 1559–1566	U
15	I. Danee, M. Jafarian, F. Forouzandeh, F. Gobal, M.G. Mahjani, *J. Phys. Chem. B* 112 (2008) 15933–15940	D, Q

Here, w_{Re} and w_{Im} are the weighting functions. They can be just unity, but other more complicated expressions have also been proposed. If the experiments are repeated many times, and if the mean and standard deviation (σ) of Z_{Re} and Z_{Im} at each E_{dc} and f is available, then the weighting function can be $w_{\text{Re}} = 1/\sigma_{\text{Re}}^2$ and $w_{\text{Im}} = 1/\sigma_{\text{Im}}^2$. This ensures that the data with more variability are given less weight. Another choice is to normalize the difference with the experimentally measured value, i.e., $w_{\text{Re}} = 1/Z_{\text{Re-exp}}^2$ and $w_{\text{Im}} = 1/Z_{\text{Im-exp}}^2$. Obviously, assigning $w_{\text{Re}} = w_{\text{Im}} = 1$ is the simplest, in terms of computational requirements, and sometimes they give more or less the same estimates for the parameters as other weighting functions do. When comparing the goodness of fit of different models, it is important to consider the number of parameters employed, and AIC_c is a better metric than RSS.

5.2.1.1 Data Form

Why should we use the real and imaginary form of the impedance data to calculate the error? Why not use the admittance values (Y_{Re} and Y_{Im}), or the phase and magnitude (θ and $|Z|$, or $\ln|Z|$)? i.e., can we use the following expression, and would we still get the same parameter estimates?

$$RSS = \sum_{\text{all } f, \text{all } E_{dc}} w_{Re}\left(Y_{Re-exp} - Y_{Re-sim}\right)^2 + \sum_{\text{all } f, \text{all } E_{dc}} w_{Im}\left(Y_{Im-exp} - Y_{Im-sim}\right)^2 \quad (5.120)$$

If the data are absolutely noise-free, and if the model is perfect, then the error would be zero, and it would not matter which form of the data was used for the CNLS analysis. In practice, the parameters estimated will differ based on which form of the data was used for error minimization. Most commercial software use Z_{Re} and Z_{Im} for the analysis. The ideal case would be to use the data in the form they are measured.

5.2.1.2 Constraint

Besides, the model predictions of the dc currents should match with the measured dc currents at these potentials. One method of including this requirement is to include the error in the dc current in the above equations, but that complicates the equations further. An alternate method is to use an *acceptable % of the variation* in the dc current and use parameter values, which would simulate a dc current matching the measured dc current (within the acceptable %, such as 2%). Thus, the error calculated in Eqs. 5.119 or 5.120 has to be minimized under the constraint that the error in dc current (i.e., the difference between the simulated current values and measured current values over a range of potential) is less than some specified value or %.

5.2.1.3 Software

If an equivalent electrical circuit is used to fit a spectrum data, then the values of the elements can be obtained quickly and accurately using commercial or noncommercial software. They can handle an arbitrary number of resistors, capacitors, and inductors and in many combinations such as series and parallel. However, to the best of our knowledge, there is no software in the public domain (commercial or otherwise) to simulate the spectrum for a general reaction mechanism and obtain the parameter fit from a set of spectra. The availability of such software will enable more users to employ mechanistic analysis of impedance data. At the minimum, the software must provide a wide range of mechanisms that the user can select to model the data. Ideally, the user should be able to specify his/her mechanism and use the software to obtain the best-fit parameters.

5.2.1.4 Direct Optimization

How do we estimate the parameter values? In theory, one can propose a mechanism, write the linearized equations, and obtain an expression for impedance. The impedance would depend on the two independent variables, viz. potential (E) and frequency (f). The parameters would be k_{i0}, b_i, Γ, R_{sol}, and C_{dl} (or a CPE with parameters Q and α). Now, any optimization program can be used to minimize the error between the model values and the experimentally measured values. Ideally, the model spectra would exactly match the measured spectra.

However, in practice, the problem is very challenging and is not amenable to the direct parameter fitting approach since the equations are highly nonlinear in f and E_{dc}. In our experience, unless the initial guesses are close to the final values, this approach does not seem to work. We have adopted a method described in Section 5.2.1.6.

5.2.1.5 Utilizing EEC Results

Here we describe a method that could be used if all the spectra, at all the E_{dc} values, exhibit similar patterns. First, we model the data using an equivalent electrical circuit and then estimate the kinetic parameters using equations relating them with the values of electrical circuit parameters. Finally, using these as initial values, we run an optimization program that would directly simulate the spectra using the kinetic parameters for all E_{dc} values and estimate the kinetic parameters, which would yield the least error.

It is relatively easy to determine the solution resistance and the double-layer capacitance (or CPE parameters Q and n, for the double-layer, discussed in Chapter 6), using a reasonable circuit model. If the spectra appear to exhibit N number of time constants, one can employ a Maxwell, ladder, or Voigt circuit with N capacitances (Figures 3.11 and 3.12). In some cases, corresponding to each potential E_{dc}, one might get different values of C_{dl} or CPE parameters $\{Q$ and $n\}$ values and slightly different R_{sol} values.

Even if inductive loops are present, one can allow for negative values of resistances and capacitances and obtain the parameter values. In our experience, the models with inductances do not converge unless good initial values are chosen, whereas, for circuits with capacitances, the optimization algorithm seems to be less sensitive to the initial values. In any case, once the parameters are estimated for a circuit with capacitances, it is easy to calculate the corresponding values for an equivalent circuit with inductances. Once the circuit fit is done, the kinetic parameters can be estimated as follows.

One should choose realistic bounds for the parameters. The charge-transfer coefficient varies between 0 and 1. Since $b_i = {\alpha_i RT}/{F}$, at room temperature, this corresponds to a value between 0 and 38 V^{-1} for the parameters b_i. Note that in our notation, the values for reverse reactions (i.e., b_{-1}, b_{-2}, and so on) would be negative, and hence they will be between −38 and 0 V^{-1}. Similarly, if any particular step involves two electrons, such as $M \xrightarrow{k_1} M^{2+}_{ads} + 2e^-$, then the value of b_1 would vary between 0 and $2RT/F$, which is approximately 76 V^{-1}. The value of Γ is also bound based on the number of atomic sites per unit area for a given material.

The relationship between the kinetic parameters and the elements of Maxwell circuits for a mechanism involving up to four adsorbed intermediates is available in Tech Notes 12 – Studies of Reaction Mechanisms by Dr. Bruno Yeum, in the tutorial section of the ZSimpWin® software. A copy of this tech note may also be available on the web. Using the relationship, we can write programs utilizing optimization libraries (e.g., *fmincon* function in MATLAB®) to obtain the best values for the parameters.

While it is relatively easy to fit the electrical circuit model to the EIS data and obtain the estimates of R, C, and L, the variability in those values (i.e., confidence levels in R, C, and L) must also be carefully noted. In our experience, impedance spectra are very sensitive to the values of a few of the electrical circuit elements. At the same time, they are practically insensitive to large changes in the values of the other electrical elements. While estimating the kinetic parameters using the equations relating the kinetic parameters to the electrical elements, this factor must also be taken into consideration. That is, one may first fit the data to electrical elements and then attempt to use the relationship between the kinetic parameters and the circuit elements, to assess the value of the kinetic parameters. However, we have encountered the following problem. Even slight changes in the values of the electrical elements may cause a large change in the spectra. Thus, if the kinetic parameters are not related *exactly* to the electrical elements, the spectra simulated from these estimates of parameters differ drastically from the spectra produced by the electrical circuits and hence from the experimental spectra. We can use these kinetic parameters as initial estimates and write programs (again utilizing optimization libraries) to obtain the best fit, by relating the kinetic parameters to the impedance data.

5.2.1.6 Grid Search

Many times, the spectra at different E_{dc} values would exhibit different patterns, and it may become difficult to use a single circuit to model the data at each E_{dc}. Then the method discussed above cannot be adopted. Instead, we adopt the second method, known as the grid search. We scan the parameter range in small steps. That is, the values of parameters k_{10}, k_{20}, b_1, b_2, and so on are incremented in small steps, and for each parameter set, the spectra for all E_{dc} are generated and compared with the experimental data. Usually, the parameters k_{i0} and Γ are incremented on the log scale, while the parameters b_is are incremented on a linear scale. The program also compares the *shape* of the simulated spectra and the measured spectra, and only those that match well are considered as valid. On a modern desktop computer, in 2020. a few million parameter combinations can be evaluated in a second. Note that if there are 9 kinetic parameters, then scanning 20 levels in each parameter would take about 10 minutes, and with a greater number of parameters, the duration increases drastically. After identifying the set of parameters that reproduce the shape of the spectra correctly and are also reasonably close to the final values, an optimization algorithm is run using these as initial guesses.

5.3 RELEVANCE OF RMA

5.3.1 The Number of Parameters. EEC vs. RMA

A common criticism of the RMA methodology is, "For a given data, EEC usually has only a few unknown values (parameters to be fit), but RMA has many more unknown values. If you have a model with many parameters, you can fit anything. Does RMA really have any value?". The assumption in the question that EEC has fewer unknowns is usually wrong, and with that, the question loses its meaning. We will illustrate this with an example. A reaction with an adsorbed intermediate can give rise to the spectra shown in Figure 5.13a. The dc current *vs.* potential is plotted in Figure 5.13b. The parameter values are given in the figure caption. We assume that the solution resistance and mass transfer resistances are negligible. We can model this using the reaction given in Eq. 5.49, i.e., $M \xrightarrow{k_1} M^{+}_{ads} + e^{-} \xrightarrow{k_2} M^{2+}_{sol} + e^{-}$. (The data were, in fact, simulated using the impedance equation corresponding to this mechanism.) In RMA, the parameters are k_{10}, b_1, k_{20}, b_2, Γ, and C_{dl} (in total 6). The values of RMA parameters are given in the figure caption. The data can also be modeled using the circuit given in Figure 5.13c. In EEC, the parameters are C_{dl}, R_1, R_2, and C_2 (four in total). The best fit values are listed in Table 5.6.

However, note carefully that the RMA model uses one parameter set (six parameters) for *all* the spectra at all the E_{dc} values. At each E_{dc} value, the EEC uses three parameters (R_1, R_2, and C_2, along with the common parameter of C_{dl}). Hence, if the data are acquired at three different E_{dc} values, then EEC uses $3 \times 3 = 9 + 1 = 10$ parameter values. If we do not carefully note this, it would *appear* that EEC uses only four parameter values. Thus, in this example, it is not correct to infer that RMA has more unknowns than EEC, and this conclusion is true for most other cases as well. As we acquire EIS data for a particular system at more E_{dc} values, EEC will need many more adjustable parameters than RMA.

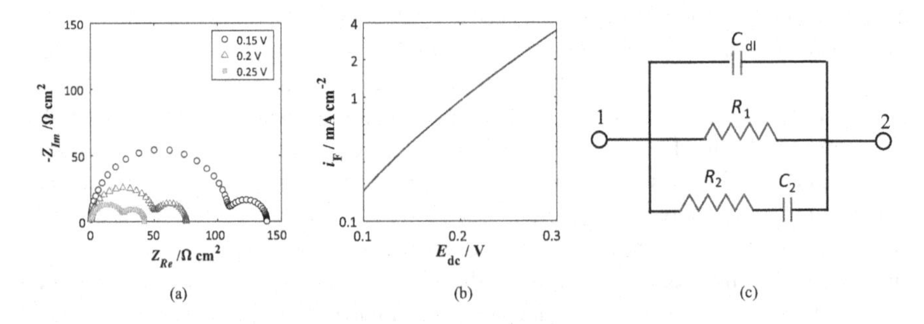

FIGURE 5.13 (a) Complex plane plots of impedance spectra generated from the reaction in Eq. 5.49 at three different dc potentials. The parameter values are $k_{10} = 10^{-10}$mol cm^{-2}s^{-1}, $b_1 = 30\,V^{-1}$, $k_{20} = 5 \times 10^{-10}$mol cm^{-2}s^{-1}, $b_2 = 12\,V^{-1}$, $\Gamma = 10^{-9}$mol cm^{-2}, and $C_{dl} = 10$ μF cm^{-2}. (b) The corresponding steady-state current potential diagram and (c) an equivalent electrical circuit that can model the data. The best-fit values are given in Table 5.6.

TABLE 5.6
Values of Electrical Elements in the Circuit in Figure 5.13c, Which Can Simulate the Data in Figure 5.13a

E_{dc} (V)	R_1 (Ω)	R_2 (Ω)	C_2 (μF)
0.15	138.5	511.7	162.5
0.2	75.5	156	139.5
0.25	42.1	67.6	77.5

C_{dl} value is 10^{-5} F.

5.3.2 Physical Interpretation

Often, the data at different E_{dc} values exhibit different patterns, and completely different circuits are employed to model the spectrum at each E_{dc}. Obtaining a unified picture of the process from the analysis of these circuits will be extremely challenging. This example also helps us understand why numerical optimization with EEC is easier. In the case of EEC, at any given instance, we model only one spectrum. So, the problem is broken into smaller pieces and solved. In the case of RMA, *all the impedance data at all the E_{dc} values* are simultaneously used for the model fit, and this is more challenging.

Another point to note is that in RMA, we also fit the dc current *vs.* potential, along with the impedance data. In EEC modeling, the dc current *vs.* potential data is not utilized at all. This fact also illustrates the fact that RMA is a more general method that can predict many sets of impedance results and multiple types of results (i.e., dc current *vs.* potential and impedance spectra) and that it is superior to EEC.

5.4 Minimum Number of Potentials (E_{DC}) Where Spectra Must Be Acquired

Consider the dissolution of a metal by the following mechanism.

$$\mathrm{M} \xrightarrow{k_1} \mathrm{M}^{+}_{\mathrm{ads}} + \mathrm{e}^- \xrightarrow{k_2} \mathrm{M}^{2+}_{\mathrm{sol}} + \mathrm{e}^- \tag{5.121}$$

with the usual notation. We have seen that the faradaic admittance can be written as

$$\frac{i_{F\text{-ac}}}{E_{\mathrm{ac}}} = F\left\{(b_1+b_2)(k_{2\text{-dc}}\theta_{\mathrm{ss}}) + (k_{2\text{-dc}} - k_{1\text{-dc}})\left(\frac{(b_1-b_2)k_{2\text{-dc}}\theta_{\mathrm{ss}}}{(k_{1\text{-dc}}+k_{2\text{-dc}}+j\omega\Gamma)}\right)\right\} \tag{5.122}$$

This can be rearranged as follows

$$\frac{1}{Z_F} = F\left\{\frac{1}{\dfrac{1}{(b_1+b_2)(k_{2\text{-dc}}\theta_{\mathrm{ss}})}} + \frac{1}{\dfrac{(k_{1\text{-dc}}+k_{2\text{-dc}}+j\omega\Gamma)}{(k_{2\text{-dc}}-k_{1\text{-dc}})(b_1-b_2)(k_{2\text{-dc}}\theta_{\mathrm{ss}})}}\right\} \tag{5.123}$$

$$\frac{1}{Z_F} = \left\{\begin{array}{l}\dfrac{1}{\left(\dfrac{1}{F(b_1+b_2)(k_{2\text{-dc}}\theta_{\mathrm{ss}})}\right)} + \\ \dfrac{1}{\left(\dfrac{k_{1\text{-dc}}+k_{2\text{-dc}}}{F(k_{2\text{-dc}}-k_{1\text{-dc}})(b_1-b_2)(k_{2\text{-dc}}\theta_{\mathrm{ss}})}\right) + j\omega\left(\dfrac{\Gamma}{F(k_{2\text{-dc}}-k_{1\text{-dc}})(b_1-b_2)(k_{2\text{-dc}}\theta_{\mathrm{ss}})}\right)}\end{array}\right\} \tag{5.124}$$

Now, consider the electrical circuit shown in Figure 5.14.

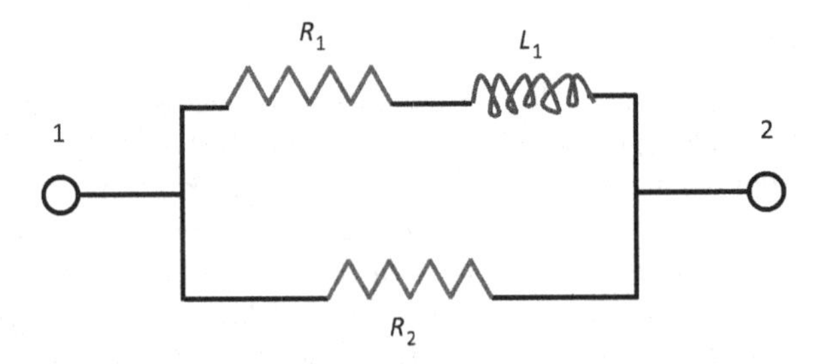

FIGURE 5.14 Equivalent electrical circuit with inductance, which can model the impedance response of an adsorbed intermediate species.

The impedance of the circuit is given by

$$\frac{1}{Z_{\text{circuit}}} = \frac{1}{R_2} + \frac{1}{R_1 + j\omega L_1} \tag{5.125}$$

A comparison of this with the previous equation indicates that the circuit can represent the faradaic process, and that Z_F can be replaced by Z_{circuit}. The following relationships exist between the electrical elements and the reaction parameters.

$$R_1 = \left(\frac{k_{1\text{-dc}} + k_{2\text{-dc}}}{F(k_{2\text{-dc}} - k_{1\text{-dc}})(b_1 - b_2)(k_{2\text{-dc}}\theta_{\text{ss}})} \right) \tag{5.126}$$

$$L_1 = \left(\frac{\Gamma}{F(k_{2\text{-dc}} - k_{1\text{-dc}})(b_1 - b_2)(k_{2\text{-dc}}\theta_{\text{ss}})} \right) \tag{5.127}$$

$$R_2 = \left(\left(\frac{1}{F(b_1 + b_2)(k_{2\text{-dc}}\theta_{\text{ss}})} \right) \right) = \frac{k_{1\text{-dc}} + k_{2\text{-dc}}}{F(b_1 + b_2) k_{1\text{-dc}} k_{2\text{-dc}}} = \frac{\frac{1}{k_{1\text{-dc}}} + \frac{1}{k_{2\text{-dc}}}}{F(b_1 + b_2)} \tag{5.128}$$

The total impedance of the electrochemical system is given by

$$Z_{\text{Total}} = R_{\text{Sol}} + \frac{1}{j\omega C_{\text{dl}} + (Z_F)^{-1}} \tag{5.129}$$

where C_{dl} is the double-layer capacitance and R_{Sol} is the solution resistance. Here we have assumed that the solution resistance is very small compared to R_1 or R_2.

An examination of these equations shows that the data at a given E_{dc} need to be acquired at five different frequencies so that all the five elements (R_{sol}, C_{dl}, R_1, L_1, and R_2) can be determined unambiguously. Usually, at a given potential E_{dc}, data are acquired at many more frequencies so that the confidence in the estimated values of the parameters R, C, and L (in general) would be good.

The equations also show that five reaction mechanism parameters, viz., k_{10}, b_1, k_{20}, b_2, and Γ are mapped onto the three electrical circuit parameters, viz., R_1, L_1, and R_2 by three equations, at a given potential E_{dc}. Hence the impedance data need to be acquired at a minimum of two different potentials. In that case, there will be six equations that map the five kinetic parameters to the three electrical circuit parameters. This will enable the determination of the kinetic parameters with some confidence. Otherwise, if the spectrum is acquired at only one E_{dc}, the system of equations will be under-defined and *will* have an infinite number of solutions.

5.4.1 Example – Multiple Solutions

We will show an example to illustrate this. Consider an electrochemical reaction following the above mechanism, with the parameter values as $k_{10} = 10^{-9}$mol cm^{-2}s^{-1}, $k_{20} = 10^{-11}$mol cm^{-2}s^{-1}, $b_1 = 10\,V^{-1}$, $b_2 = 20\,V^{-1}$, and $\Gamma = 10^{-9}$mol cm^{-2}. At $E_{dc} = 0.2$ V, the rate constants are $k_{1\text{-}dc} = 7.39 \times 10^{-9}$mol cm^{-2}s^{-1}, $k_{2\text{-}dc} = 5.46 \times 10^{-10}$mol cm^{-2}s^{-1}, and $\theta_{SS} = 0.9312$. The corresponding electrical parameter values are $R_1 = 2363.9\ \Omega$, $L_1 = 297.9$ H, and $R_2 = 679.5\ \Omega$. The spectrum generated in the frequency range of 1 kHz to 100 mHz is given as a complex plane plot in Figure 5.15.

It is easy to use commercial software such as ZSimpWin® and fit the model to a circuit and to estimate the R_{sol}, C_{dl}, R_1, L_1, and R_2 values. However, by using the three equations, we cannot uniquely determine all the five kinetic parameters because there are infinite numbers of solutions to the equations relating the kinetic parameters to the circuit parameters. As an example, the reader can verify that in addition to the values above, each one of the following sets of kinetic parameters (Table 5.7) also satisfy the three equations and result in the same spectrum. Infinite numbers of such solution sets are possible.

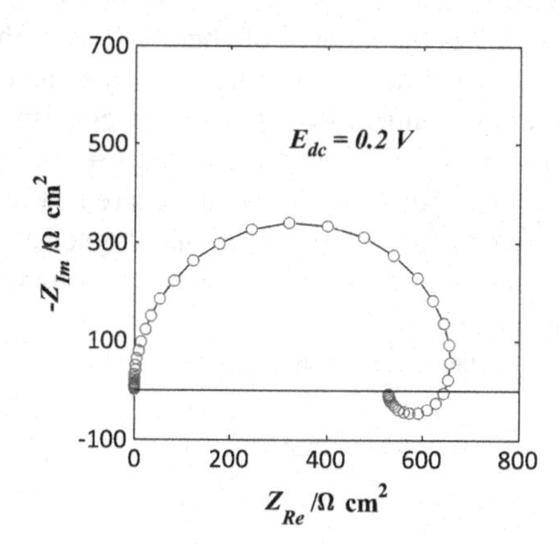

FIGURE 5.15 Complex plane plot of impedance spectrum for a reaction with an adsorbed intermediate. The kinetic parameters are $k_{10} = 10^{-9}$mol cm^{-2}s^{-1}, $k_{20} = 10^{-11}$mol cm^{-2}s^{-1}, $b_1 = 10\,V^{-1}$, $b_2 = 20\,V^{-1}$,and $\Gamma = 10^{-9}$mol cm^{-2}. The dc potential is $E_{dc} = 0.2$ V and the double-layer capacitance is $C_{dl} = 10^{-5}$ F cm^{-2}. EEC values are $R_1 = 2363.9\ \Omega$, $L_1 = 297.9$ H, and $R_2 = 679.5\ \Omega$. The area unit (cm^2) is omitted here.

TABLE 5.7
Example of Two Sets of Kinetic Parameters (Reaction in Eq. 5.121), Which Can Result in the Spectra Shown in Figure 5.15

Set	k_{10} (mol cm^{-2}s^{-1})	b_1 (V^{-1})	k_{20} (mol cm^{-2}s^{-1})	b_2 (V^{-1})	Γ (mol cm^{-2})
1	1.1364×10^{-9}	4	2.6098×10^{-11}	18	4.3908×10^{-10}
2	2.0153×10^{-9}	4	1.0980×10^{-10}	12	7.1780×10^{-10}

However, if the spectra are acquired at two or more E_{dc} values, then for this mechanism, there is less likelihood of obtaining multiple sets of kinetic parameters simulating data that correctly match the experimental results. If the data are acquired at a greater number of potentials, then the confidence level in the parameter estimates would be better.

5.4.2 Example – Calculation of Minimum Number of E_{DC} Where EIS Data Must Be Acquired

If the proposed mechanism contains a greater number of rate constants, the impedance would need to be measured at correspondingly a greater number of potentials to determine the parameters uniquely. The analysis of mechanisms with N number of intermediate adsorbed species leads to equivalent circuits with one charge-transfer resistance and N blocks of resistance in series with an inductance (Reference Technical Note 12, ZSimpWin®), i.e., the total number of equivalent electrical elements is $(2N+1)$. In the mechanistic representation, the number of steps for a mechanism with N intermediate species is at least $(N+1)$. Among the steps, if P is the number of electrochemical steps and Q is the number of chemical steps, then $(P+Q) \geq (N+1)$. Note that any electrochemical reaction must have at least one electrochemical step, i.e., $P>0$, but the number of chemical steps (Q) in a mechanism may be zero.

Now, the total number of kinetic parameters is $(2P+Q+1)$, and this is always $>(2N+1)$. The minimum number of dc potentials where EIS data must be acquired so that the kinetic parameters can be estimated unambiguously is given by the ratio $(2P+Q+1)/(2N+1)$, rounded up to the next integer, and this value is always two or more. An example is given below.

Consider the following reaction mechanism.

$$M \underset{k_{-1}}{\overset{k_1}{\rightleftarrows}} M^+_{ads} + e^- \tag{5.130}$$

$$M^+_{ads} \underset{k_{-2}}{\overset{k_2}{\rightleftarrows}} M^+_{sol} \tag{5.131}$$

The dissolution reaction occurs via the intermediate species M^+_{ads}. Note that $k_2 = k_{20}$ and $k_{-2} = k_{-20}$, since the forward and reverse reactions in the second step do not involve electron transfer. There are seven kinetic parameters, viz., k_{10}, b_1, k_{-10}, b_{-1}, k_{20}, k_{-20}, and Γ. The unsteady-state mass balance results in

$$\Gamma\frac{d\theta}{dt} = k_1(1-\theta) - k_{-1}\theta - k_2\theta + k_{-2}C_{M^+_{sol}}(1-\theta) \tag{5.132}$$

The steady-state surface coverage is given by

$$\theta_{SS} = \frac{k_{1dc} + k_{-2}C_{M^+_{sol}}}{k_{1dc} + k_{-1dc} + k_2 + k_{-2}C_{M^+_{sol}}} \tag{5.133}$$

After linearization of Eq. 5.128 and rearrangement of the terms, we can arrive at the following equation

$$\frac{d\theta}{dE} = \frac{\left(b_1k_{1dc}(1-\theta_{SS}) - b_{-1}k_{-1dc}\theta_{SS}\right)}{k_{1dc} + k_{-1dc} + k_2 + k_{-2}C_{M^+_{sol}} + j\omega\Gamma} \tag{5.134}$$

The current is given by

$$i = F\left[k_1(1-\theta) - k_{-1}\theta\right] \tag{5.135}$$

The faradaic admittance Y_F is obtained by expanding Eq. 5.135 in the Taylor series and substituting from Eq. 5.134,

$$Y_F = \frac{1}{Z_F} = \frac{\delta i_F}{\delta E} = F\left[b_1k_{1dc}(1-\theta_{SS}) - b_{-1}k_{-1dc}\theta_{SS}\right] - F\left(k_{1dc} - k_{-1dc}\right)\frac{d\theta}{dE} = \frac{1}{R_2} + \frac{1}{R_1 + j\omega L_1} \tag{5.136}$$

where

$$R_2 = \frac{1}{F\left[b_1k_1(1-\theta_{SS}) - b_{-1}k_{-1}\theta_{SS}\right]} \tag{5.137}$$

$$R_1 = \frac{k_{1dc} + k_{-1dc} + k_2 + k_{-2}C_{M^+_{sol}}}{-F\left(k_{1dc} - k_{-1dc}\right)\left(b_1k_{1dc}(1-\theta_{SS}) - b_{-1}k_{-1dc}\theta_{SS}\right)} \tag{5.138}$$

$$L_1 = \frac{\Gamma}{-F\left(k_{1dc} - k_{-1dc}\right)\left(b_1k_{1dc}(1-\theta_{SS}) - b_{-1}k_{-1dc}\theta_{SS}\right)} \tag{5.139}$$

Thus, at each potential E_{dc}, there are three equations relating the kinetic parameters to the electrical circuit parameters, as shown in Eqs. 5.137–139. Since there are seven kinetic parameters, and since each potential will result in three equations, the impedance must be acquired at three or more potentials to determine the kinetic parameters uniquely. We have two electrochemical steps ($P = 2$), two chemical steps ($Q = 2$), and one intermediate species ($N = 1$). The ratio, $(2P + Q + 1)/(2N + 1) = 7/3$, is rounded up to three. Thus, EIS must be acquired at three or more dc potentials to enable proper estimation of the parameter values.

If we have an impedance spectrum acquired at one *dc* potential, then we have seven unknowns and three equations, which would yield an infinite number of solutions. If we have the spectra at two *dc* potentials, there will be seven unknowns and six equations. This would again yield an infinite number of solutions. But with spectra at three or more *dc* potentials, there will be seven unknowns and nine or more equations relating them. Now, nonlinear least square regression can be employed to determine the kinetic parameters. This does not guarantee that the solution will be unique, but there is better confidence in the model identification and in determining the parameter set.

5.4.3 Why Is not the Kinetic Parameter Set Unique?

In the above example, when we use spectra at three or more *dc* potentials and fit the model, why can we not *guarantee* that the parameter set is unique? This is because the equations relating the kinetic parameters to the circuit parameters are nonlinear. If we have a linear equation, then the solution will be unique, but a nonlinear equation *may* have more than one valid solution. For example, the equation $4x+1=9$ is a linear equation with one unknown (x), and the solution is $x = 2$. The equation $4x^2+1=9$ is a nonlinear equation with one unknown (x), and it has two possible solutions, viz., $x=+\sqrt{2}$ or $x=-\sqrt{2}$. In the case of RMA, the equations are highly nonlinear in potential and frequency, and it is very difficult to estimate the number of possible solutions.

5.4.4 Frequency Intervals, Frequency Range, and dc Potential

One must acquire data at fine frequency intervals (seven or more frequencies per decade, logarithmically spaced), in a wide frequency range (10^5 Hz–1 mHz or lower), and at as many *dc* potentials as possible (at least 3). Now, it takes a longer time to acquire data at very fine frequency intervals, at more potentials or a wider frequency range. The lowest frequency that can be used in probing the system is often limited by the long duration needed to acquire the data and the associated noise and stability problems. If the data at low frequencies are noisy, then they may be discarded, and modeling can be limited to the data, which are repeatable and which can be validated.

5.5 LIMITATIONS OF THE RMA METHODOLOGY

5.5.1 Software Availability

Although the RMA of impedance data gives much better insight than equivalent circuit analysis, it is worth noting the limitations and not be carried away. At present, the greatest challenge in analyzing EIS data using RMA is that there is no commercial or noncommercial software available in public, to perform the fit. For each mechanism evaluated, the user has to write a program himself/herself and analyze the data. The equations are highly nonlinear, and unless the initial values are chosen carefully, the model will not converge. In our experience, once a custom program is written, it is relatively easy to fit the model using one impedance spectrum and polarization data. On the other hand, when data at many dc potentials are employed, it is very difficult to model all the data. In the case of EEC, there are several free software and commercial software available, and convergence is usually not an issue. One must also note that, in the case of EEC, only one impedance spectrum is fit at a given time, and in the case of RMA, all spectra are fit simultaneously, making it a lot more challenging problem. Also, in EEC analysis, the user does not fit the polarization data at all.

5.5.2 Unambiguous Mechanism Identification

From the impedance data, using RMA, we can propose a *plausible* mechanism and set of parameters that can match the experimentally observed data. It does not mean that we have *determined* the exact mechanism. There are likely many other reaction mechanisms that can satisfactorily explain the data. Certainly, a superset of the proposed mechanism (i.e., a mechanism that contains all the steps of the proposed mechanism and a few more steps) can be used to model the data equally well or better. Other phenomena, such as the formation of a film on the surface, may explain the data equally well. Unless there is corroborative evidence (such as other spectroscopic data), it is difficult to pinpoint the mechanism or the intermediate species.

However, RMA does help in eliminating certain mechanisms. As an example, if the complex plane plots show three loops, we cannot eliminate the film formation process or conclude that the reaction will have exactly two intermediate species. But we can say that a reaction with single adsorbed species without a film cannot be used to describe the data.

5.6 SUMMARY

First, we learned that usually, multiple equivalent electrical circuits can model a given data and that the choice has to be based on a physical understanding of the system. Next, we learned about the (quasi)potentiostatic and (quasi)galvanostatic mode of acquiring EIS data. The methodology of linearization of the current equation and obtaining the impedance under small amplitude potential perturbation was illustrated with a variety of examples, viz., electron transfer reaction, reactions with one or two adsorbed intermediates, a catalytic reaction, and an electron-electroadsorption reaction. These examples show how the impedance spectra can be simulated if a reaction and the corresponding kinetic parameter values are given.

In most cases, the user would collect impedance data and want to find a suitable reaction mechanism that can model the data. The user would also want to find the values of kinetic parameters that would result in the best fit. We show with an example that it may be possible to eliminate a few reactions. However, it is not possible to show that only one reaction would fit the data; at the minimum, any reaction, which is a superset of the proposed reaction, will fit the data. A methodology for estimating the parameters from the impedance and polarization data is outlined. The minimum number of dc potentials, where the EIS data must be acquired to minimize the possibility of multiple solutions, is specified. Some of the limitations of the RMA methodology that the authors are aware of, are listed.

5.7 EXERCISE – MECHANISTIC ANALYSIS

Q5.1 (a) For a simple electron transfer reaction given in Eq. 5.2, calculate the steady-state polarization current when the potential is varied from −0.1 to +0.1 V *vs.* OCP. Given that $k_{f0} = k_{r0} = 3 \times 10^{-6}$m s^{-1} and $\alpha = 0.5$ for a particular electrode and [Fe^{2+}] = [Fe^{3+}] = 1 mM each, and temperature = 25°C. Assume that solution resistance and mass transfer resistance are negligible. (b) Calculate the impedance spectra at $E_{dc} = 0$, +0.1, and −0.1 V where the potentials are measured with respect to the equilibrium potential. The double-layer capacitance is 18 μF cm^{-2}. The frequency range is 100 kHz–1 mHz, with seven points per decade and logarithmically spaced. Plot the spectra as complex plane plots as well as Bode plots.

Q5.2 For the reaction given in Eqs. 5.18 and 5.19, the variation of θ with E is given in Eq. 5.46. Plot the magnitude and phase of $\frac{d\theta}{dE}$ *vs.* frequency, using the set of parameters given in Figure 5.5. Explore the effect of varying the parameters.

Q5.3 From the equation for the faradaic admittance Eq. 5.47, derive the expression for polarization resistance for this mechanism.

Q5.4 Generate Figure 5.5 at $E_{dc} = 0.25$ V *vs.* the OCP. Fit the EIS data to the circuit shown in Figure 4.2. Compare with the values obtained at other potentials (Table 5.1). What inferences can we draw?

Q5.5 Fit the data in the above problem, as well as in Figure 5.5, using the Voigt circuit (Figure 4.7b) to represent the faradaic impedance.

Q5.6 In Figure 5.6a, generate the plot at other dc potentials. Do we get any pattern other than a semicircle? Why?

Q5.7 Fit the Randles circuit (Figure 4.1b) to the above data. What inference can we draw about the number of adsorbed intermediates of the reaction mechanism that gave rise to this data?

Q5.8 From Eq. 5.56, derive the expression for polarization resistance.

Q5.9 How does $\frac{d\theta}{dE}$, given in Eq. 5.53, vary with frequency? Generate Bode plots (i.e., magnitude and phase as a function of frequency). Use the parameter sets employed to generate the data in Figure 5.6.

Q5.10 Fit the data in Figure 5.7 using a Voigt circuit to represent the faradaic impedance (e.g., Figure 4.7b).

Q5.11 Generate Figure 5.7 with $E_{dc} = 0.15$, 0.2, 0.25, and 0.35 V *vs.* the OCP.

Q5.12 Derive the expression for Z_F for the mechanism given in Eq. 5.65. Generate complex plane plots of impedance at a few E_{dc}. For the kinetic parameter sets, you can select values based on those used in this chapter. Explore the effect of varying these values.

Q5.13 Generate the data in Figure 5.8. Model it using the circuit shown in Figure 4.11a.

Q5.14 Repeat the above problem using ladder (e.g., Figure 4.11b) and Voigt (e.g., Figure 4.11c) circuits to represent the faradaic impedance.

Q5.15 In the above two questions, for a given set of kinetic parameters, (a) vary E_{dc} and plot the results in a complex plane. (b) Fit a Maxwell-based circuit (Figure 4.11a) to each of this data set. (c) At a given E_{dc}, if we swap (R_2, C_2) with (R_3, C_3), will the resulting spectrum change? (d) Do the values of elements (R_2, C_2, R_3, C_3) exhibit any trend with E_{dc}? (e) In light of Q5.15c and d, explain the challenge in drawing inferences based on circuit-based analysis of EIS data.

Q5.16 For the mechanism in Eqs. 5.84–5.86, derive the expression for the faradaic impedance. Generate EIS data for a set of kinetic parameters and E_{dc} and show the results as a complex plane plot as well as Bode plots. How many *loops* are visible to the naked eye in the complex plane plot?

Q5.17 Fit Maxwell, ladder, and Voigt-based circuit to the above data.

Q5.18 Consider the impedance data shown in Figure 5.10b–d. Fit an equivalent circuit to the data. Use (a) Maxwell, (b) ladder/Voigt element-based circuit to model the faradaic impedance.

Q5.19 Using Eq. 5.114 and the kinetic parameters in Figure 5.11, determine the value of $\frac{d\theta}{dE}$ as a function of frequency.

Q5.20 Derive the expression for polarization resistance from Eqs. 5.14 and 4.16, for the mechanism given in Eqs. 5.99–5.101.

Q5.21 Browse through recent literature showing impedance data and organize them based on the patterns given in Figure 5.12.

Q5.22 Write a program to fit a kinetic model and obtain best-fit parameters from impedance data. You can use the following procedure.

a. Use some of the data sets generated in the earlier questions. Choose a mechanism (e.g., Eq. 5.49) and a set of kinetic parameters (e.g., in Figure 5.6). In a potential window, generate polarization curve (i_F *vs.* E_{dc}) and, at a few dc potentials, generate EIS data. Consider these as 'ideal data,' where we know the mechanism and the kinetic parameter values.

b. Pretending that we do not know the kinetic parameters (but we know R_{sol} and C_{dl}), write an optimization program that will vary the kinetic parameters, compare the predicted EIS data with 'ideal data,' and minimize the difference. You do not need to use polarization data in this process. Use EIS data at one dc potential only, and run the program. What are the results? What is the effect of changing the initial value?

Q5.23 Repeat Q5.22 using EIS at two DC potentials.

Q5.24 Repeat Q5.22 using EIS at three DC potentials.

Q5.25 Repeat Q5.22, with EIS at one DC potential, but with an additional constraint that the model polarization data should not deviate from the experimental data by more than 0.5 mA cm^{-2}.

Q5.26 Repeat Q5.25, using EIS at three DC potentials.

Q5.27 What inferences can we draw from the results of Q5.22–Q5.26?

Q5.28 In the mechanism given below, at how many dc potentials should one acquire EIS data so that the kinetic parameters can be estimated with some confidence? Assume that the polarization data are not utilized in the analysis.

a.
$$M \underset{k_{-1}}{\overset{k_1}{\rightleftarrows}} M_{ads}^{+} + e^{-}$$
$$M_{ads}^{+} \underset{k_{-2}}{\overset{k_2}{\rightleftarrows}} M_{ads}^{2+} + e^{-}$$
$$M_{ads}^{2+} \xrightarrow{k_3} M_{sol}^{2+}$$

b.
$$M \underset{k_{-1}}{\overset{k_1}{\rightleftarrows}} M_{ads}^{+} + e^{-}$$
$$M_{ads}^{+} \xrightarrow{k_2} M_{sol}^{+}$$
$$M + M_{ads}^{+} \xrightarrow{k_3} M_{ads}^{+} + M_{sol}^{+} + e^{-}$$

c.
$$P \underset{k_{-1}}{\overset{k_1}{\rightleftarrows}} Q + e^{-}$$
$$M + A_{sol}^{-} \underset{k_{-2}}{\overset{k_2}{\rightleftarrows}} M - A_{ads} + e^{-}$$

6 EIS – Other Physical Phenomena

The important thing is not to stop questioning. Curiosity has its own reason for existing.

—**Albert Einstein**

Apart from the reactions, many other physical phenomena affect the impedance spectra. For example, the electrode surface may not be uniform, and this can cause the electrode–electrolyte interface to deviate from the ideal behavior of a simple capacitor. If the mass transfer is slow, then the diffusion of the relevant species in the solution plays an important role in determining the current, and hence the impedance. The electrode may be porous, and the impedance spectrum would be quite different from that of a planar surface. The electrode may have a thin film on the electrode surface. In that case, the film thickness and other properties would alter the impedance spectra. In this chapter, we examine these four phenomena and their effects on the EIS response.

6.1 CONSTANT PHASE ELEMENTS (CPE)

In 1932, **Kenneth S. Cole** studied the electrical impedance of suspension of spheres and showed that the complex plane plot of the imaginary *vs.* real component of dielectric constants was a depressed semicircle.

An electrode–electrolyte double-layer interface is modeled as a capacitor. In the EIS spectrum, it is expected to give rise to a capacitance loop in the high-frequency (kHz) range. The admittance of a capacitor is given by $Y = j\omega C$. However, in many real systems, we find that a capacitor cannot model the admittance data. Instead, we must use a modified equation to represent the admittance (Y) and impedance (Z) respectively as

$$Y = Q(jw)^n \tag{6.1}$$

$$Z = Q^{-1}(jw)^{-n} \tag{6.2}$$

This representation is called a *constant phase element* (CPE). Often in the literature, the exponent is denoted by α. However, since we use α to denote the transfer coefficient, we employ n for the exponent. Likewise, the pre-exponent is sometimes denoted by Y_0, and since we denote admittance by Y, we employ Q to represent the pre-exponent.

We first see an example of data where a capacitor in EEC is not sufficient to model it, but a CPE is. A graphical representation of the data will help us understand the origin of the name 'constant phase element.' We also will see the flexibility offered by CPE in modeling. Then we present a brief survey of experimental and theoretical publications focusing on the origin of CPE behavior. Finally, we present the expression proposed in the literature to extract an equivalent capacitance from the CPE element values.

When the EIS data of a system represented by an equivalent electrical circuit (Figure 1.18) is plotted in a complex plane, it would appear as a semicircle, as shown in Figure 6.1a. However, if the capacitance is not ideal, but is replaced by a CPE, then it would appear as shown in Figure 6.1b. In this example, the high-frequency limit is marked by *X*, while the low-frequency limit touching the abscissa is marked by *Y*. For an ideal capacitance behavior, the spectra will appear as a semicircle with its center (*O*) on the real axis, as shown in Figure 6.1a. For a CPE, the arc will appear, as shown in Figure 6.1b. This arc is also a part of a circle, but when we draw a line, starting at the high-frequency impedance (*X*) and passing through the center of the circle (*O*), it makes a non-zero angle with the abscissa ($\angle OXY$). In other words, for an ideal capacitance, the complex plane plot would be half of a circle with the center (*O*) *on* the real axis, whereas, for a CPE, it is a smaller part of a circle with the center (*O*) *below* the real axis. It is a small part of a circle, whose center has been pushed *down*. This trend is often referred to as a *depressed semicircle*.

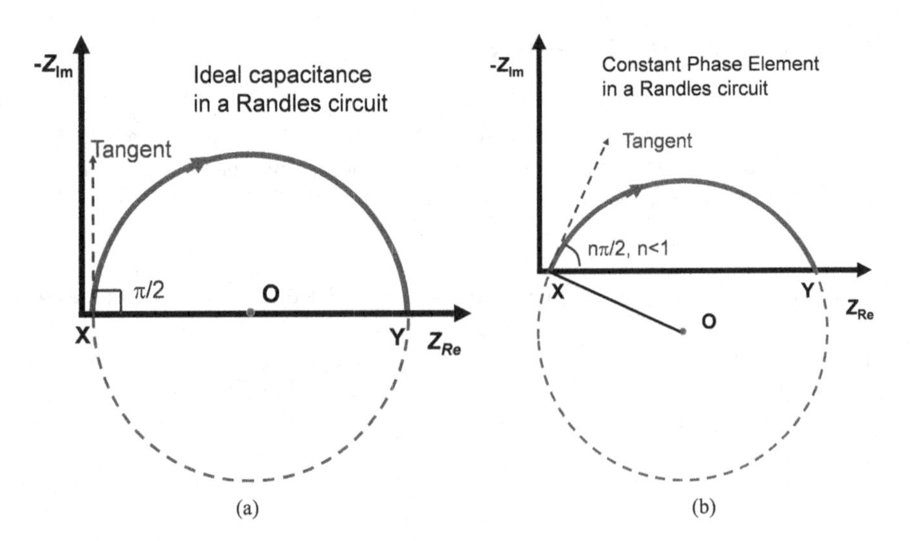

FIGURE 6.1 Complex plane plot of a simple equivalent electrical circuit with (a) an ideal capacitance or (b) a constant phase element. The angle between the tangent and the real axis is given in radians. The high-frequency limit is marked by *X*, while the low-frequency limit touching the abscissa is marked by *Y*. The arrow marks in the spectra show the direction of decreasing frequency. The dotted curved line is the extension of the arc to show the complete circle. The tangent at the point of intersection of the spectra with the real axis is related to the value of parameter *n*.

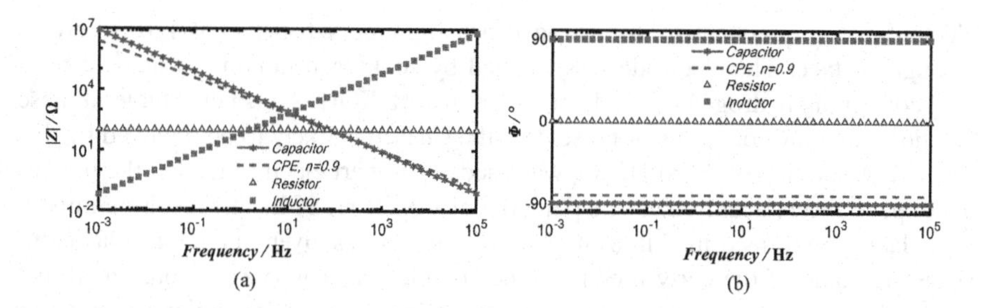

FIGURE 6.2 Example Bode plots, (a) magnitude and (b) phase *vs.* frequency, of the impedance of a capacitor, CPE, resistor, and inductor. Note that the phase is constant for each one of these circuit elements.

The 'data' in Figure 6.1a can be modeled using the circuit in Figure 1.18, which employs a capacitor in parallel with a resistor. However, the data in Figure 6.1b can be modeled adequately only if the capacitor is replaced by a CPE, with the exponent 'n' less than 1. When the exponent 'n' is 1, the CPE, of course, becomes a capacitor. Besides, when $n = 0.5$, the CPE behaves as Warburg impedance (discussed in Section 6.2), when $n = 0$, the CPE becomes a resistor, and when $n = -1$, the CPE becomes an inductor. The impedance values of a capacitor, resistor, inductor, and a CPE with exponent values of 0.95 and 0.5 are plotted in the Bode format (Figure 6.2). While the magnitude of the impedance of a CPE can change with frequencies, its phase value is a constant at all frequencies, which explains its name *constant phase* element. A capacitor, a resistor, and an inductor can be thought of as a special case of CPE. By varying the value of the exponent 'n' of the CPE, we can simulate a rather wide spectrum of impedance data. Thus, CPE is a very flexible element that can be used in EEC analysis. However, in electrochemical literature, the element CPE usually refers to the non-ideal behavior of the double-layer with the exponent n varying between 0.8 and 1.

CPE behavior was called *frequency dispersion of capacitance* because it appears as if the double-layer is behaving as a capacitor whose capacitance depends on the frequency. Now the general agreement in the research community is that CPE behavior is due to variation of current density and potential across the electrode surface. This variation occurs due to inhomogeneity on the surface, but it is not related to the macroscopic (micron or larger level) roughness (Kerner and Pajkossy 1998, 2000). Rather, it is due to microscopic inhomogeneity, such as the orientation of the crystal face or the energy levels on the electrode surface. Besides, the variation may be caused by inhomogeneities across the surface or normal to the surface, such as an inhomogeneous film (Hirschorn et al. 2010b). The following sections provide a more detailed description of this topic.

6.1.1 Experimental Results

In this section, we present a summary of published experimental results of research work focusing mainly on CPE behavior. To determine the origin of CPE behavior in rough electrodes, Kerner and Pajkossy (1998) fabricated a polycrystalline gold electrode on a quartz plate and abraded it using emery paper (grit 120) to create a rough surface. The electrode was cleaned by soaking it in concentrated sulfuric acid containing hydrogen peroxide and then in 10% KOH. Au is not soluble in these solutions, but any contaminant present on the surface is likely to be removed by this process. A solution of 0.1 M $HClO_4$ was used as an inert electrolyte, and the impedance spectra were acquired at a *dc* potential of −0.2 V *vs.* SCE. The resulting impedance data were transformed into *capacitance spectra,* as given in the formula below. (Note: In some of the early literature, the complex capacitance was presented, but this is no longer a common practice.) At any frequency ω, the *complex capacitance* *C(ω)* is defined as

$$C(\omega)=\frac{1}{\left[Z(\omega)-Z(\omega\to\infty)\right]\times(j\omega)\times A_g} \tag{6.3}$$

Here, A_g is the geometrical area of the electrode. Since *C(ω)* is a complex quantity, its magnitude and phase can be plotted against the frequency, in the Bode format. If the magnitude of the complex capacitance is independent of the frequency and if the phase is 0° for all the frequencies, then we can conclude that the electrode exhibits an ideal capacitance behavior. If the phase is greater than 0°, it exhibits a CPE behavior. The magnitude of the complex capacitance would also then vary with the frequency. The phase angle θ, expressed in degrees, is related to the CPE exponent n by the equation

$$\theta=90(1-n) \tag{6.4}$$

The roughened polycrystalline Au electrode exhibited a CPE behavior, and the phase angle θ varied between 5° and 10°, depending on the frequency. In contrast, a single crystal electrode, where the electrode surface is homogeneous, exhibited an ideal capacitance behavior. When a CPE circuit was used to fit the data, the CPE exponent was 0.99. Next, the polycrystalline electrodes were annealed in a bunsen burner for approximately 5 seconds and quenched in ultrapure water. The impedance spectrum was measured again in the same system, and the phase angle decreased, i.e., the electrode behavior moved toward that of an ideal capacitor. Longer annealing (1 minute) resulted in more decrease of the phase angle. However, annealing, for an even longer time, did not cause much change in the phase angle. The electrodes were then anodically etched in $CaCl_2$ solution. This resulted in removing the rough upper layer and gave a bright surface. The impedance spectra of this electrode in the same solution indicated that the phase angle decreased further, and the electrode behavior was very close to that of an ideal electrode whose double-layer can be represented by a capacitor. While the individual spectra were not very reproducible, the trends (i.e., annealing by chemicals or chemically etching removing the surface decreases the phase angle) were reproducible.

Later, the same research group (Kerner and Pajkossy 2000) made a very rough electrode of Au, Ag, and Pt and acquired impedance spectra. The electrode was roughened by a steel file or by emery paper (Grit 100) and then cleaned with chemicals. The impedance spectra of the rough electrodes exhibited CPE behavior. Then the electrodes were annealed to reduce the atomic level inhomogeneities without decreasing the macroscopic roughness, and impedance spectra were acquired again. Annealing was done at <500°C using a flame for several seconds, and optical and scanning electron microscopic (SEM) analysis showed that the macroscopic roughness did not decrease with annealing. An analysis of the resulting spectra showed that the frequency dispersion of capacitance decreased with annealing, i.e., after annealing, the double-layer behavior was closer to that of an ideal capacitor. Hence, they concluded that macroscopic roughness (irregular geometry) is not the cause of CPE behavior, but atomic level inhomogeneity, which results in surface energy differences, is the root cause.

6.1.2 Models to Explain the Origin of CPE

There are various explanations offered to describe CPE behavior. In general, if a particular property such as the capacitance or charge-transfer resistance varies from point to point on the electrode, then we can expect the system to exhibit CPE behavior. This is often described as 'dispersion' of capacitance or charge-transfer resistance. In the earlier days, it was believed that rough electrodes with macroscopic roughness (of the order of mm), with the corresponding variation in double-layer capacitance and solution resistance, result in CPE behavior. However, later analyses indicate that atomic level heterogeneity or micro-level variations are likely to cause CPE behavior (Pajkossy, Wandlowski, and Kolb 1996, Pajkossy 1997, 2005). For example, even a single crystal Au electrode can exhibit CPE behavior if some of the surface sites are modified by the presence of adsorbed species such as Br^- (Pajkossy, Wandlowski, and Kolb 1996, Pajkossy 1997). There is no macro-level roughness in the single crystal electrode, and yet CPE behavior was reported. Another report also suggests that energetic heterogeneity, and not geometric heterogeneity, causes CPE behavior (Cordoba-Torres et al. 2012). These types of variations that occur across the surface are grouped as surface variations.

Another type of variation can occur in the direction normal to the surface, e.g., if there is a thin film present on the surface and if the film thickness or properties such as dielectric constant vary from location to location, then again, CPE behavior can arise in that situation (Hirschorn et al. 2010a, b).

6.1.3 Equations to Relate CPE to the Effective Capacitance

When a CPE is needed to model an electrode surface, is it possible or even meaningful to extract an effective capacitance from the data? This is not settled yet, and at the same time, a few formulas are proposed in the literature to relate the effective capacitance from the CPE parameters. They also include resistance values, as described below. The relationship between the CPE parameter Q and the effective double-layer capacitance is proposed as

$$C_{\text{dl-effective}} = Q^{1/n}\left(R_{\text{sol}}^{-1} + R_t^{-1}\right)^{\left(\frac{n-1}{n}\right)} \tag{6.5}$$

by Brug (1984). Another relationship proposed by Hsu and Mansfield (2001) is

$$C_{\text{dl-effective}} = Q^{1/n} R_t^{(1-n)/n} \tag{6.6}$$

Later it was shown (Hirschorn et al. 2010b) that the Brug formula can be obtained by equating the characteristic frequency maximizing the imaginary part of the data in the admittance form for a CPE and an ideal capacitor. Similarly, the formula by Hsu and Mansfield can be obtained by equating the characteristic frequency maximizing the imaginary part of the data in the impedance form. Equation 6.5 is appropriate if the frequency dispersion arises due to variation along the surface as shown in Figure 6.3a. In this example, the activity of the electrode varies across the surface and this is referred to as two-dimensional heterogeneity. This can be represented by the equivalent circuit shown in Figure 6.3c. On the other hand, it is possible to have a film on the surface whose properties vary across the surface as well as in the direction perpendicular to the surface (Figure 6.3b). In that case, the equivalent circuit shown in Figure 6.4d has to be employed to model that electrode.

The parameters Q and n are not exactly independent. By analyzing the EIS data of polycrystalline iron in 1 M NaOH + 0.1 M $K_3[Fe(CN)_6]$ under cathodic polarization, it is shown that Q and n are coupled together and as n decreases, Q increases (Cordoba-Torres et al. 2012). This coupling (Figure 6.4) is seen in other experimental results as well (Prasad, Kumar, and Ramanathan 2009, Venkatesh and Ramanathan 2010b) and appears to be applicable in general.

The equation relating Q and n can be obtained by rearranging Eq. 6.5 and is written as

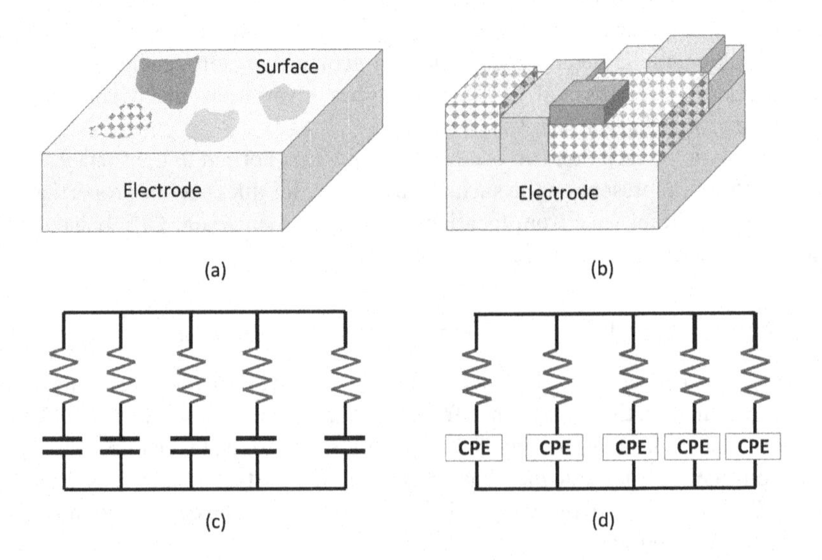

FIGURE 6.3 Pictorials of (a) two-dimensional heterogeneity, (b) three-dimensional heterogeneity, (c) equivalent circuit to represent the two-dimensional heterogeneity, and (d) equivalent circuit to represent the three-dimensional heterogeneity.

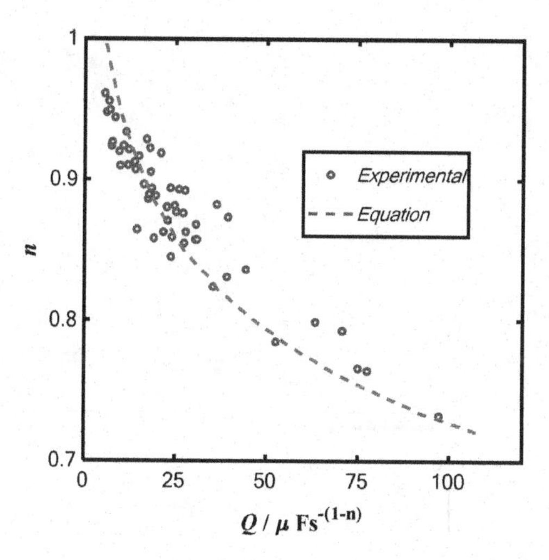

FIGURE 6.4 Correlation between CPE exponent 'n' and pre-exponent 'Q.' (Adapted from Electrochimica Acta, 72, P. Córdoba-Torres, T.J. Mesquita, O. Devos, B. Tribollet, V. Roche, R.P. Nogueira, On the intrinsic coupling between constant-phase element parameters n and Q in electrochemical impedance spectroscopy, pp. 172–178, Copyright (2012), with permission from Elsevier.)

$$n_2 = n_1 \times \frac{\left[\ln(Q_2) - \ln\left(R_{\text{sol2}}^{-1} + R_{t2}^{-1}\right)\right]}{\left[\ln(Q_1) - \ln\left(R_{\text{sol1}}^{-1} + R_{t1}^{-1}\right)\right] + n_1 \times \left[\ln\left(R_{\text{sol1}}^{-1} + R_{t1}^{-1}\right) - \ln\left(R_{\text{sol2}}^{-1} + R_{t2}^{-1}\right)\right]} \tag{6.7}$$

where the subscripts 1 and 2 refer to two slightly different conditions in a given system.

The particular case of $n_1 = 1$, $Q_1 = C_0$, $R_{\text{sol1}} = R_{\text{sol2}}$, and $R_{t1} = R_{t2}$ can be written as

$$n = \frac{\log\left(R_{\text{sol}}^{-1} + R_t^{-1}\right) - \log(Q)}{\log\left(R_{\text{sol}}^{-1} + R_t^{-1}\right) - \log(C_0)} \tag{6.8}$$

6.2 DIFFUSION EFFECTS

If the reaction rate is fast, but the diffusion of the species from (to) the solution to (from) the electrode is slow, then diffusion will also influence the net reaction rate and the current. In these cases, the impedance spectra can exhibit a completely different behavior. Theoretical analysis of this case has been developed, and experiments match the theoretical results well. In a simplified analysis, the relevant differential equations and boundary conditions are developed as follows; for the domain under consideration, one boundary is at the electrode ($x = 0$). The other boundary may lie at a finite distance ($x = \delta$), as shown in the cartoon Figure 6.5. The solution beyond that boundary is assumed to be well mixed and of large volume, and the concentration is uniform in the bulk. In the case of rotating disc electrodes, this boundary layer thickness can be estimated using the **Levich** diffusion layer thickness equation

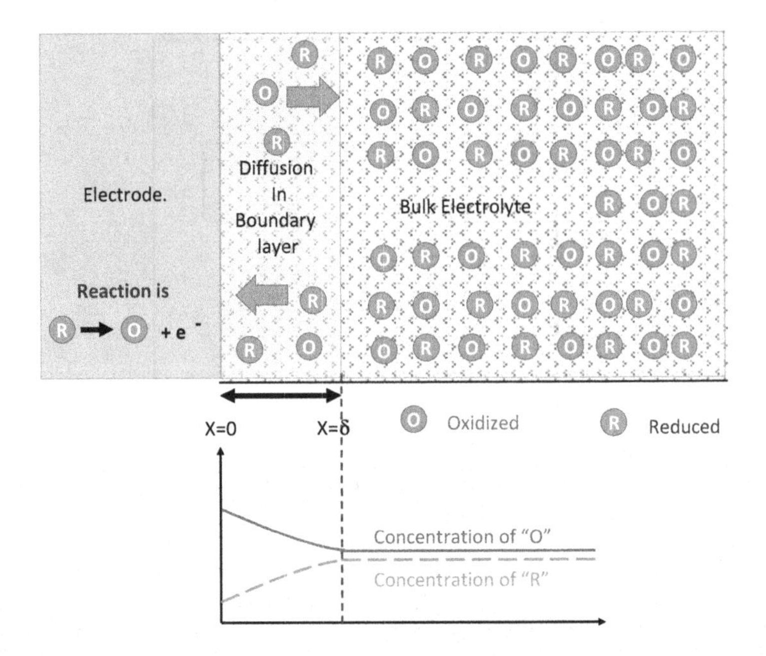

FIGURE 6.5 Cartoon illustrating diffusion of solution species toward and away from the electrode. On the electrodes, species 'R' is consumed, and the species 'O' is produced. The concentration of 'R' at the interface is lower compared to the bulk concentration, and 'R' diffuses from the bulk to the surface. Species 'O' diffuses from the surface to the bulk. Note that the double-layer is not explicitly shown, for the sake of simplicity.

$$\delta = 1.61 \times D_0^{1/3} \psi^{-1/2} \gamma^{1/6} \tag{6.9}$$

where δ is in cm, D_0 is diffusivity in cm^2s^{-1}, ψ is rotational speed in rad s^{-1}, and γ is kinematic viscosity in cm^2s^{-1}. Using nominal values of $\gamma = 0.01\,cm^2s^{-1}$, $D_0 = 10^{-5}cm^2s^{-1}$, and rotations speeds of 100–1000 rpm, the diffusion layer thickness is estimated to be in the range of 50–16 μm, with higher rpm corresponding to smaller boundary layer thickness. If the solution is unstirred, and the electrode is stationary, the boundary layer thickness is infinite ($x = \infty$), and this is denoted as *semi-infinite boundary conditions.* In this section, we consider only one-dimensional diffusion, i.e., perpendicular to the electrode surface.

6.2.1 Finite Boundary Conditions

In 1899, **Emil Warburg** developed an expression for the current response of a non-polarizable electrode, to sinusoidal potential, when the mass transfer is slow and Fick's second law of diffusion is applicable. The dependence of the response current on the frequency and the phase difference of 45° with potential are identified.

In 1947, **J.E.B. Randles** developed an expression for the impedance of an electrochemical system for which the Butler Volmer equation is applicable. An equivalent circuit combining the solution resistance, double-layer capacitance, and faradaic impedance including the diffusion effect was proposed, although the faradaic impedance was described as a resistor in series with a capacitor both of which are *frequency dependent.*

Consider an elementary redox reaction

$$A + ne^- \underset{k_{-1}}{\overset{k_1}{\rightleftharpoons}} B \tag{6.10}$$

The rate constants are given by $k_1 = k_{10}e^{\frac{\alpha F}{RT}(E)}$ and $k_{-1} = k_{-10}e^{\frac{-(1-\alpha)F}{RT}(E)}$, where the potential E is measured with respect to the equilibrium potential (which is also the open-circuit potential (OCP)). At equilibrium, the net rate is zero, and the concentration of A and B on the surface is the same as the concentration in the bulk. The concentrations of A and B in the bulk are denoted by $C_{A\text{-bulk}}$ and $C_{B\text{-bulk}}$, respectively, and these values are assumed to be constants; i.e., there is a large reservoir of electrolyte, and the reactions occurring at the electrode do not change the bulk concentration to a significant extent.

Note that k_{10} and k_{-10} are not independent, but are related by the equation

$$k_{10}C_{A\text{-bulk}} = k_{-10}C_{B\text{-bulk}} \tag{6.11}$$

This is because, at equilibrium, the steady-state value of the net current is zero.

The movement of any species is governed by the Nernst–Planck equation, which, in one dimension, can be written as

$$\frac{\partial C_i}{\partial t} = D_i\left(\frac{\partial^2 C_i}{\partial x^2}\right) - v_y\frac{\partial C_i}{\partial x} + \frac{D_i ne}{k_B T}C_i\left(\frac{\partial \varphi}{\partial x}\right) \tag{6.12}$$

where i is A or B, D_i is the diffusivity of species i, v_y is the velocity of the fluid in the direction perpendicular to the electrode, φ is the electric potential measured with respect to point-of-zero-charge, e is the elementary charge, k_B is the Boltzmann constant, and T is the temperature. Within the boundary layer, the concentration can change with location, depending on the reaction rate, and hence the applied potential. The concentration is written with two indices as $C_A(x, t)$, with the first one denoting the position and the second one denoting the time. The velocity v_y (in cm s^{-1}) can be obtained using the Von Karman and Cochran equation (Bard and Faulkner 1980)

$$v_y = -0.51\psi^{3/2}\gamma^{-1/2}y^2 \tag{6.13}$$

where y is the distance from the surface in the direction perpendicular to the electrode, ψ is the rotational speed in rad s^{-1}, and γ is the kinematic viscosity in cm^{2}s^{-1}.

In Eq. 6.12, the first term on the right side describes the mass transfer by Fickian diffusion. The second term describes the mass transfer by convection, while the last term accounts for the movement of charged species in an electric field. If we assume that convection effects are negligible, and all the potential drop occurs at the electrode–electrolyte interface (i.e., solution resistance is negligible), then the equation can be simplified to Fick's second law of diffusion, i.e.,

$$\frac{\partial C_i}{\partial t} = D_i\left(\frac{\partial^2 C_i}{\partial x^2}\right) \tag{6.14}$$

The faradaic current is given by

$$i_F = nF\left(k_{10}e^{\frac{\alpha F}{RT}(E)}C_A(0,t) - k_{-10}e^{\frac{-(1-\alpha)F}{RT}(E)}C_B(0,t)\right) \tag{6.15}$$

Steady-state conditions:

If a dc potential E_{dc} is applied, the steady-state concentrations of species A and B at the surface (C_{AS} and C_{BS} respectively) can be found using Eq. 6.14, along with the following boundary conditions.

At $x=0$, the flux at the surface is related to the current as

$$D_A\left.\frac{\partial C_A}{\partial x}\right|_{0,t} = \frac{i_F}{nF} = -D_B\left.\frac{\partial C_B}{\partial x}\right|_{0,t} \tag{6.16}$$

$$\text{At}\, x=\delta, C_A = C_{A\text{-bulk}} C_B = C_{B\text{-bulk}} \tag{6.17}$$

The steady-state concentration of A and B on the surface is given by

$$C_A(x=0) = \frac{D_A\frac{(C_{A\text{-bulk}})}{\delta} + k_{-1dc}C_{B\text{-bulk}} + k_{-1dc}\frac{D_A}{D_B}C_{A\text{-bulk}}}{\left[k_{1dc} + \frac{D_A}{\delta} + k_{-1dc}\frac{D_A}{D_B}\right]} \tag{6.18}$$

$$C_B(x=0) = C_{B\text{-bulk}} + \frac{D_A}{D_B}\left(C_{A\text{-bulk}} - C_A(x=0)\right) \tag{6.19}$$

Here, k_{1dc} and k_{-1dc} are the rate constants evaluated at E_{dc}. The steady-state current can be obtained by substituting Eqs. 6.18 and 6.19 in Eq. 6.15. For a particular set of variables and parameters, the current and concentrations on the surface are given in Figure 6.6a and b, respectively.

6.2.1.1 Unsteady-State Conditions

Here we follow the methodology illustrated by Lasia (2014). When a small amplitude sinusoidal potential perturbation is applied, the current can be obtained by solving

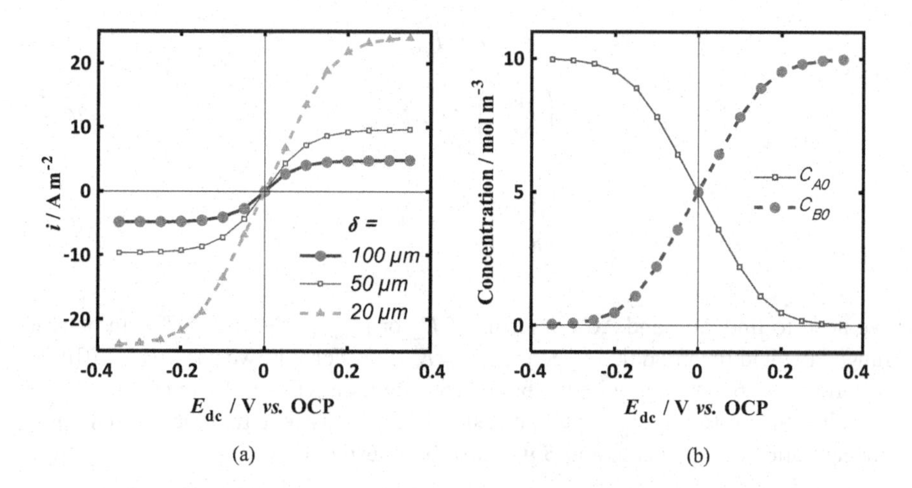

FIGURE 6.6 (a) Potentiodynamic polarization curves for a simple electron transfer reaction at various boundary layer thickness values. (b) Surface concentrations of A and B as a function of *dc* potential, at $\delta = 50$ μm. Here, $k_{10} = k_{-10} = 10^{-4}$m s^{-1}, $D_A = D_B = 10^{-9}$m^{2}s^{-1}, $C_{A\text{-bulk}} = C_{B\text{-bulk}} = 5$ mM, $\alpha = 0.5$, and $T = 300$ K.

Eq. 6.14, along with Eq. 6.15. Expanding the rate constants in the Taylor series and truncating after the first-order term, we get $k_i \approx k_{i\text{-dc}}(1 + b_i E_{ac})$. We can also write the concentration at any location as

$$C_A(x,t) = C_A(x,\text{SS}) + \frac{\text{d}C_A(x,t)}{\text{d}E} E_{ac} \tag{6.20}$$

and substitute in the current Eq. 6.15. We will use a slightly different notation. Note that E, C_A, and C_B are independent variables. The concentrations of A and B in the boundary layer, for example, can be varied by changing the rotational speed or the concentrations in the bulk. Hence, when a small change in potential and concentrations are applied, Eq. 6.15 can be written as

$$i_F = i_{F\text{-dc}} + i_{F\text{-ac}} = i_{F\text{-dc}} + \Delta i_F \tag{6.21}$$

where

$$\Delta i_F = i_{F\text{-ac}} = \left(\frac{\partial i_F}{\partial E}\right)_{C_A,C_B} \Delta E + \left(\frac{\partial i_F}{\partial C_A}\right)_{E,C_B} \Delta C_A + \left(\frac{\partial i_F}{\partial C_B}\right)_{E,C_A} \Delta C_B \tag{6.22}$$

Here,

$$\left(\frac{\partial i_F}{\partial E}\right)_{C_A,C_B} = F\left(b_1 k_{1\text{-dc}} C_A\big|_{x=0,ss} - b_{-1} k_{-1\text{-dc}} C_B\big|_{x=0,ss}\right) \tag{6.23}$$

$$\Delta E = E_{\text{ac}} \tag{6.24}$$

$$\left(\frac{\partial i_F}{\partial C_A}\right)_{E,C_B} = Fk_{1\text{dc}} \tag{6.25}$$

$$\left(\frac{\partial i_F}{\partial C_B}\right)_{E,C_A} = -Fk_{-1\text{dc}} \tag{6.26}$$

We need to find ΔC_A and ΔC_B in terms of E_{ac} or $i_{F\text{-ac}}$, so that Eq. 6.22 can be rearranged to yield the faradaic impedance $Z_F = E_{\text{ac}}/i_{F\text{-ac}}$. For this, we use Fick's diffusion equation (Eq. 6.14) along with the boundary conditions (Eqs. 6.16 and 6.17).

When the perturbation amplitude is small, we can assume that the oscillations of concentrations of species A and B will also be sinusoidal, i.e.,

$$C_A(x,t) = C_{A\text{-dc}} + C_{A\text{-ac}} = C_{A\text{-dc}}(x) + C_{A\text{-ac}0}(x)\sin(\omega t + \phi_A(x)) \tag{6.27}$$

$$C_B(x,t) = C_{B\text{-dc}} + C_{B\text{-ac}} = C_{B\text{-dc}}(x) + C_{B\text{-ac}0}(x)\sin(\omega t + \phi_B(x)) \tag{6.28}$$

This can be written in complex notation as

$$C_A = C_{A\text{-dc}} + C_{A\text{-ac}0}\mathrm{e}^{j(\omega t + \phi_A)} = C_{A\text{-dc}} + C_{A\text{-ac}} = C_{A\text{-dc}} + \Delta C_A \tag{6.29}$$

$$C_B = C_{B\text{-dc}} + C_{B\text{-ac}0}\mathrm{e}^{j(\omega t + \phi_B)} = C_{B\text{-dc}} + C_{B\text{-ac}} = C_{B\text{-dc}} + \Delta C_B \tag{6.30}$$

with the understanding that $C_{A\text{-dc}}$ is a function of location (x), while $C_{A\text{-ac}0}$ and ϕ_A are functions of x and frequency (ω). $C_{A\text{-ac}0}$ is expected to be proportional to $E_{\text{ac}0}$, whereas ϕ_A will be independent of $E_{\text{ac}0}$, as long as $E_{\text{ac}0}$ is small enough. This substitution simplifies the partial differential equation (Eq. 6.14) to an ordinary differential equation, as shown below.

Equation 6.14 can be written as

$$\frac{\partial C_{A\text{-ac}}}{\partial t} = D_A\left(\frac{\partial^2 C_{A\text{-ac}}}{\partial x^2}\right) \tag{6.31}$$

$$\frac{C_{A\text{-ac}0}\mathrm{e}^{j\phi_A}\,\partial\left[\mathrm{e}^{j\omega t}\right]}{\partial t} = D_A \mathrm{e}^{j\omega t}\left(\frac{\partial^2\left[C_{A\text{-ac}0}\mathrm{e}^{j\phi_A}\right]}{\partial x^2}\right) \tag{6.32}$$

$$j\omega C_{A\text{-ac}0}\mathrm{e}^{j\phi_A}\mathrm{e}^{j\omega t} = D_A \mathrm{e}^{j\omega t}\left(\frac{\partial^2\left[C_{A\text{-ac}0}\mathrm{e}^{j\phi_A}\right]}{\partial x^2}\right) \tag{6.33}$$

Denoting $\widetilde{C_A} = C_{A\text{-ac}0}\mathrm{e}^{j\phi_A}$, we can write

$$\frac{\mathrm{d}^2\widetilde{C_A}}{\mathrm{d}x^2} = \frac{j\omega}{D_A}\widetilde{C_A} \tag{6.34}$$

The problem is then reduced to solving this ordinary differential equation, the solution of which is given by

$$\widetilde{C_A} = A1 \times e^{mx} + A2 \times e^{-mx}, \tag{6.35}$$

where $m = \sqrt{\frac{j\omega}{D_A}}$ and $A1$ and $A2$ are integration constants.

The boundary conditions 6.16 can be written as

$$\left.\frac{\partial \widetilde{C_A}}{\partial x}\right|_{0,t} = \frac{\widetilde{i_F}}{nFD_A} \tag{6.36}$$

$$\text{Here } \widetilde{i_F} = i_{F\text{-ac}0} e^{j\omega\phi_i} = \frac{i_{F\text{-ac}}}{e^{j\omega t}} \tag{6.37}$$

At $x = \delta$, we can write $\widetilde{C_A} = 0$, since the concentrations of A and B are fixed and are equal to the respective bulk values.

We can evaluate the constants $A1$ and $A2$ using the boundary conditions and show that

$$A1 = \frac{\widetilde{i_F} e^{-\left(\sqrt{\frac{j\omega}{D_A}}\right)\delta}}{D_A F \sqrt{\frac{j\omega}{D_A}} \left[e^{-\left(\sqrt{\frac{j\omega}{D_A}}\right)\delta} + e^{\left(\sqrt{\frac{j\omega}{D_A}}\right)\delta} \right]} \tag{6.38}$$

$$A2 = \frac{-\widetilde{i_F} e^{\left(\sqrt{\frac{j\omega}{D_A}}\right)\delta}}{D_A F \sqrt{\frac{j\omega}{D_A}} \left[e^{-\left(\sqrt{\frac{j\omega}{D_A}}\right)\delta} + e^{\left(\sqrt{\frac{j\omega}{D_A}}\right)\delta} \right]} \tag{6.39}$$

The concentration fluctuations at any location within the boundary layer are given by

$$\widetilde{C_A}(x) = \frac{\widetilde{i_F} \left[e^{\left(\sqrt{\frac{j\omega}{D_A}}\right)(x-\delta)} - e^{-\left(\sqrt{\frac{j\omega}{D_A}}\right)(x-\delta)} \right]}{F\sqrt{j\omega D_A} \left[e^{-\left(\sqrt{\frac{j\omega}{D_A}}\right)\delta} + e^{\left(\sqrt{\frac{j\omega}{D_A}}\right)\delta} \right]} = \frac{-\widetilde{i_F} \sin h\left(\left(\sqrt{\frac{j\omega}{D_A}}\right)(\delta - x)\right)}{F\sqrt{j\omega D_A} \cos h\left(\left(\sqrt{\frac{j\omega}{D_A}}\right)\delta\right)} \tag{6.40}$$

At the surface ($x = 0$), they are given by

$$\widetilde{C_A}(x) = \frac{-\widetilde{i_F} \sin h\left(\left(\sqrt{\frac{j\omega}{D_A}}\right)\delta\right)}{F\sqrt{j\omega D_A} \cos h\left(\left(\sqrt{\frac{j\omega}{D_A}}\right)\delta\right)} \tag{6.41}$$

The corresponding values for species B can be written as

$$\widetilde{C_B} = \frac{\widetilde{i_F} \tan h\left(\left(\sqrt{\frac{j\omega}{D_B}}\right)\delta\right)}{F\sqrt{j\omega D_B}} \tag{6.42}$$

By substituting Eqs. 6.23–26, 6.41, and 6.42, in Eq. 6.22 and rearranging, we get

$$Y_F = \frac{i_{F\text{-ac}}}{E_{\text{ac}}} = \frac{\left\{F\left(b_1 k_{1\text{dc}}\, C_A|_{x=0} - b_{-1}k_{-1\text{dc}}\, C_B|_{x=0}\right)\right\}}{\left(1 + k_{1\text{dc}}\frac{\tan h\left(\left(\sqrt{\frac{j\omega}{D_A}}\right)\delta\right)}{\sqrt{j\omega D_A}} + k_{-1\text{dc}}\frac{\tan h\left(\left(\sqrt{\frac{j\omega}{D_B}}\right)\delta\right)}{\sqrt{j\omega D_B}}\right)} \tag{6.43}$$

and the faradaic impedance is

$$Z_F = R_t + \left(Z_{\text{BW},A} + Z_{\text{BW},B}\right) = \frac{\left(1 + k_{1\text{dc}}\frac{\tan h\left(\left(\sqrt{\frac{j\omega}{D_A}}\right)\delta\right)}{\sqrt{j\omega D_A}} + k_{-1\text{dc}}\frac{\tan h\left(\left(\sqrt{\frac{j\omega}{D_B}}\right)\delta\right)}{\sqrt{j\omega D_B}}\right)}{\left\{F\left(b_1 k_{1\text{dc}}\, C_A|_{x=0} - b_{-1}k_{-1\text{dc}}\, C_B|_{x=0}\right)\right\}} \tag{6.44}$$

$$= R_t + Z_{\text{BW}} \tag{6.45}$$

The impedance can be represented by the equivalent circuit shown in Figure 6.7.

In Eq. 6.44, the charge-transfer resistance is given by

$$(R_t)^{-1} = nF\left(b_1 k_{1\text{dc}}\, C_A|_{x=0,\text{SS}} - b_{-1}k_{-1\text{dc}}\, C_B|_{x=0,\text{SS}}\right) \tag{6.46}$$

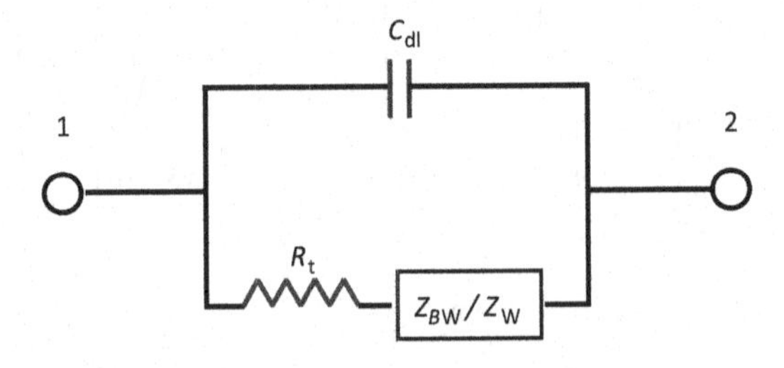

FIGURE 6.7 Schematic of an equivalent electrical circuit to represent a simple electron transfer reaction under diffusion and kinetics control. Z_{BW} is the bounded Warburg impedance, as shown in Eq. 6.44, and is applicable for a boundary layer of a finite thickness δ (rotating disc electrode). Z_W is the semi-infinite Warburg impedance, shown later in Eq. 6.75, and is applicable when the boundary layer thickness is infinity (stationary electrode).

The other components, often referred to as mass-transfer impedance, are given by

$$Z_{\mathrm{BW},A} = \frac{1}{nF\sqrt{j\omega D_A}}\left[\frac{k_{1\mathrm{dc}}}{b_1 k_{1\mathrm{dc}}\, C_A\big|_{x=0,\mathrm{SS}} - b_{-1}k_{-1\mathrm{dc}}\, C_B\big|_{x=0,\mathrm{SS}}}\right]\tanh\left(\delta\sqrt{\frac{j\omega}{D_A}}\right) \tag{6.47}$$

$$Z_{\mathrm{BW},B} = \frac{1}{nF\sqrt{j\omega D_B}}\left[\frac{k_{-1\mathrm{dc}}}{b_1 k_{1\mathrm{dc}}\, C_A\big|_{x=0,\mathrm{SS}} - b_{-1}k_{-1\mathrm{dc}}\, C_B\big|_{x=0,\mathrm{SS}}}\right]\tanh\left(\delta\sqrt{\frac{j\omega}{D_B}}\right) \tag{6.48}$$

This is also called *finite Warburg impedance* or *Nernst diffusion impedance* or *open finite length diffusion impedance*. The boundary conditions are sometimes referred to as *transmittive boundary conditions*. These impedance expressions can also be written as

$$Z_{\mathrm{BW},A} = \frac{\sigma'_A}{\sqrt{2\omega}}\tan h\left(\delta\left[\frac{j\omega}{D_A}\right]^{1/2}\right)(1-j) = \frac{\sigma'_A}{\sqrt{j\omega}}\tan h\left(\delta\left[\frac{j\omega}{D_A}\right]^{1/2}\right) \tag{6.49}$$

$$Z_{BW,B} = \frac{\sigma'_B}{\sqrt{2\omega}}\tan h\left(\delta\left[\frac{j\omega}{D_B}\right]^{1/2}\right)(1-j) = \frac{\sigma'_B}{\sqrt{j\omega}}\tan h\left(\delta\left[\frac{j\omega}{D_B}\right]^{1/2}\right) \tag{6.50}$$

where σ'_A and σ'_B are given by

$$\sigma'_A = \frac{1}{nF\left(b_1 k_{1\mathrm{dc}} C_{\mathrm{AS}} - b_{-1}k_{-1\mathrm{dc}} C_{\mathrm{BS}}\right)}\left[\frac{k_{1\mathrm{dc}}}{\sqrt{D_A}}\right] \tag{6.51}$$

$$\sigma'_B = \frac{1}{nF\left(b_1 k_{1\mathrm{dc}} C_{\mathrm{AS}} - b_{-1}k_{-1\mathrm{dc}} C_{\mathrm{BS}}\right)}\left[\frac{k_{-1\mathrm{dc}}}{\sqrt{D_B}}\right] \tag{6.52}$$

Figure 6.8 shows a complex plane plot and a Bode plot of impedance spectra of a simple electron transfer reaction. Although a cursory look of the complex plane plot may indicate that the finite Warburg impedance introduces a second 'capacitive' loop at low frequencies, there is a marked difference between a capacitive loop and a finite Warburg impedance. A capacitor in parallel with a resistor will result in a semicircle in a complex plane plot; a CPE in parallel with a resistor will yield a depressed semicircle, and a finite Warburg impedance will yield a 45° line at the high-frequency range and a circular part at the low-frequency range.

Under equilibrium conditions (i.e., when $E_{\mathrm{dc}} = 0\,\mathrm{V}$ *vs.* OCP), Eqs. 6.47 and 6.48 can be simplified as follows. When $E_{\mathrm{dc}} = 0\,\mathrm{V}$ *vs.* OCP, $C_{A\text{-dc}}\,(x=0) = C_{A\text{-bulk}}$ and $C_{B\text{-dc}}\,(x=0) = C_{B\text{-bulk}}$. In addition, $k_{1\mathrm{dc}} = k_{10}$ and $k_{-1\mathrm{dc}} = k_{-10}$. Remembering that $k_{10}C_{A\text{-bulk}} = k_{-10}C_{B\text{-bulk}}$ as given in Eq. 6.11, we can write

$$b_1 k_{1\mathrm{dc}}\, C_A\big|_{x=0} - b_{-1}k_{-1\mathrm{dc}}\, C_B\big|_{x=0} = \frac{\alpha F}{RT}k_{10}C_{A\text{-bulk}} - \left[\frac{-(1-\alpha)F}{RT}\right]k_{-10}C_{B\text{-bulk}} = k_{10}C_{A\text{-bulk}}\frac{F}{RT} \tag{6.53}$$

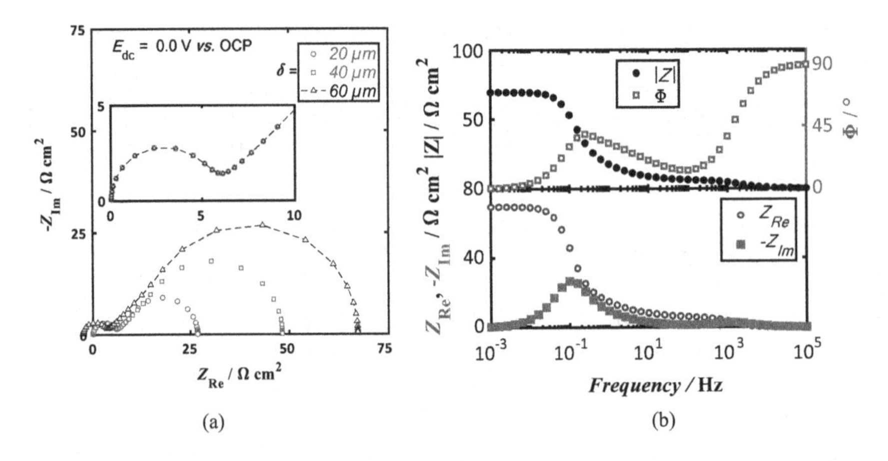

FIGURE 6.8 (a) Complex plane plot of finite Warburg impedance data, at various δ. The inset shows the high and mid frequency data, expanded for clarity. In this, the data for all three δ overlap and are practically indistinguishable. Values used are $k_{10}=k_{-10}=10^{-4}$m s^{-1}, $E_{dc}=0$ V *vs.* OCP, $D_A=D_B=10^{-9}$m^{2}s^{-1}, $C_{A\text{-bulk}}=C_{B\text{-bulk}}=5$ mM, $\alpha=0.5$, $T=300$ K, and $C_{dl}=20$ μF cm^{-2}. (b) Same data presented in the Bode format.

Remembering

$$Z_F = R_t + Z_{\text{BW},A} + Z_{\text{BW},B} \tag{6.54}$$

we can write

$$R_t = \frac{RT}{F^2 k_{10} C_{A\text{-bulk}}} \tag{6.55}$$

$$Z_{\text{BW},A} = \frac{RT}{n^2 F^2 C_{A\text{-bulk}} \sqrt{j\omega D_A}} \tan h\left(\delta\sqrt{\frac{j\omega}{D_A}}\right) \tag{6.56}$$

$$Z_{\text{BW},B} = \frac{RT}{n^2 F^2 C_{B\text{-bulk}} \sqrt{j\omega D_B}} \tan h\left(\delta\sqrt{\frac{j\omega}{D_B}}\right) \tag{6.57}$$

6.2.1.2 Zero dc Bias

Note that Eqs. 6.55–6.57 are valid only at equilibrium, whereas Eqs. 6.44–6.48 are valid even if E_{dc} is non-zero. At equilibrium (i.e., $E_{dc}=0$ V *vs.* OCP), the bounded Warburg impedance (Z_{BW}) can be written as

$$Z_{\text{BW}} = \left\{\frac{\sigma_A}{\sqrt{\omega}} \tan h\left(\delta\left[\frac{j\omega}{D_A}\right]^{1/2}\right) + \frac{\sigma_B}{\sqrt{\omega}} \tan h\left(\delta\left[\frac{j\omega}{D_B}\right]^{1/2}\right)\right\}(1-j) \tag{6.58}$$

$$= \left\{\sigma_A \tan h\left(\delta\left[\frac{j\omega}{D_A}\right]^{1/2}\right) + \sigma_B \tan h\left(\delta\left[\frac{j\omega}{D_B}\right]^{1/2}\right)\right\}\frac{\sqrt{2}}{\sqrt{j\omega}} \tag{6.59}$$

where σ_A and σ_B are given by

$$\sigma_A = \frac{RT}{n^2F^2\sqrt{2D_A}}\left[\frac{1}{C_{A\text{-bulk}}}\right] \tag{6.60}$$

$$\sigma_B = \frac{RT}{n^2F^2\sqrt{2D_B}}\left[\frac{1}{C_{B\text{-bulk}}}\right] \tag{6.61}$$

In case $D_A = D_B$ and $C_{A\text{-bulk}} = C_{B\text{-bulk}}$, one can write

$$\sigma = \sigma_A + \sigma_B = \frac{RT}{n^2F^2\sqrt{D}}\left[\frac{\sqrt{2}}{C_{\text{bulk}}}\right] \tag{6.62}$$

At this point, a few remarks are in order. At equilibrium, $C_{AS} = C_{A\text{-bulk}}$ and $C_{BS} = C_{B\text{-bulk}}$. Under that condition, it is seen that R_t, given in Eq. 6.55, is independent of the boundary layer thickness δ. This is shown in simulated values of impedance, shown as complex plane plots in Figure 6.8a. Besides, at equilibrium, Z_{BW}, as given in Eq. 6.58–6.61, does not depend on the rate constant but has a contribution from the diffusivities. The R_t and σ values are shown in Table 6.1.

It is important, however, to realize that the circuit elements R_t and Z_{BW} shown in Figure 6.8 *do not* represent a clear separation of kinetic and mass-transfer resistances. Recall that when E_{dc} is non-zero, then C_{AS} and C_{BS} depend on δ, as described in Eqs. 6.18 and 6.19, and R_t depends on kinetics as well as mass transfer. Likewise, when E_{dc} is non-zero, Z_{BW} as shown in Eq. 6.44 depends on kinetic as well as mass-transfer parameters. When the electrode is held at a non-zero dc bias, the effect of varying the boundary layer thickness on the impedance spectra is shown in Figure 6.9.

TABLE 6.1
Variation of Charge-Transfer Resistance (R_t) and Warburg parameter σ, on Boundary Layer Thickness and E_{dc}

Source	Boundary Layer Thickness (μm)	E_{dc} (V vs. OCP)	R_t (Ω cm²)	σ (Ω cm² s$^{0.5}$)
Figure 6.8	20	0.0	5.4	16.9
	40			
	60			
Figure 6.9	20	0.1	10.8	235.4
	40		13.6	297.4
	60		15.0	327.4
Figure 6.10	20	0	5.4	16.9
		0.05	6.8	56.8
		0.1	10.8	235.4

The parameters employed in the simulation are listed in the captions of Figures 6.8–6.10.

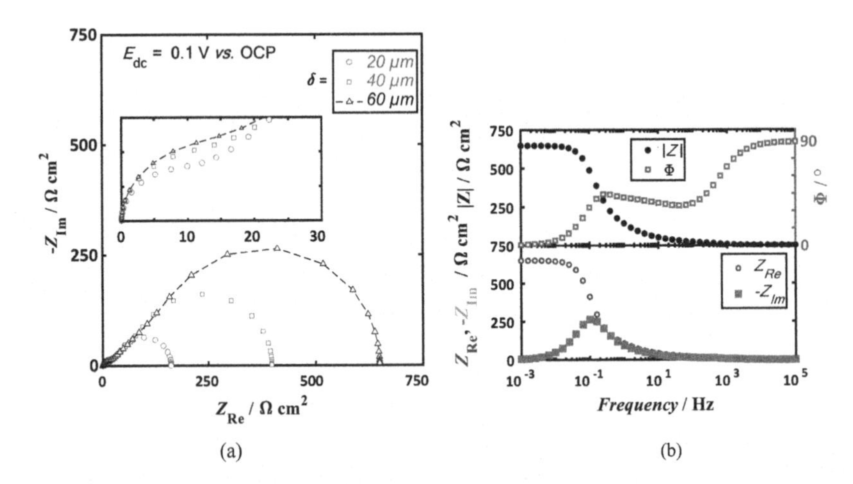

FIGURE 6.9 (a) Complex plane plot of finite Warburg impedance data, at various δ. The inset shows the high and mid frequency data, expanded for clarity. In this, the data for all three δ overlap and are practically indistinguishable. Values used are $k_{10}=k_{-10}=10^{-4}$m s^{-1}, $E_{dc}=0.1$ V *vs.* OCP, $D_A=D_B=10^{-9}$m^{2}s^{-1}, $C_{A\text{-bulk}}=C_{B\text{-bulk}}=5$ mM, $\alpha=0.5$, $T=300$ K, and $C_{dl}=20$ μF cm^{-2}. (b) Same data presented in the Bode format.

R_t depends on boundary layer thickness and thus is not independent of the mass-transfer resistance. The effect of varying E_{dc} is shown in the complex plane plots presented in Figure 6.10. The corresponding R_t and σ values are also listed in Table 6.1. We thus conclude that, in general, both R_t and Z_{BW} depend on kinetics as well as mass-transfer properties, and these two effects cannot always be separated.

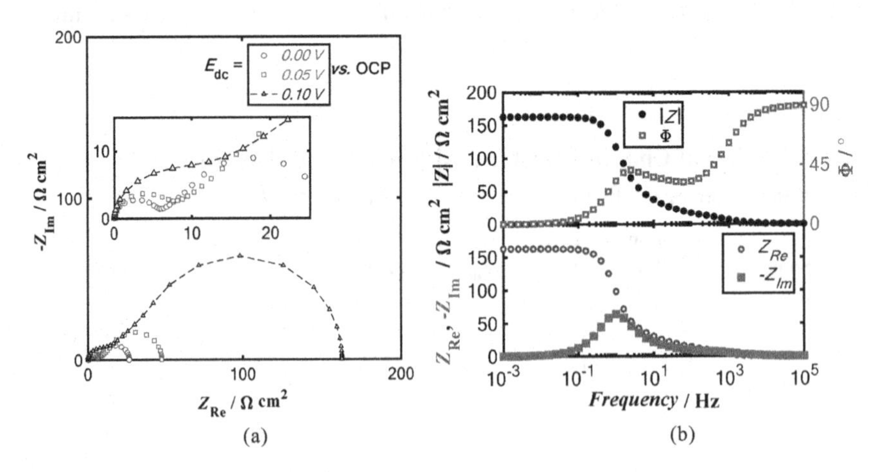

FIGURE 6.10 Effect of changing E_{dc} on the impedance spectra. The boundary layer thickness is maintained constant ($\delta=20$ μm). The values of all other variables used are the same as those in Figure 6.8. (a) A complex plane plot of finite Warburg impedance data, at various E_{dc}. The inset shows the high and mid-frequency data, expanded for clarity. (b) Same data presented in the Bode format.

6.2.2 Semi-Infinite Boundary Conditions

If the solution is not stirred and if the electrode is not rotated, then when an ac perturbation is applied, the boundary layer thickness keeps increasing with time, and this is referred to as a moving boundary. After some time, when the boundary layer thickness is very large, the resulting current will be more or less steady periodic. Mathematically, the boundary layer thickness can be considered infinite.

The boundary conditions on the electrode surface remain the same, i.e.,

$$\left.\frac{\partial \widetilde{C_A}}{\partial x}\right|_{0,t} = \frac{\widetilde{i_F}}{nFD_A} \tag{6.63}$$

where

$$\widetilde{i_F} = i_{F\text{-ac}0} e^{j\omega\phi_i} = \frac{i_{F\text{-ac}}}{e^{j\omega t}} \tag{6.64}$$

On the other hand, the other boundary condition is different. At $x \to \infty$, we can write $\widetilde{C_A} = 0$, since the concentrations of A and B are fixed and are equal to the respective bulk values.

Recall the general solution (Eq. 6.35)

$$\widetilde{C_A} = A1 \times e^{mx} + A2 \times e^{-mx} \tag{6.65}$$

where $m = \sqrt{\frac{j\omega}{D_A}}$ and $A1$ and $A2$ are integration constants.

By substituting the boundary conditions in the solution (6.65), we can show that $A1 = 0$ and

$$\left.\frac{\partial \widetilde{C_A}}{\partial x}\right|_{0,t} = \frac{\widetilde{i_F}}{nFD_A} = -A2\sqrt{\frac{j\omega}{D_A}} \tag{6.66}$$

i.e.,

$$A2 = \frac{\widetilde{-i_F}}{nF\sqrt{j\omega D_A}} \tag{6.67}$$

Hence,

$$\widetilde{C_A} = \frac{\widetilde{-i_F}}{nF\sqrt{j\omega D_A}} e^{-\sqrt{\frac{j\omega}{D_A}}x} \tag{6.68}$$

On the electrode surface, we can write

$$\Delta C_A\big|_{x=0} = \widetilde{C_A}\Big|_{x=0} e^{j\omega t} = \frac{\widetilde{-i_F}}{nF\sqrt{j\omega D_A}} e^{j\omega t} = \frac{-\Delta i_F}{nF\sqrt{j\omega D_A}} \tag{6.69}$$

Similarly,

$$\Delta C_B\big|_{x=0} = \frac{+\Delta i_F}{nF\sqrt{j\omega D_B}} \tag{6.70}$$

Therefore, Eq. 6.22 becomes

$$\Delta i_F = \left\{ F\left(b_1 k_{10} C_{A\text{-bulk}} - b_{-1} k_{-10} C_{B\text{-bulk}}\right) E_{\text{ac}} + F k_{1\text{dc}} \left(\frac{-\Delta i_F}{F\sqrt{j\omega D_A}} \right) - F k_{-1\text{dc}} \left(\frac{\Delta i_F}{F\sqrt{j\omega D_B}} \right) \right\} \tag{6.71}$$

The faradaic admittance can be written as

$$Y_F = \frac{\Delta i_F}{E_{\text{ac}}} = \frac{F\left(b_1 k_{10} C_{A\text{-bulk}} - b_{-1} k_{-10} C_{B\text{-bulk}}\right)}{\left(1 + k_{1\text{dc}} \left(\frac{1}{\sqrt{j\omega D_A}} \right) + k_{-1\text{dc}} \left(\frac{1}{\sqrt{j\omega D_B}} \right) \right)} \tag{6.72}$$

The faradaic impedance can be written as

$$Z_F = \left(Y_F\right)^{-1} = \frac{\left(1 + \left(\frac{k_{10}}{\sqrt{j\omega D_A}} \right) + \left(\frac{k_{-10}}{\sqrt{j\omega D_B}} \right) \right)}{F\left(b_1 k_{10} C_{A\text{-bulk}} - b_{-1} k_{-10} C_{B\text{-bulk}}\right)} \tag{6.73}$$

Note that under semi-infinite boundary conditions, a (pseudo)steady-state solution is possible only when the dc potential is at equilibrium ($E_{\text{dc}} = 0\,\text{V}$ *vs.* OCP). Hence, we can employ the simplification used in Eq. 6.39 and write

$$Z_F = R_t + Z_{W,A} + Z_{W,B} = \frac{RT}{F^2 k_{10} C_{A\text{-bulk}}} + \frac{RT}{F^2 C_{A\text{-bulk}} \sqrt{j\omega D_A}} + \frac{RT}{F^2 C_{B\text{-bulk}} \sqrt{j\omega D_B}} \tag{6.74}$$

The second and third parts of Eq. 6.74 can be rewritten as (Bard and Faulkner 1980)

$$Z_W = Z_{W,A} + Z_{W,B} = \frac{\sqrt{2}\sigma_W}{\sqrt{j\omega}} = \sigma_W \frac{(1-j)}{\sqrt{\omega}} \tag{6.75}$$

where

$$\sigma_W = \frac{RT}{n^2 F^2 \sqrt{2}} \left[\frac{1}{\sqrt{D_A} C_{A\text{-bulk}}} + \frac{1}{\sqrt{D_B} C_{B\text{-bulk}}} \right] \tag{6.76}$$

This is called *semi-infinite Warburg impedance* or simply, *Warburg impedance.*

In the complex plane plots, the Warburg impedance will appear as a 45° line, as shown in Figure 6.11a. Remember, the scales must be equal in the abscissa and the ordinate. The corresponding Bode plots are shown in Figure 6.11b.

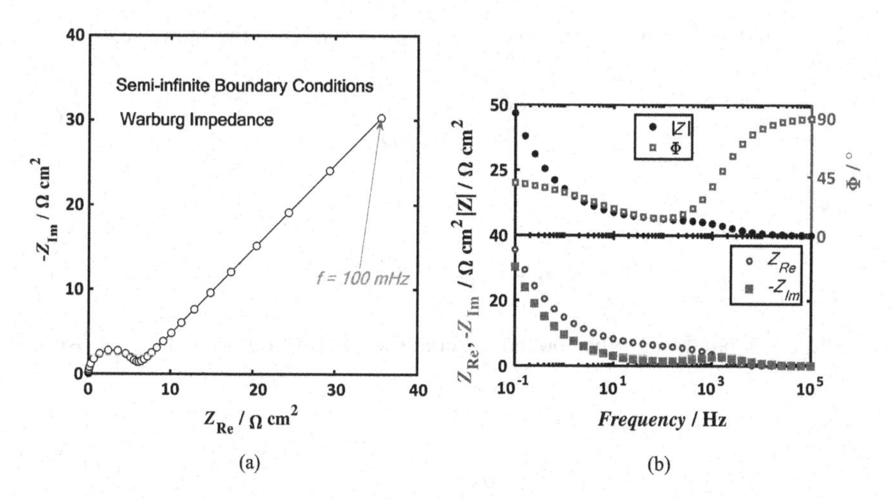

FIGURE 6.11 EIS of a simple electron transfer reaction, with diffusion limitation, at OCP – semi-infinite boundary conditions. Here, $k_{10}=k_{-10}=10^{-4}$m s^{-1}, $E_{dc}=0$V *vs.* OCP, $D_A=D_B=10^{-9}$m^{2}s^{-1}, $C_{A\text{-bulk}}=C_{B\text{-bulk}}=5$mM, $\alpha=0.5$, $T=300$ K, and $C_{dl}=20$ μF cm^{-2}. (a) Complex plane plot. (b) Bode plots.

At very low frequencies, the semi-infinite diffusion boundary condition predicts that the magnitude of impedance will tend to infinity. In contrast, the finite boundary condition predicts that the magnitude of impedance tends to a finite value (Lasia 2014). Besides, when $\omega \to \infty$ or when $\delta \to \infty$, the expression for finite Warburg impedance Z_{BW} tends toward the expression for infinite Warburg impedance Z_W, as expected.

6.2.3 Blocking Boundary Conditions

Another type of boundary conditions that may occur in certain electrochemical systems is called *finite length diffusion with reflecting boundary conditions.* This is sometimes called as *finite length diffusion with blocking boundary conditions.* In this case, an active species diffuses in a direction, perhaps due to electrical force. However, it is blocked at a boundary, which means the flux across the boundary, given by $\partial C/\partial x$, will be zero. This condition is also satisfied if the concentration profiles on either side of the boundary are mirror images. Hence the name *reflecting boundary conditions.* It is used to explain the impedance plots obtained in secondary Li-ion batteries and also in some transpassive dissolution.

The general solution obtained earlier will be used with the appropriate boundary conditions.

$$\widetilde{C_A} = A1 \times e^{mx} + A2 \times e^{-mx} \tag{6.77}$$

where $m = \sqrt{\dfrac{j\omega}{D_A}}$.

The boundary conditions on the electrode surface remain the same, i.e.,

$$\left.\frac{\partial \widetilde{C_A}}{\partial x}\right|_{0,t} = \frac{\widetilde{i_F}}{nFD_A} \tag{6.78}$$

where

$$\widetilde{i_F} = i_{F\text{-ac}0}\mathrm{e}^{j\omega\phi_i} = \frac{i_{F\text{-ac}}}{\mathrm{e}^{j\omega t}} \tag{6.79}$$

On the other hand, the other boundary condition is different, i.e., at $x=\delta$, we can write

$$\left.\frac{\partial \widetilde{C_A}}{\partial x}\right|_{\delta,t} = 0, \tag{6.80}$$

This leads to

$$\left.\frac{\partial \widetilde{C_A}}{\partial x}\right|_{\delta,t} = 0 = m\times\left[A1\times\mathrm{e}^{m\delta} - A2\times\mathrm{e}^{-m\delta}\right] \tag{6.81}$$

$$A2 = A1\times\mathrm{e}^{2m\delta} \tag{6.82}$$

Therefore,

$$\left.\frac{\partial \widetilde{C_A}}{\partial x}\right|_{0,t} = \frac{\widetilde{i_F}}{nFD_A} = m\times[A1 - A2] = mA1\times\left(1-\mathrm{e}^{2m\delta}\right) \tag{6.83}$$

Rearranging, we get the values of $A1$ and $A2$ as

$$A1 = \frac{\widetilde{i_F}}{mnFD_A\left(1-\mathrm{e}^{2m\delta}\right)} = \frac{\mathrm{e}^{-m\delta}\,\widetilde{i_F}}{mnFD_A\left(\mathrm{e}^{-m\delta}-\mathrm{e}^{m\delta}\right)} = \frac{-\mathrm{e}^{-m\delta}\,\widetilde{i_F}}{mnFD_A\left(\mathrm{e}^{m\delta}-\mathrm{e}^{-m\delta}\right)} \tag{6.84}$$

$$A2 == \frac{-\mathrm{e}^{m\delta}\,\widetilde{i_F}}{mnFD_A\left(\mathrm{e}^{m\delta}-\mathrm{e}^{-m\delta}\right)} \tag{6.85}$$

Now, the concentration oscillations can be written as

$$\widetilde{C_A} = \frac{-\mathrm{e}^{-m\delta}\,\widetilde{i_F}}{mnFD_A\left(\mathrm{e}^{m\delta}-\mathrm{e}^{-m\delta}\right)}\times\mathrm{e}^{mx} + \frac{-\mathrm{e}^{m\delta}\,\widetilde{i_F}}{mnFD_A\left(\mathrm{e}^{m\delta}-\mathrm{e}^{-m\delta}\right)}\times\mathrm{e}^{-mx} \tag{6.86}$$

$$\widetilde{C_A} = \frac{-\mathrm{e}^{-m(\delta-x)}\,\widetilde{i_F}}{mnFD_A\left(\mathrm{e}^{m\delta}-\mathrm{e}^{-m\delta}\right)} + \frac{-\mathrm{e}^{m(\delta-x)}\,\widetilde{i_F}}{mnFD_A\left(\mathrm{e}^{m\delta}-\mathrm{e}^{-m\delta}\right)} \tag{6.87}$$

$$\widetilde{C_A} = \frac{-\widetilde{i_F}}{mnFD_A}\frac{e^{m(\delta-x)}+e^{-m(\delta-x)}}{\left(e^{m\delta}-e^{-m\delta}\right)} = \frac{-\widetilde{i_F}}{mnFD_A}\frac{\cosh\left(m(\delta-x)\right)}{\sinh\left(m\delta\right)} \tag{6.88}$$

Thus,

$$\widetilde{C_A}\Big|_{x=0} = \frac{-\widetilde{i_F}\cot h\left(m\delta\right)}{mnFD_A} \tag{6.89}$$

and similarly,

$$\widetilde{C_B}\Big|_{x=0} = \frac{+\widetilde{i_F}\cot h\left(m\delta\right)}{mnFD_A} \tag{6.90}$$

The impedance of a system with finite length diffusion with blocking boundary conditions is given by

$$Z_{\text{Reflective}} = \frac{RT}{n^2F^2DC_A}\frac{\cot h\left(\delta\sqrt{j\omega/D}\right)}{\left(\delta\sqrt{j\omega/D}\right)} \tag{6.91}$$

A sample spectrum is shown in Figure 6.12 in the complex plane format and Bode format. Note, first, that at high frequencies, the reflecting boundary conditions yield

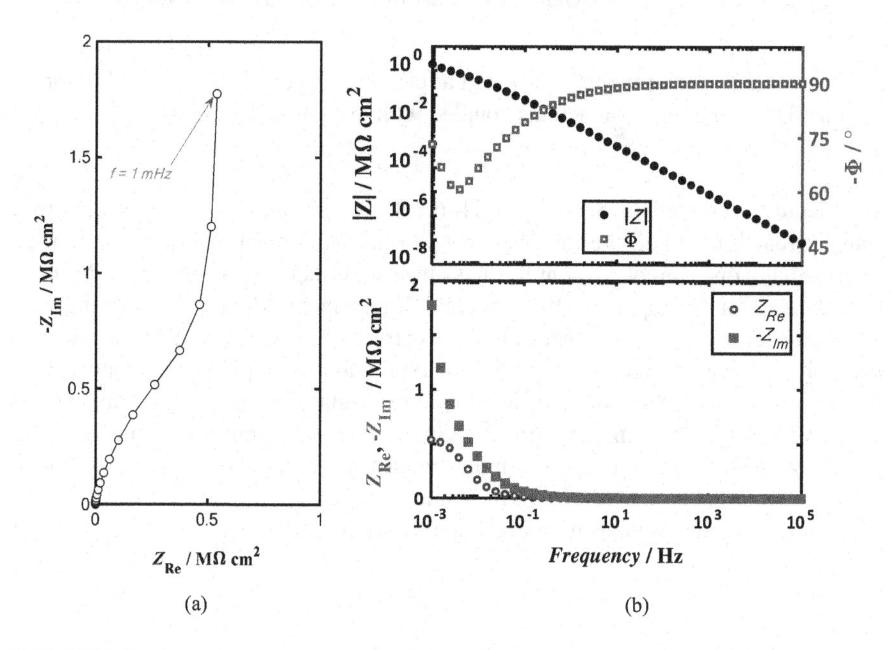

FIGURE 6.12 Impedance spectrum of a simple reaction, with blocking boundary conditions. Here, $k_{10}=k_{-10}=10^{-6}$m s^{-1}, $E_{dc}=0$V *vs.* OCP, $D_A=D_B=10^{-10}$m^{2}s^{-1}, $C_{A\text{-bulk}}=C_{B\text{-bulk}}=10$mM, $\alpha=0.5$, $T=300$ K, $C_{dl}=20$ μF cm^{-2}, and $\delta=20$nm. (a) Complex plane plot. (b) Bode plot.

impedance that is similar to semi-infinite Warburg impedance. At low frequencies, the resulting impedance is similar to a capacitive line. In the example shown in Figure 6.12, the values are chosen such that the semicircle corresponding to the double-layer capacitance and the charge-transfer resistance appears at high frequencies in the complex plane plot. At intermediate frequencies, the Warburg impedance-like behavior of the $Z_{\text{Reflective}}$ manifests and at low frequencies, the capacitance-like behavior of the $Z_{\text{Reflective}}$ manifests. It is quite possible that, for a given system and a frequency range, only the capacitive line may show up in the low frequencies and/or only the Warburg-like line may be seen in the low frequencies. The reader is encouraged to generate these plots and play with the kinetic parameter values and observe their effect on the complex plane and Bode plots.

The main difference between the *reflective* and *transmittive boundary* conditions is that at low frequencies, the former appears like a capacitance. Meanwhile, the latter exhibits a characteristic loop, with a 45° slope at high frequencies and an arc at low frequencies. Note that the discussion assumes an ideal case of diffusion in one dimension under steady-state conditions. In complex systems, mass transfer in 2D, and possibly in 3D, has to be taken into account, and the reader can visualize many such scenarios with different geometries and boundary conditions. In many specific cases, analytical expressions of the impedance corresponding to those diffusions are available, and the interested reader is referred to *Handbook of Electrochemical Impedance Spectroscopy* (Diard, Le Gorrec, and Montella 2012).

6.2.4 Reactions with Adsorbed Intermediates, Coupled with Diffusion

In 1974, **S.K. Rangarajan** developed a formalism to derive an expression for impedance of complex reactions coupled with mass-transfer effects.

The details presented in Sections 6.2.1–6.2.3 are all valid for a simple electron transfer reaction. When the mechanism is a multi-step reaction with an adsorbed intermediate, one cannot account for mass-transfer effects by just adding a Warburg impedance (for example) to the Faradaic impedance calculated assuming that mass transfer is rapid. The kinetics and mass transfer are coupled at the electrode–electrolyte interface, and we can derive the expression for impedance accounting for the mass transfer; the derivation is lengthy but straightforward, and the final expressions, when expressed in terms of the kinetic and mass-transfer parameters, are unwieldy. However, using a program and intermediate variables, one can simulate the impedance spectra, as shown in the example below.

Consider the electrodeposition of a metal from the solution.

$$M_{sol}^{2+} + e^- \underset{k_{-1}}{\overset{k_1}{\rightleftharpoons}} M_{ads}^{+} \tag{6.92}$$

$$M_{ads}^{+} + e^- \underset{k_{-2}}{\overset{k_2}{\rightleftharpoons}} M \tag{6.93}$$

Let us denote the rate of each step as r_1, r_2, and so on, i.e.,

$$r_1 = k_1 C_{\text{surface}}(1-\theta) \tag{6.94}$$

$$r_{-1} = k_{-1}\theta \tag{6.95}$$

$$r_2 = k_2\theta \tag{6.96}$$

$$r_{-2} = k_{-2}(1-\theta) \tag{6.97}$$

where C_{surface} is the concentration of M^{2+}_{sol} at the interface.

The diffusion of M^{2+}_{sol} from the bulk to the surface is not necessarily rapid. Assuming that convection and electromigration effects can be neglected, we can write

$$\frac{\partial C}{\partial t} = D\frac{\partial^2 C}{\partial x^2} \tag{6.98}$$

where C is the concentration of M^{2+}_{sol} in the boundary layer.

The boundary conditions are given as

$$\text{At } x = 0, D\left.\frac{\partial C}{\partial x}\right|_{x=0} = r_1 - r_{-1} \tag{6.99}$$

$$\text{At } x = \delta, \text{ the boundary condition is } C = C_{\text{bulk}}. \tag{6.100}$$

The mass balance equation for the adsorbed intermediate is given by

$$\Gamma\frac{d\theta}{dt} = r_1 - r_{-1} - r_2 + r_{-2} \tag{6.101}$$

The faradaic current is given by

$$i_F = F(r_1 - r_{-1} + r_2 - r_{-2}) \tag{6.102}$$

Under steady-state conditions, the concentration profile within the boundary layer is linear, and hence Eq. 6.99 can be written as

$$D\left.\frac{\partial C}{\partial x}\right|_{x=0} = k_{1\text{dc}}(C_{\text{ss-surface}})(1-\theta_{\text{ss}}) - k_{-1\text{dc}}\theta_{\text{ss}} = D\frac{(C_{\text{bulk}} - (C_{\text{ss-surface}}))}{\delta} \tag{6.103}$$

The steady-state surface concentration $(C_{\text{ss-surface}})$ and fractional surface coverage value (θ_{ss}) can be obtained as follows. First, we set Eq. 6.101 to zero.

$$k_{1\text{dc}}C_{\text{ss-surface}}(1-\theta_{\text{ss}}) - k_{-1\text{dc}}\theta_{\text{ss}} - k_{2\text{dc}}\theta_{\text{ss}} + k_{-2\text{dc}}(1-\theta_{\text{ss}}) = 0 \tag{6.104}$$

Solving Eqs. 6.103 and 6.104 together, we get a quadratic equation in θ_{ss}. Only one of the two solutions will be physically meaningful, i.e., between 0 and 1. This can be substituted in Eqs. 6.103 or 6.104 to get $C_{ss\text{-surface}}$.

Using these values in Eq. 6.102, the steady-state faradaic current can be calculated.

To calculate the impedance response, we can apply a sinusoidal perturbation E_{ac}, superimposed on E_{dc}. Following the procedure adopted earlier in Section 6.2.1, we further assume that the concentration of M^{2+}_{sol} in the boundary layer can be written as

$$C = C_{ss} + C_{ac} = C_{ss} + \tilde{C}\ E_{ac0} e^{j\omega t} = C_{ss} + \tilde{C} E_{ac} \tag{6.105}$$

$$\text{where } \tilde{C} = \left({C_{ac0}}/{E_{ac0}} \right) e^{j\phi}. \tag{6.106}$$

Note that both C_{ss} and $\tilde{C}$ are functions of location (x) within the boundary layer.

Then Eq. 6.101 can be written as

$$j\omega\Gamma \frac{d\theta}{dE} = r_1' - r_{-1}' - r_2' + r_{-2}' \tag{6.107}$$

where

$$r_1' = b_1 k_{1dc} C_{ss\text{-surface}} (1-\theta_{ss}) + k_{1dc}(1-\theta_{ss})\tilde{C} - k_{1dc} C_{ss\text{-surface}} \frac{d\theta}{dE} \tag{6.108}$$

$$r_{-1}' = b_{-1} k_{-1dc} \theta_{ss} + k_{-1dc} \frac{d\theta}{dE} \tag{6.109}$$

$$r_2' = b_2 k_{2dc} \theta_{ss} + k_{2dc} \frac{d\theta}{dE} \tag{6.110}$$

$$r_{-2}' = b_{-2} k_{-2dc} (1-\theta_{ss}) - k_{-2dc} \frac{d\theta}{dE} \tag{6.111}$$

Substitution of Eq. 6.105 in Eq. 6.98 transforms the partial differential equation to a second-order ordinary differential equation with constant coefficients. The general solution of that *ode* is given by

$$\tilde{C} = A1 \times e^{mx} + A2 \times e^{-mx} \tag{6.112}$$

where $m = \sqrt{\frac{j\omega}{D}}$

The boundary condition in Eq. 6.100 yields

$$A2 = -A1 \times e^{2m\delta} \tag{6.113}$$

and the other boundary condition in Eq. 6.99 can be written as

$$mD(A1 - A2) = mD\left(1 + e^{2m\delta}\right) A1 = r_1' - r_{-1}' \tag{6.114}$$

By substituting Eqs. 6.108 and 6.109 in 6.114 and rearranging, we get

$$A1 = \frac{\left[b_1 k_{1dc} C_{ss\text{-surface}}(1-\theta_{ss}) - b_{-1}k_{-1dc}\theta_{ss}\right] + k_{1dc}(1-\theta_{ss})\left(\tilde{C}\Big|_{x=0}\right) - \left(k_{1dc}C_{ss\text{-surface}} + k_{-1dc}\right)\dfrac{d\theta}{dE}}{mD\left(1+e^{2m\delta}\right)} \quad (6.115)$$

Now, from Eq. 6.112, we can write the value of $\tilde{C}$ at the surface as

$$\tilde{C}\Big|_{x=0} = A1 + A2 = \left(1 - e^{2m\delta}\right)A1 \quad (6.116)$$

$$= \frac{\left(1-e^{2m\delta}\right)}{\left(1+e^{2m\delta}\right)} \frac{\left[b_1 k_{1dc} C_{ss\text{-surface}}(1-\theta_{ss}) - b_{-1}k_{-1dc}\theta_{ss}\right] + k_{1dc}(1-\theta_{ss})\left(\tilde{C}\Big|_{x=0}\right) - \left(k_{1dc}C_{ss\text{-surface}} + k_{-1dc}\right)\dfrac{d\theta}{dE}}{mD} \quad (6.117)$$

$$= -\tan h(m\delta) \times \frac{\left[b_1 k_{1dc} C_{ss\text{-surface}}(1-\theta_{ss}) - b_{-1}k_{-1dc}\theta_{ss}\right] + k_{1dc}(1-\theta_{ss})\left(\tilde{C}\Big|_{x=0}\right) - \left(k_{1dc}C_{ss\text{-surface}} + k_{-1dc}\right)\dfrac{d\theta}{dE}}{mD} \quad (6.118)$$

This can be rearranged as

$$\left(\tilde{C}\Big|_{x=0}\right) = -\frac{\left[b_1 k_{1dc} C_{ss\text{-surface}}(1-\theta_{ss}) - b_{-1}k_{-1dc}\theta_{ss}\right] - \left(k_{1dc}C_{ss\text{-surface}} + k_{-1dc}\right)\dfrac{d\theta}{dE}}{\left[k_{1dc}(1-\theta_{ss}) + mD \times \cot h(m\delta)\right]} \quad (6.119)$$

By substituting Eqs. 6.108–6.111 and 6.119 in Eq. 6.107 and rearranging, we can find the expression for $\frac{d\theta}{dE}$ in terms of known quantities. Note that Eqs. 6.112 and 6.113 are independent of the mechanism and only the boundary condition at the surface, i.e., Eq. 6.114 depends on the detailed mechanism.

The faradaic admittance is given by expanding Eq. 6.102 in the Taylor series, truncating after the first-order term and rearranging the equation as

$$(Z_F)^{-1} = Y_F = F\left(r_1' - r_{-1}' + r_2' - r_{-2}'\right) \quad (6.120)$$

Note that the faradaic impedance cannot be written in terms of the simple Warburg impedance and charge-transfer resistance, as was the case for a simple electron transfer reaction. A sample polarization curve and complex plane plot of impedance spectra are given in Figure 6.13a and b, respectively. The results for the kinetic-limited case (i.e., assuming mass transfer is rapid) are also shown for comparison. Note that the cathodic overpotential and currents are shown to be positive. When the boundary layer thickness was set at 0.1 μm, the results overlap with the kinetic-limited case. When the boundary layer thickness is increased, Figure 6.13a shows that the current saturates at large cathodic overpotentials and that the maximum current value decreases with the boundary layer thickness.

Figure 6.13b shows that the mass-transfer effects manifest at mid and low frequencies, and when the boundary layer thickness is increased, even the patterns

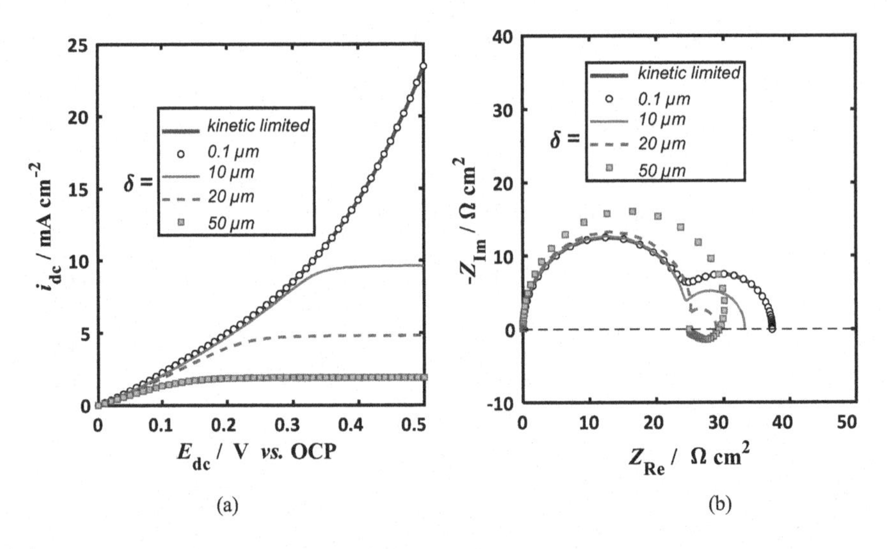

FIGURE 6.13 Response of a reaction given in Eqs. 6.92 and 6.93 accounting for mass-transfer effects. Here, $k_{10}=10^{-2}$cm s^{-1}, $k_{-10}=3\times10^{-9}$cm^{2}s^{-1}, $k_{20}=10^{-8}$cm s^{-1}, $b_1=15$ V^{-1}, $b_{-1}=-10$ V^{-1}, $b_2=5$ V^{-1}, $b_{-2}=-9$ V^{-1}, $\Gamma=5\times10^{-9}$cm^{2}s^{-1}, $D=10^{-5}$cm^{2}s^{-1}, $C_{bulk}=5$ mM, and $C_{dl}=20$ μF cm^{-2}. The results for a kinetically limited case are also shown for comparison. *Here, cathodic overpotential and cathodic currents are taken to be positive.* (a) Polarization plot. (b) Complex plane plot of impedance spectra, $E_{dc}=0.15$ V *vs.* OCP (cathodic polarization).

change dramatically. In the kinetic-limited case, two capacitive loops appear, with the high-frequency loop corresponding to the double-layer and charge-transfer resistance, and the low-frequency loop corresponding to the adsorbed intermediate. At 50 μm boundary layer thickness, mid and low-frequency loops appear in the fourth quadrant of the complex plane plot, appearing as inductive loops. The coupling of mass transfer and kinetics via the boundary condition can lead to very complex patterns (Molina Concha et al. 2013).

6.3 POROUS ELECTRODES

In 1967, **Robert de Levie** described the impedance of a porous electrode.

So far, we have assumed that the electrode surface is two-dimensional – mostly planar surface. In the case of a rough surface, the roughness considered is either microscopic or at the atomic level. However, electrodes can have significant porosity. For example, when silicon is electrochemically oxidized under certain conditions, pores are formed due to the selective dissolution of silicon. After some time, the planar electrode is essentially converted to a porous silicon electrode. In fuel cells, the electrodes are deliberately made porous so that the active surface area is large. The impedance response of the porous electrode differs from that of a planar electrode. The basic

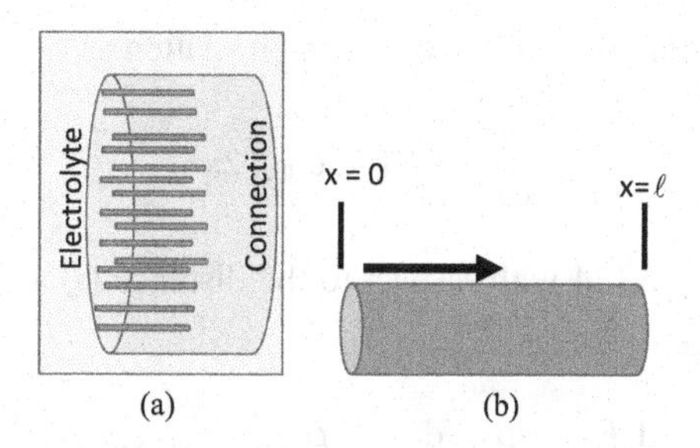

FIGURE 6.14 (a) Pictorial of a porous electrode. The electrode material is a conductor and has many parallel long cylindrical pores, filled with electrolyte (b) a single pore, with $x = 0$ at the electrode surface and $x = \ell$ at the other end of the pore.

model used to describe the porous electrode is the **de Levie** model. In this, the pores are assumed to be cylindrical. Here, we consider a metal electrode only. (In contrast, porous silicon is a porous semiconductor electrode, and the development of equations for such an electrode is not included here.) In addition, the potential is assumed to be the same everywhere in the pore wall, and axial diffusion effects are neglected.

A pictorial of a porous electrode is given in Figure 6.14a. An expanded version of a single pore is given in Figure 6.14b. Note that if ideal capacitance behavior can be assumed in the small element of a pore, then the specific impedance of the pore wall–electrolyte interface is given by

$$Z_0 = \left(\left(Z_f \right)^{-1} + j\omega C_{\mathrm{dl}} \right)^{-1} \tag{6.121}$$

where Z_f is the Faradaic impedance. Furthermore, we assume that there is no reaction, i.e., the faradaic impedance is infinite. Now, if ρ is the specific resistivity of the electrolyte, ℓ is the length of a pore, and r is the radius of the pore, then the potential drop across a short distance Δx in the pore is given by

$$\Delta E = -i \times R \tag{6.122}$$

$$R = \frac{\rho \Delta x}{\pi r^2} \tag{6.123}$$

Therefore, the potential drop per unit length along the pore length can be written as

$$\frac{\Delta E}{\Delta x} = \frac{\mathrm{d}E}{\mathrm{d}x} = \frac{-\rho}{\pi r^2} \times i \tag{6.124}$$

On the other hand, the current (Δi) passing through the cylindrical sidewall, which is of length Δx and radius r, is given by

$$\Delta i = \frac{-E}{Z_{\mathrm{dl}}} (2\pi r \Delta x) \tag{6.125}$$

At the limit of Δx going to zero, Eq. 6.96 can be written as,

$$\frac{\mathrm{d}i}{\mathrm{d}x} = -E\left(j\omega C_{\mathrm{dl}}\right)(2\pi r) \tag{6.126}$$

By differentiating Eq. 6.124 with respect to the distance (x) and substituting for $\frac{\mathrm{d}i}{\mathrm{d}x}$ from Eq. 6.126, we get

$$\frac{\mathrm{d}^2E}{\mathrm{d}x^2} = \frac{-\rho}{\pi r^2} \times \frac{\mathrm{d}i}{\mathrm{d}x} = \frac{+\rho}{\pi r^2} E\left(j\omega C_{\mathrm{dl}}\right)(2\pi r) \tag{6.127}$$

which can be simplified as

$$\frac{\mathrm{d}^2E}{\mathrm{d}x^2} = E\left(\frac{2\rho j\omega C_{\mathrm{dl}}}{r}\right) \tag{6.128}$$

The solution of this differential equation is

$$E = C_1 \times \mathrm{e}^{mx} + C_2 \times \mathrm{e}^{-mx} \tag{6.129}$$

where

$$m = \sqrt{j\omega\left(\frac{2\rho C_{\mathrm{dl}}}{r}\right)} \tag{6.130}$$

The boundary conditions are

$$\text{At } x = 0, E = E_{\mathrm{ac0}} \tag{6.131}$$

$$\text{At } x = \ell, \frac{\mathrm{d}E}{\mathrm{d}x} = 0 \tag{6.132}$$

By substituting Eqs. 6.131 and 6.132 in Eq. 6.129, we can show that

$$C_2 = C_1 \times \mathrm{e}^{2m\ell} \tag{6.133}$$

$$C_1 = \frac{E_{\mathrm{ac0}}}{\left(1+\mathrm{e}^{2m\ell}\right)} = \frac{E_{\mathrm{ac0}}}{\mathrm{e}^{m\ell}\left(\mathrm{e}^{m\ell}+e^{-m\ell}\right)} \tag{6.134}$$

Thus, the potential inside the pore can be written as

$$E = E_{\mathrm{ac0}} \frac{\cos h\left(m[\ell - x]\right)}{\cos h(m\ell)} \tag{6.135}$$

Moreover, the potential gradient at any location inside the pore is

$$\frac{\mathrm{d}E}{\mathrm{d}x} = -mE_{\mathrm{ac0}} \frac{\sin h\left(m[\ell - x]\right)}{\cos h(m\ell)} \tag{6.136}$$

At the surface, the potential gradient is

$$\left.\frac{dE}{dx}\right|_{x=0} = -mE_{ac0}\tan h(m\ell) \tag{6.137}$$

By combining Eqs. 6.124 and 6.137, we get

$$\frac{dE}{dx} = \frac{-\rho}{\pi r^2}\times i = -mE_{ac0}\tan h(m\ell) \tag{6.138}$$

Now, the impedance of the pore can be written as

$$Z_{pore} = \frac{E_{ac0}}{i} = \frac{\rho}{\pi r^2}\frac{\cot h(m\ell)}{m} = \left[\frac{\rho\ell}{\pi r^2}\right]\frac{\cot h\left(\Lambda^{1/2}\right)}{\Lambda^{1/2}} = \frac{R_{\Omega,p}}{\Lambda^{1/2}}\cot h\left(\Lambda^{1/2}\right) \tag{6.139}$$

where

$$R_{\Omega,p} = \rho\ell/\pi r^2 \tag{6.140}$$

is the resistance per unit length of the pore and Λ is the dimensionless admittance given by

$$\Lambda = m^2\ell^2 = j\omega C_{dl}\left(\frac{2\rho}{r}\right)\ell^2 = \left(Y_{C_{dl}}2\pi r\ell\right)\frac{\rho\ell}{\pi r^2} \tag{6.141}$$

which is the ratio of 'admittance of the double-layer capacitor' to the 'admittance of the solution resistor.'

When n such pores are present on the surface, the total resistance is given by

$$Z_T = R_{sol} + Z_{pore}/n \tag{6.142}$$

In complex plane plots, Z_T would appear as a 45° line at the high-frequency limit and as a capacitive line at the low-frequency limit, as shown in Figure 6.15. A more rigorous treatment accounting for the change of potential along the pore wall, axial diffusion, and non-cylindrical geometries is available (Lasia 2002), but the expressions are more complex, and the *de Levie* model is more commonly employed.

Figure 6.15 shows the complex plane plot and Bode plot of the impedance spectra of two porous electrodes, with identical features, but with different pore densities, i.e., the number of pores per unit area. The pore radius is measured in μm, and a pore density of 10^6 pore cm^{-2} means that there is a pore of 1 μm *radius* for every '10 μm × 10 μm *square*'. The total *pore sidewall area* is 31.4 cm^2 per "cm^2 of the geometric area of the electrode". Here, a rather limited frequency range, i.e., from 100 kHz to 10 Hz is employed. This is because the magnitude of the impedance of this system, at lower frequencies, is very high, and if those values are included, the major features of the spectra would not be visible. We have assumed that there are no reactions. Hence, if the electrode had a flat and non-porous surface, then we would

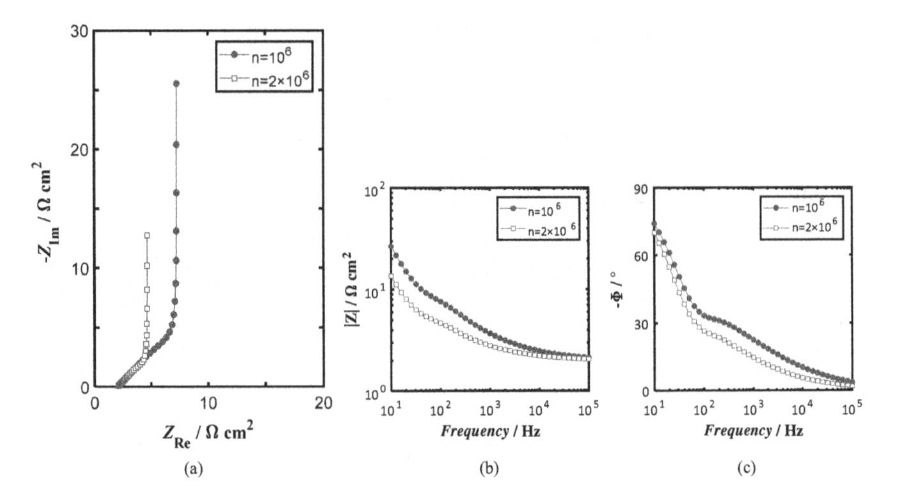

FIGURE 6.15 (a) Complex plane plot of impedance spectra of a porous electrode, with different pore densities (n). The corresponding Bode (b) magnitude and (c) phase plots of the impedance spectra. $n = 10^6$ or 2×10^6 pores (/cm²), $r = 1$ μm, $\ell = 500$ μm, $C_{dl} = 20$ μF cm⁻², $\rho = 10$ Ω cm, and $R_{sol} = 2$ Ω cm².

expect to see the impedance response of an ideal capacitor in series with an ideal resistor, i.e., in the complex plane plot, a vertical line with a non-zero offset. Since the electrode is porous, the impedance behavior is much more complex. When the number of pores is increased, the impedance magnitude decreases, but mainly at mid and low frequencies. The effect of changing the pore size is shown in Figure 6.16. Not surprisingly, when the pore radius decreases, the impedance magnitude increases, in particular at mid and low frequencies.

The literature on a porous electrode with pores of other shapes, pores with a size distribution, electrochemical reactions in porous electrodes, and so on is available, and the interested reader is referred to Lasia (2014).

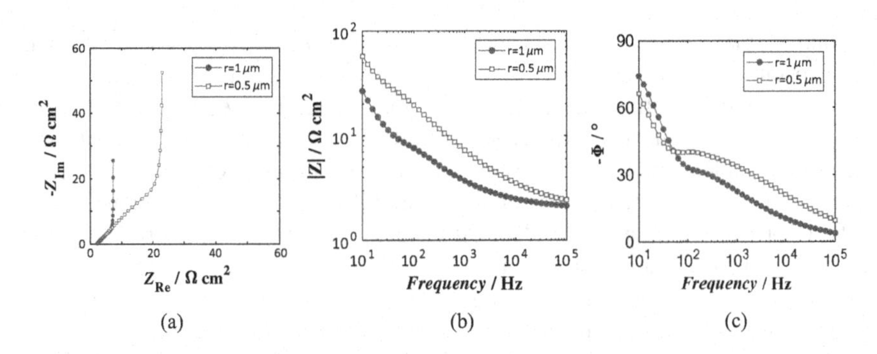

FIGURE 6.16 (a) Complex plane plot of impedance spectra of a porous electrode, with different pore densities (n). The corresponding Bode (b) magnitude and (c) phase plots of the impedance spectra. $n = 10^6$ pores (/cm²), $r = 1$ or 0.5 μm radius, $\ell = 500$ μm, frequency range is 100 kHz–10 Hz, $C_{dl} = 20$ μF cm⁻², and $\rho = 10$ Ω cm.

6.4 FILM FORMATION AND PASSIVATION

On many occasions, a film is formed on the electrode surface and will change the EIS response of the electrode. The nature of the film strongly depends on the film content and the method of formation, and no single model can represent all the electrode–film–solution interfaces. Films can also be made of organic chemicals, either inadvertently adsorbed from the solution, or deliberately introduced. Films can be inorganic, e.g., metal oxides formed by the reaction of the electrode with the electrolyte. Sometimes, organic films are coated on the electrode surface to protect the surface from corrosion. In biosensor fabrication, self-assembled monolayers (SAM) are created by treating the electrodes with certain organic chemicals, and a quasi-film is formed on the surface.

When a metal is exposed to water, it may or may not react with water, e.g., noble metals such as Au do not react with water while base metals such as Fe will react. When a base metal reacts with water, the reaction product is likely to be an oxide, which is insoluble in water. The oxide will form a film on the metal surface. The film may be an insulator or a semiconductor. Based on the morphology, the film can be classified as porous or non-porous. If a material forms a tenacious, chemically inert, non-porous film on the surface, then the film is called passive. Apart from these, a film can be formed on the surface of certain metals in some specific electrolytes by anodically polarizing the metal for a given time, and these are referred to as *anodic films*.

Now let us presume that a film is present on the metal surface. The impedance response of the electrode with a film is of interest to us. Figure 6.17 shows pictorials of a few models proposed to describe a film on the electrode surface. The corresponding equivalent electrical circuits are also presented and are described here.

6.4.1 Models Employed

1. A film can be a perfect insulator, and if it is non-porous, it will completely block the electrode surface. This is given in Figure 6.17a. The film can be described as a capacitor that is in series with the double-layer capacitor. The charge-transfer resistance is either very high or, in an extreme case, it would be infinity. In this case, the EIS will be that of a resistance (corresponding to solution resistance) in series with a capacitance (corresponding to the film and double-layer taken together). A complex plane plot of the EIS of this system would appear as a vertical line, especially in the low frequencies. If the film is heterogeneous, the film capacitor would be replaced by a CPE. Likewise, if the film–solution interface is heterogeneous, then the double-layer capacitor would be replaced by a CPE.
2. A film can be thought of as an insulator with some porosity, which allows electrolytes to pass, and thus some current will pass through the film. Then, a capacitor with a resistor in parallel can be used to model the film (MacDonald and Andreas 2014), as shown in Figure 6.17b. The capacitance depends on the dielectric constant of the film and its thickness.

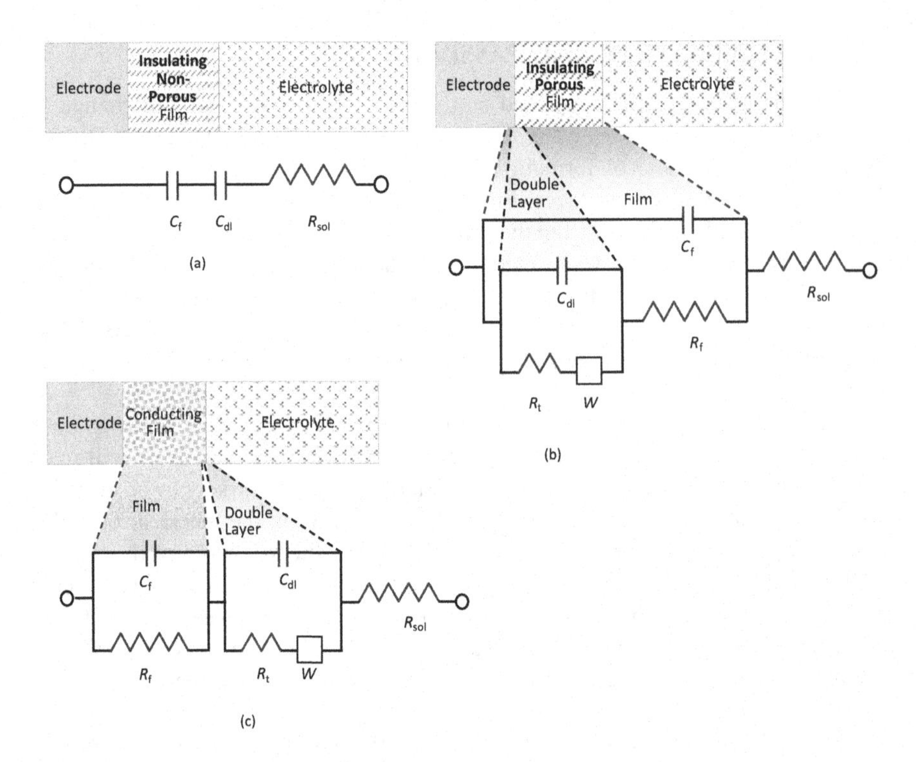

FIGURE 6.17 Pictorials of models proposed to describe a film on the electrode surface, and the corresponding equivalent electrical circuits. (a) A perfectly insulating film without any porosity. (b) An electrically insulating film with some porosity. (c) An electrically conducting film.

Some current will pass through the electrolyte filled in the pores. The double-layer is in series with the film resistance. Here, the film resistance is due to the restricted electrolyte movement in the pores. A complex plane plot of the EIS of such a system can have two capacitive loops. One loop would arise from the double-layer, and the second loop would arise from the film. In case the mass transfer of a species in solution is limiting, a Warburg impedance is added to the charge-transfer resistance in the double-layer. Besides, any heterogeneity in the film or the electrode–electrolyte interfaces can lead to the replacement of the corresponding capacitor with a CPE.

3. The film may be electrically conducting and can have a moderate dielectric constant. It is assumed to form without any porosity. The film is described by a capacitor that is in parallel with a resistance, and the double-layer is in series with the film (MacDonald and Andreas 2014). The double-layer is described by a capacitor that is in parallel with a charge-transfer resistor. A Warburg impedance may be incorporated here to account for mass-transfer limitations. This is shown in Figure 6.17c.
4. An electrically conducting film with porosity would be considered simply as a porous electrode.

When a film is formed by electrode oxidation, if the film is continuous (i.e., non-porous) and does not dissolve in the electrolyte, then it will offer excellent passivation. However, many such films lose passivity in the presence of chloride ions, by a phenomenon known as *pitting corrosion*. This is of major importance in any industry where chloride ions are present. In the beginning, a few *pits* would develop on the surface, but later, they become larger and can cause severe damage and, ultimately, failure of the structure. Pitting corrosion is much more pronounced in the presence of fluoride ions, and almost all metal oxides are attacked by F^- species. Unlike uniform corrosion, pitting corrosion is not easily detectable, and hence is extremely dangerous.

6.4.2 Point Defect Model

In 1981, **C. Y. Chao, L. F. Lin, and D. D. Macdonald** published the derivation of the first-generation point defect model to describe an anodic passive film.

1. To describe pitting corrosion, a model of the film, which is a lot more sophisticated than those depicted in Figure 6.16, was proposed by Digby D. Macdonald's group in 1985, and it was further upgraded over some time. It is known as the point defect model (PDM), and its salient features are qualitatively described below. For the detailed derivation of the relevant equations and model, the reader is referred to the classical papers (Chao, Lin, and Macdonald 1981, Lin, Chao, and Macdonald 1981, Chao, Lin, and Macdonald 1982, Macdonald 1992, 2011).
2. A physical description of the processes occurring at the metal–film–electrolyte solution volume is given below. A film covers the electrode surface, and the film is made of the metal oxide. The film is a lattice, which is not perfect and can have cation and anion vacancies. The film can also have cation interstitials. Note that anions are large and are not expected to move as interstitials in the film. In total, seven processes are considered.
   a. *Cation Interstitial Generation:* A metal atom (M) goes from the metal into the film as a cation interstitial ($M_i^{\chi+}$). This results in a vacancy in the metal v_m and the release of χ electrons, as depicted in Figure 6.18.

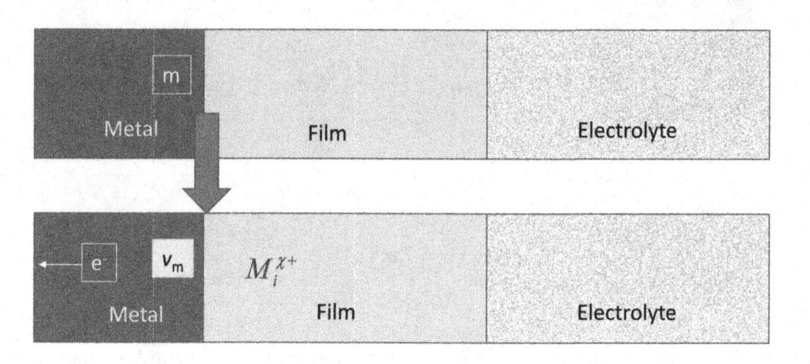

FIGURE 6.18 Generation of cation interstitials at the metal–film interface.

This step contributes to the current but does not result in increasing or decreasing the film thickness, i.e., film formation or destruction. Hence, this is a lattice-conserving reaction. The reaction is written as $m \rightarrow M_i^{\chi+} + v_m + \chi e^-$. It is assumed that the metal vacancies (v_m) quickly diffuse into the bulk metal and are consumed in a large sink, i.e., there will not be any accumulation of v_m near the metal–film interface and all the v_m disappear fast. If the cation vacancy diffusion in the metal is slower than that in the film, then pitting can occur. This will be described in detail later.

Initial development of PDM (Chao, Lin, and Macdonald 1981, Lin, Chao, and Macdonald 1981, Chao, Lin, and Macdonald 1982) did not incorporate cation interstitial generation or consumption, but the next generation model (Macdonald 1999) included them.

b. *Cation Interstitial Consumption:* The cation interstitials diffuse through the film and reach the film solution interface. Here, the interstitials cross the interface and move into the solution either at the same oxidation state (χ+) or possibly at a higher oxidation state (δ+), as shown in Figure 6.19. If the dissolution is oxidative, then a current is generated. The film thickness is not changed during this process, and hence this is also a lattice conserving reaction. The reaction is written as $M_i^{\chi+} \rightarrow M_{sol}^{\delta+} + (\delta - \chi)e^-$.
c. *Cation Vacancy Generation:* The movement of a cation from the film into the electrolyte at the film–solution interface results in the generation of a cation vacancy in the film, as depicted in Figure 6.20. The cation in the film may oxidize further (from the oxidation state of χ+ to δ+) when it moves into the solution, and this would result in a current. In certain cases, the oxidation state may not change when the cation moves into the solution, and in those cases, there will not be any current associated with this step. In general, the reaction is written as $M_M^{\chi+} \rightarrow M_{sol}^{\delta+} + V_M^{\chi+} + (\delta - \chi)e^-$. Since the film thickness does not change during this process, it is a lattice conserving reaction.
d. *Cation Vacancy Consumption:* The cation vacancies will diffuse through the film and arrive at the metal–film interface where they are consumed

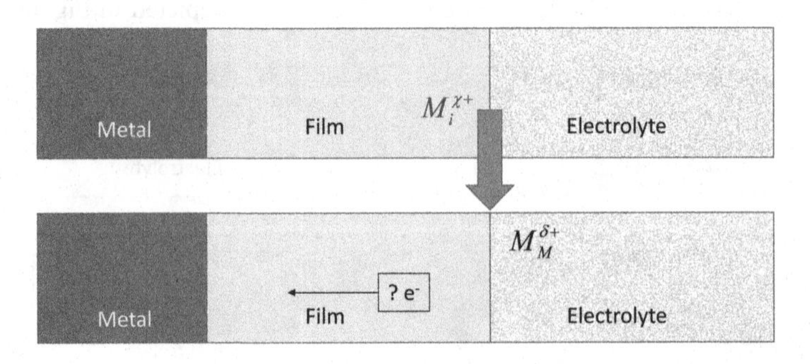

FIGURE 6.19 Consumption of cation interstitials at the film–solution interface. Lattice conserving reaction.

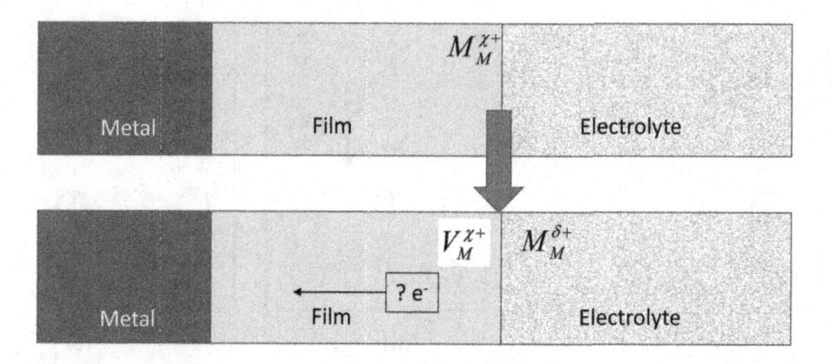

FIGURE 6.20 Generation of cation vacancies at the film–solution interface. Lattice conserving reaction.

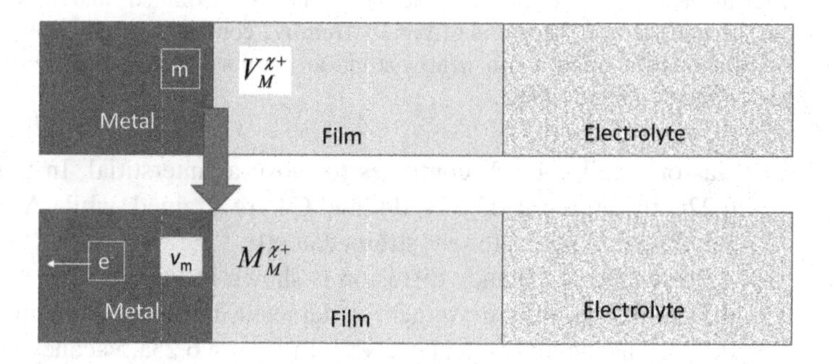

FIGURE 6.21 Consumption of cation vacancies at the metal–film interface. Lattice conserving reaction.

by the following reaction $m + V_M^{\chi+} \rightarrow M_M^{\chi+} + v_m + \chi e^-$. This is shown in Figure 6.21. Here again, it is assumed that the metal vacancies (v_m) quickly diffuse into the bulk metal and are consumed in a large sink,

i. *Difference between an Interstitial Movement and Vacancy Movement:* There is a considerable difference between a cation-interstitial movement and a cation (or anion) vacancy movement. If a cation diffuses from location 'A' to 'B,' then we assume that one single ion moves from 'A' to 'B' by taking many individual steps. If a cation vacancy moves from B to A, then it means that many atoms have moved one step each, so that the vacancy appears in 'A.' This is illustrated in Figures 6.22 and 6.23. The schematics show the movement of *atoms* in a two-dimensional lattice, but the same idea can be applied to a crystal made of *anions* and *cations*. In Figure 6.22, one atom makes two jumps, and at the end (Figure 6.22c) is in the process of making the third jump. All other atoms remain in their nominal lattice position, with only minor adjustments. In Figure 6.22b, the atoms at A3, B3, A4, and B4 are 'strained,' i.e., moved slightly away from their original lattice position. At other locations, the atoms are strained less (or not strained at all). The atom, which

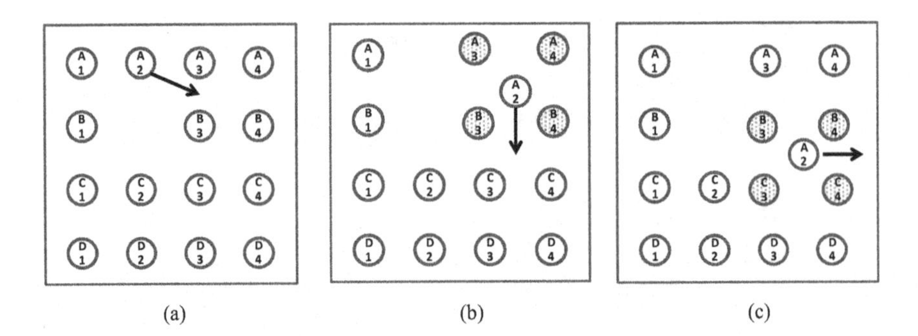

FIGURE 6.22 Point defect model – schematics. Atomic diffusion through a two-dimensional lattice of atoms. (a) Note the vacancy at B2. The atom at A2 moves 'between the lattice sites,' i.e., interstitial movement. (b) Atoms near the interstitials are 'strained' and atoms at farther locations are not strained. The atom originally from A2 continues to move interstitially. (c) Interstitial at another location; again, nearby atoms are strained. The atom originally from A2 continues its interstitial movement.

was originally at A2, continues to move as interstitial. In Figure 6.22c, the atoms at B3, C3, B4, and C4 are strained, while A3 and A4 are strained less (or not strained at all).

ii. A schematic of vacancy diffusion is shown in Figure 6.23. Again, the atomic vacancy movement is depicted, but the idea applies to the ion vacancy movement as well. In Figure 6.23a, vacancy at B2 is shown. In Figure 6.23b, the atom at C2 moves up to the vacant site B2, and this is considered as the movement of the vacancy from B2 to C2. In Figure 6.22c, the next atom at D3 moves diagonally up to the vacant site (which was the original location of the atom marked C2), i.e., vacancy at C2 moves diagonally down to D3. In Figure 6.23d, the atom at E2 moves diagonally up to the vacant site (which was originally occupied by the atom marked D3), i.e., vacancy at D3 moves diagonally down to E2. In Figure 6.23e, all the above three movements are shown. One can visualize this as "three different atoms executing one jump each in a particular sequence," as shown by continuous black arrows, or "an atomic vacancy executing three sequential jumps," as shown by brown dashed arrows.

e. *Anion Vacancy Generation:* A metal atom, near the metal–film interface, oxidizes and becomes a cation. This can be visualized as the formation of a metal oxide, but with an anion vacancy generation, as depicted in Figure 6.24. This step will contribute to the current. It will also increase the film thickness and hence is a lattice non-conserving reaction. The reaction is written as $m \rightarrow M_M^{\chi+} + \chi' e^- = MO_{\chi/2} + \frac{\chi}{2} V_{O^{\cdot\cdot}} + \chi' e^-$

f. *Anion Vacancy Consumption:* The anion vacancies generated at the metal–film interface diffuse through the film and reach the film–solution interface. Here, they react with water and are filled with the oxide anion, essentially consuming the anion vacancy. Two hydrogen

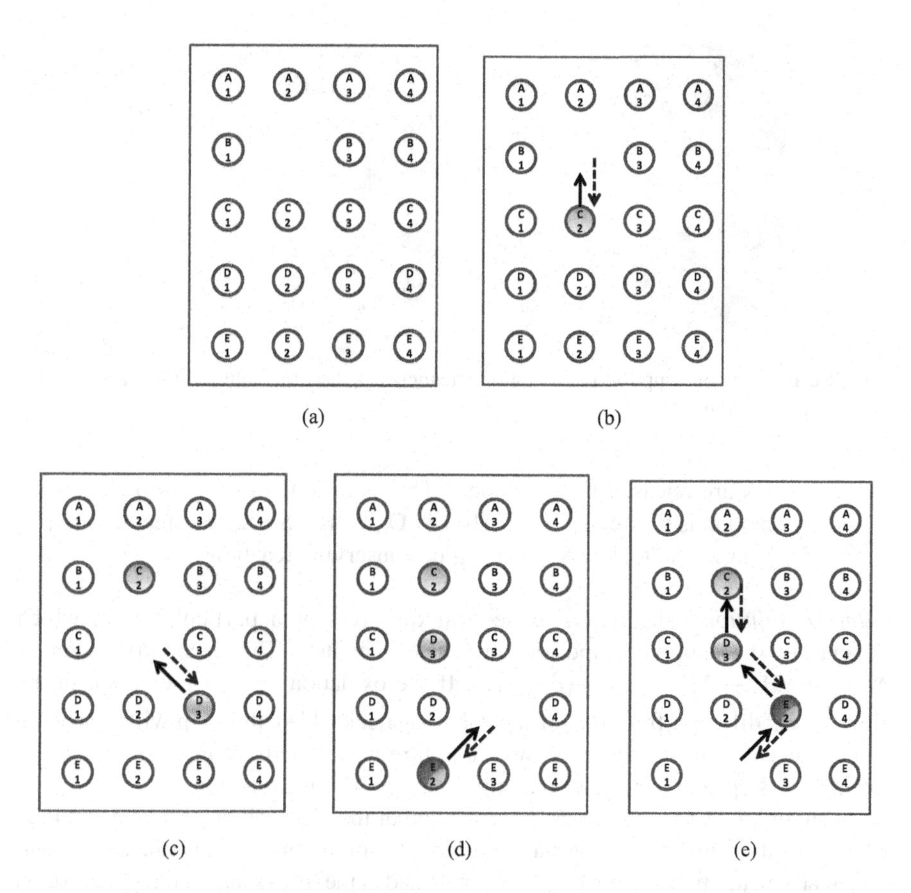

FIGURE 6.23 Vacancy diffusion through a two-dimensional lattice of atoms. (a) Note the vacancy at B2. (b) Single vacancy movement from B2 to C2. (c) Vacancy movement from C2 to D3. (d) Vacancy movement from D3 to E2. (e) All three vacancy movements. Atom movements are shown as continuous black arrows, and the corresponding vacancy movements are shown as brown dashed arrows.

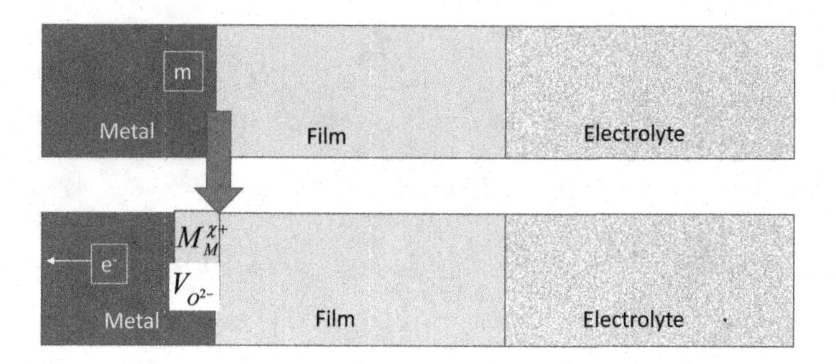

FIGURE 6.24 Generation of anion vacancies at the metal–film interface. Lattice non-conserving reaction.

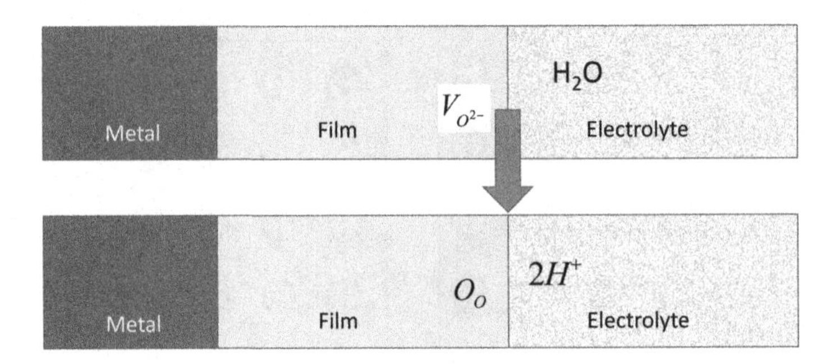

FIGURE 6.25 Consumption of the anion vacancies at the film–solution interface. Lattice conserving reaction.

ions are released as byproducts. This is depicted in Figure 6.25. The reaction is written as $V_{O^{2-}} + H_2O \rightarrow O_O + 2H^+$. Since this step does not change film thickness, it is a lattice conserving reaction.

Film Dissolution: The metal oxide film dissolves at a particular rate, which depends on the solution composition and the film content. The reaction is written as $MO_{\chi/2} + \chi H^+ \rightarrow M^{\delta+} + H_2O + (\delta - \chi)e^-$. If the oxidation state of the cation in the film (χ+) is different from that of the solution species (δ+), this step will contribute to the current. This process is shown in Figure 6.26. The film thickness decreases during this step, and hence this is a lattice non-conserving reaction.

In summary, cation vacancies are generated at the film–solution interface and are consumed at the metal–film interface. Cation interstitials and anion vacancies are generated at a metal–film interface and are consumed at the film–solution interface. Anion vacancy generation leads to film formation. Film dissolution at the film–solution interface leads to film destruction. The key steps of PDM are listed in Table 6.2.

Note that we neglect the possibility of a cation interstitial filling in (i.e., consuming) a cation vacancy in the film or the possibility of an anion vacancy and cation vacancy combining to contribute to film consumption. All of the three species, viz., cation vacancy or anion vacancy or cation interstitial, can carry the current, but usually,

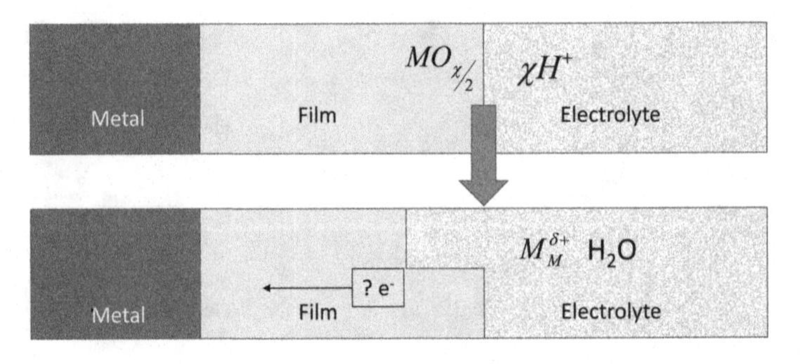

FIGURE 6.26 Dissolution of the film at the film–solution interface. Lattice non-conserving reaction.

TABLE 6.2
List of Key Steps in the Point Defect Model

Sl. No.	Step	Occurs at	Lattice Conserving?	Contributes to Current?	Comment
1	Cation interstitial generation	Metal/film	Yes	Yes	
2	Cation interstitial consumption	Film/solution	Yes	Yes, if dissolution is oxidative	
3	Cation vacancy generation	Film/solution	Yes	Yes, if dissolution is oxidative	
4	Cation vacancy consumption	Metal/film	Yes	Yes	
5	Anion vacancy generation	Metal/film	No, film thickness increases	Yes	
6	Anion vacancy consumption	Film/solution	Yes	No	H_2O is consumed and H^+ is generated
7	Film dissolution	Film/solution	No, film thickness decreases	Yes, if dissolution is oxidative	H^+ is consumed, and H_2O is generated

only one of them is a dominant species in a given system, and neglecting the interaction between these species is justified.

If the film is a good electrical conductor, then electrons and holes can also travel through the film easily and conduct electricity. Typically, the diffusivities of electrons and holes in the film are much larger than those of cations or anions. Then, the total film admittance may be approximated by the sum of the admittance corresponding to the electrons and the holes, and the film would be resistive. In that case, the movement of cations and anions in the film can be ignored. However, this can arise only in a very few cases.

Under steady-state conditions, the film thickness is a constant, since the rate of film formation and that of film dissolution are equal. When diffusion of a cation interstitial or a cation vacancy or an anion vacancy is the rate-limiting step, then the impedance response is qualitatively similar to the Warburg impedance. The film thickness and the ionic diffusivities in the film can, of course, be vastly different compared to the diffusion layer thickness and diffusivities of soluble species in the liquid phase.

The total current through the film (i_{Total}) can be written as the sum of current carried by electrons (i_e), holes (i_h), cation vacancies ($i_{V_M^{\chi+}}$), anion vacancies ($i_{V_{O^{2-}}}$), and interstitials ($i_{M_i^{\chi+}}$). The diffusion of the species was described using the Nernst–Planck equation, i.e., Fick's law modified for the effect of the electric field on the movement of charged species. The potential drop across the film–solution interface depends on the solution pH and the applied potential. In contrast, the potential drop across the metal–film interface depends on the solution pH, applied potential, and the film thickness (Engelhardt, Case, and Macdonald 2016). The model predicts the film thickness as a function of potential, the impedance response of a passive film,

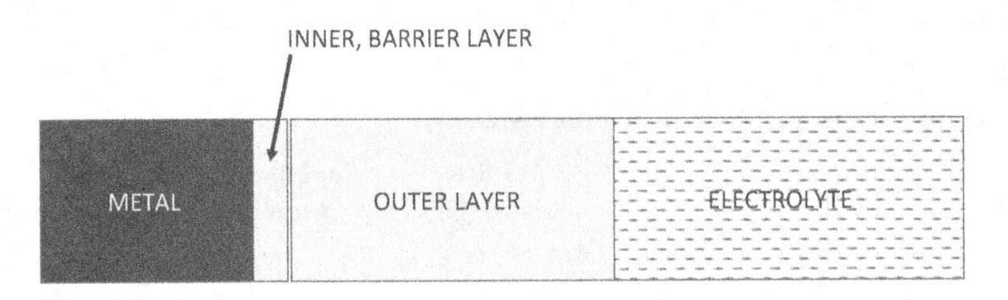

FIGURE 6.27 Schematic of PDM with a bilayer film.

and also a film breakdown by pitting corrosion. If anion vacancy transmission is the dominant mechanism of conduction in the film, then the film is an n-type semiconductor (e.g., W/WO_3). If cation vacancy transmission is the dominant mechanism, then the film is a p-type semiconductor (e.g., Ni/NiO). If a cation-interstitial movement is the dominant mechanism of conduction in the film, then the film is again an n-type semiconductor (Fe/FeO_x).

In an updated version of the model (Macdonald 2011), the film is visualized as a bi-layer (Figure 6.27), with the inner barrier layer offering protection and growing into the metal. The outer layer is formed when the metal ion reacts with solution species and precipitates. Typically, the inner layer is compact and very thin, while the outer layer may be porous and may even contain the electrolyte. In some materials, the outer layer can be very resistive and offers protection (e.g., valve metals like Ta) and it is used to determine the impedance response.

6.4.2.1 Pitting Corrosion

Chloride ions are known to cause pitting corrosion in many metals. When a metal, which is normally passive in a chloride-free solution, is immersed in a chloride-containing solution, there is an incubation time before the breakdown of passivity. Unlike uniform corrosion, pitting corrosion occurs in specific locations only. PDM explains the occurrence of pitting corrosion as follows. The generation of cation vacancies (Figure 6.20) is accelerated by chloride ions present at the film–solution interface. This leads to a higher rate of cation vacancies diffusing and arriving at the metal–film interface. The rate of consumption of cation vacancies at the metal–film interface is a finite number. If the rate of arrival of cation vacancies at the metal–film interface is more than the rate of consumption of the cation vacancies through the metal, then cation vacancies would accumulate, and a void is generated, at the metal–film interface (Macdonald 1992). Right at this location, the film thickness decreases, and eventually, the film breaks down, causing pits. This explains why there is an incubation time before pitting corrosion is seen and also why pitting corrosion occurs at only certain locations (where chloride ions accelerate the generation of cation vacancies) rather than at all locations. Note that, in this description, chloride ion does *not* travel inside the film and reach the metal–film interface.

6.4.3 Surface Charge Approach

The surface charge approach (SCA) can be viewed as an extension of PDM. In PDM, the possibility of accumulation of the ion or vacancies inside the film, near

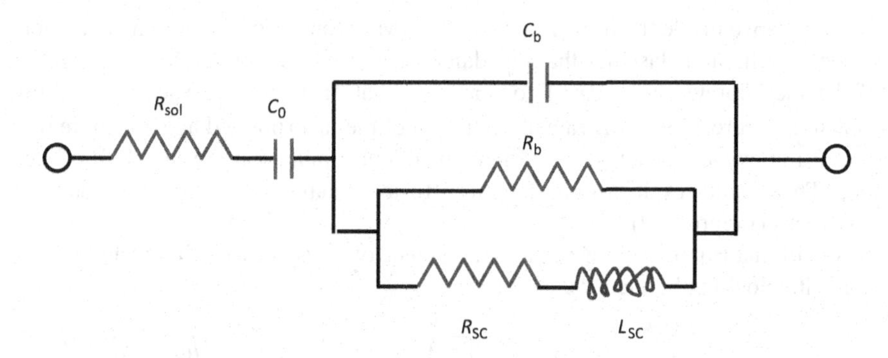

FIGURE 6.28 Equivalent electrical circuit to represent an electrode with a barrier film using a surface charge approach. (Cattarin, S., M. Musiani, and B. Tribollet, Nb electrodissolution in acid fluoride medium steady-state and impedance investigations. *Journal of The Electrochemical Society*, 2002. 149(10): pp. B457–B464. © The Electrochemcial Society. Adapted with permission from IOP publishing. All rights reserved.)

the metal–film interface or film–solution interface, was not considered. The original PDM uses the Nernst–Planck equation to describe the diffusion of the ions in the film, under an electric potential. The Nernst–Planck equation is meant for the movement of charged species in solution, under concentration gradient and electrostatic potential. A modification to this was effected by Bojinov when he employed the Fromhold model, which was specifically derived to describe the diffusion of charged species in a discrete lattice at high electric fields (Bojinov 1997a).

SCA leads to the prediction of a mid-frequency inductive loop in the complex plane plot of the impedance spectrum. Detailed mechanistic analysis of SCA has been used to describe the anodic dissolution of W (Bojinov, Karastoyanov, and Tzvetkov 2010), Mo (Bojinov, Betova, and Raicheff 1996), Nb (Bojinov et al. 2003), and so on in various solutions. For a given system, at a given dc potential, an electrical circuit incorporating resistors, capacitors, and inductors (Figure 6.28) can be used to represent SCA and model the impedance spectrum. Here, C_0 is denoted as the faradaic pseudocapacitance, C_b is the barrier layer capacitance, R_b is the resistance to migration, and R_{sc} and L_{sc} are the elements to represent accumulated negative charge in the film near the film–solution interface. Note that the circuit predicts that at very low frequencies, the system behavior will be that of a simple capacitor in series with a resistor, and the circuit-based results often do not match with the experimental data in a low-frequency regime. The reader is referred to the relevant journal articles for a more detailed description of the SCA model (Bojinov 1997a, Bojinov, Karastoyanov, and Tzvetkov 2010, Bojinov et al. 2003, Bojinov, Betova, and Raicheff 1996).

6.4.4 Anion Incorporation Model

An anion incorporation model (AIM) can be viewed as another extension of PDM. In PDM, it is assumed that anions cannot travel, like interstitials, inside the film, since anions tend to be large. SCA assumes that any anions present as an interstitial in the film would accumulate near the film–solution interface. However, in the case of small anions (such as fluoride), they may be incorporated in the bulk film as well and may move inside the film (Kong 2010). AIM accounts for the presence of anions inside the film and classifies the impedance response based on the rate of migration of the incorporated anion (r_{A^-}) compared to that of an oxygen vacancy, ($r_{O^{2-}}$) as shown in Figure 6.29. If the rate of migration of the incorporated anion is more than that of the oxygen vacancy, then anions would accumulate at the metal–film interface. This would result in finite-length diffusion behavior with blocking boundary conditions (Figure 6.12).

Recall that Eq. 6.91 describes the impedance of a system with finite length diffusion with blocking boundary conditions,

$$Z_D = \frac{RT}{n^2F^2DC_A}\frac{\cot h\left(\delta\sqrt{j\omega/D}\right)}{\left(\delta\sqrt{j\omega/D}\right)} = \frac{RT}{n^2F^2DC_A}\frac{\cot h\left(\sqrt{j\omega/\omega_0}\right)}{\left(\sqrt{j\omega/\omega_0}\right)} \tag{6.143}$$

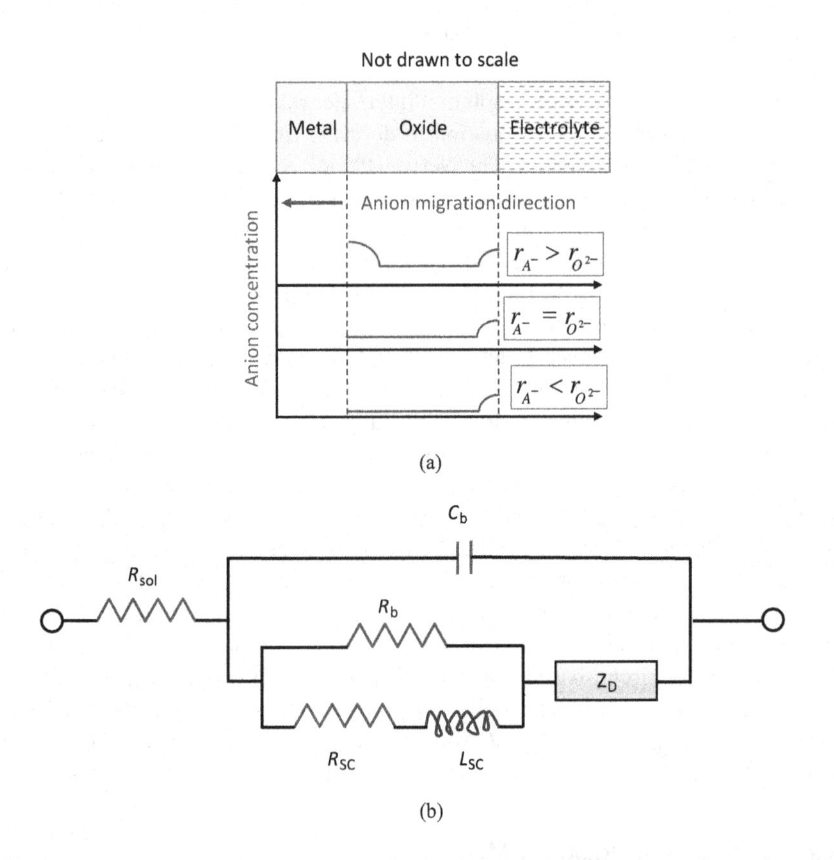

FIGURE 6.29 (a) Pictorial of anion distribution in the film in various scenarios, (b) Equivalent electrical circuit used to describe an anion incorporation model. (Adapted with permission from *Langmuir*, 26, De-Sheng Kong, Anion-Incorporation Model Proposed for Interpreting the Interfacial Physical Origin of the Faradaic Pseudocapacitance Observed on Anodized Valve Metals—with Anodized Titanium in Fluoride-Containing Perchloric Acid as an Example, pp. 4880–4891, Copyright (2010) American Chemical Society.)

where D is the diffusivity of species A, δ is the film thickness, C_A is the concentration of the particular ion (A) responsible for this impedance, and $\omega_0 = \frac{D}{\delta^2}$.

At high frequencies ($\omega \gg \omega_0$), Eq. 6.143 simplifies to

$$Z_D \simeq \frac{RT}{n^2F^2DC_A}\frac{1}{\sqrt{j\omega\delta^2/D}} = \frac{R_D}{\sqrt{\left(2\omega/\omega_0\right)}} - j\frac{R_D}{\sqrt{\left(2\omega/\omega_0\right)}} \tag{6.144}$$

which is similar to Warburg impedance. Here,

$$R_D = \frac{RT}{n^2F^2DC_A} \tag{6.145}$$

At low frequencies ($\omega << \omega_0$), Eq. 6.143 simplifies to

$$Z_D \simeq \frac{RT}{3n^2F^2DC_A} - j\frac{RT}{n^2F^2C_A\delta^2\omega} = \frac{R_D}{3} + \frac{1}{(j\omega C_0)} \tag{6.146}$$

where $C_0 = \frac{1}{R_D\omega_0}$. AIM was used to model the impedance spectrum of anodized Ti in acidic fluoride media in the transpassive dissolution region (Kong 2010).

6.5 EXERCISE – IMPEDANCE OF CPE, DIFFUSION, AND FILM

Q6.1 In the circuit given in Figure 6.26, let $R_1 = 10\ \Omega$ and $R_2 = 100\ \Omega$. Draw the impedance as a complex plane plot and Bode plots, when (a) $Q = 10\ \mu F$ and $n = 1$ (b), $Q = 20\ \Omega^{-1} s^n$ and $n = 0.9$, and (c) $Q = 40\ \Omega^{-1} s^n$ and $n = 0.8$. The frequency range can be 10^{-1}–10^5 Hz. List the inferences.

Q6.2 Derive Eq. 6.7 from Eq. 6.5.

Q6.3 Select a few publications in the literature, where the electrode–electrolyte interface is modeled using a CPE instead of an ideal capacitor.

(Example 1: The impedance spectra of Cu in 1 vol% H_2O_2, 0.1 M NH_4OH, and 0.1 M Na_2SO_4 at various anodic potentials were reported by Venkatesh and Ramanathan (2010a).

Example 2: Impedance spectra of Nb dissolving in 500 mM HF at various anodic potentials were presented by Mandula and Srinivasan (2017).)

a. What inference can one draw from the values of the CPE exponent n?
b. Is there a correlation between Y_0 and n?
c. Assuming that Brug's formula is applicable, what is the effective capacitance of the double-layer at each potential?

Q6.4 Determine the boundary layer thickness as given by the Levich equation, for several typical inorganic and organic electrolytes, as a function of rpm. The electrode rotational speed may vary from 100 to 10,000 rpm. Plot the boundary layer thickness as a function of rpm for each electrolyte in the same figure. What inferences can we draw from these results?

Q6.5 The diffusivity of the H^+ ion in the aqueous electrolyte can be ten times higher than that of a few large ions. What is the effect of the species diffusivity on the boundary layer thickness, as given by the Levich equation?

Q6.6 For a typical electrolyte, using Eq. 6.13, plot the value of v_y as a function of rpm, when $y = \delta/2$ and when $y = \delta$.

Q6.7 From Eqs. 6.14–6.17, show that the steady-state concentrations of A and B can be written using Eqs. 6.18 and 6.19.

Q6.8 Generate the plots given in Figure 6.6. Explore the effect of varying the kinetic parameters.

Q6.9 In Figure 6.6, explore the effect of varying D_A while keeping D_B fixed.

Q6.10 For the given set of parameters in Figure 6.6, plot $C_{A\text{-surface}}$ and $C_{B\text{-surface}}$ *vs.* E_{dc}, at various delta values.

Q6.11 The concentration oscillations of A and B are given in Eqs. 6.41 and 6.42. Plot the magnitude of the concentration oscillations *vs.* frequency, given that $\delta = 20\ \mu m$, $D_A = D_B = 10^{-9}\, m^2 s^{-1}$, and i_{ac} = (a) 100 nA cm^{-2} and (b) 100 µA cm^{-2}.

Q6.12 In Figure 6.8, what is the effect of changing D_A (while keeping $D_A = D_B$)? What is the effect of changing C_{dl} from 10 to 30 µF cm^{-2}? The boundary layer thickness can be taken as 40 µm.

Q6.13 Repeat the above for a non-zero dc bias (e.g., Figure 6.9).

Q6.14 Using the parameter set employed in generating Figure 6.9, explore the effect of changing dc bias at a fixed δ. For example, set $E_{dc} = 0$ or -0.1 or $+0.1$ V *vs.* OCP. Compare the results in complex plane plots and Bode plots. List the inferences.

Q6.15 Repeat the above, while keeping $D_A \neq D_B$.

Q6.16 Under semi-infinite boundary conditions, explain why a non-zero dc bias will not yield steady-state results.

Q6.17 Show that Eq. 6.44 reduces to Eq. 6.73, when $E_{dc} = 0$ and $\delta \rightarrow \infty$

Q6.18 Using any circuit fitting software, fit the data that you generated in the above questions. Do we get the same parameter set that we used to generate the data?

Q6.19 Generate Figure 6.12 and explore the effect of change in (a) diffusivity and (b) boundary layer thickness δ.

Q6.20 Generate the spectrum in Figure 6.13. Explore the effect of changes in diffusivity, boundary layer thickness, the concentration of A in the bulk, and the dc bias, E_{dc}.

Q6.21 Plot $C_{A\text{-ss}}$ *vs.* E_{dc} and θ_{1ss} *vs.* E_{dc}, for a few values of δ.

Q6.22 Plot the magnitude of the concentration oscillations *vs.* frequency for the given set of parameter values and $E_{ac0} = 10$ mV.

Q6.23 For a few other reaction mechanisms (chosen either from Chapter 5 or from the literature), incorporate the mass-transfer effect of one species in solution and derive the expressions for impedance. For example take the hydrogen evolution reaction in acidic solutions. Note: This will require considerable time and effort to complete.

Q6.24 Recalculate and plot Figure 6.15, by varying the solution conductivity (ρ).

Q6.25 Explore the effect of change in pore length on the impedance (Figure 6.15).

Q6.26 Using the circuits given in Figure 6.17 and the element values from the literature, generate the EIS data and present them as a complex plane plot and Bode plots.

Q6.27 In the point defect model, in the literature, what are the typical values of diffusivity of cation interstitials and ionic vacancies employed? How do they compare with diffusivities of ions in liquids?

Q6.28 Using the circuit given in Figure 6.28, and using the typical value of the elements from the literature (e.g., Cattarin, Musiani, and Tribollet (2002)), generate the EIS data and present them as complex plane plots and Bode plots.

Q6.29 Using the circuit given in Figure 6.29b, and using the typical value of the elements from the literature (e.g., Kong (2010)), generate the EIS data and present them as complex plane plots and Bode plots.

7 Applications – A Few Examples

I was taught the way of progress is neither swift nor easy.

—**Marie Curie**

There are thousands of research articles published each year on electrochemical impedance spectroscopy (EIS) and its applications. The number of articles that contain EIS in the main fields, viz., title, abstract, or keywords, increased exponentially over time (Figure 7.1). A significant part of the application of EIS is in the field of corrosion, while relatively fewer articles are on the application of EIS in the energy field (battery, fuel cells, and supercapacitors) and sensors. In this chapter, a few examples of applications of EIS reported in the literature are illustrated. First, we describe an application of EIS to corroborate the results of other electrochemical techniques to characterize corrosion. Next, mechanistic analysis of the anodic dissolution of a metal in an acid using EIS is presented. Third, a couple of examples of EIS applications in biosensing are described. Finally, the use of EIS to determine the cycle life and state of charge of a battery, and to understand the effect of temperature on the physical processes, is illustrated.

7.1 CORROSION

7.1.1 Corrosion of Valve Metals in Acidic Fluoride Media

Corrosion of metal in aqueous solutions consists of at least two electrochemical reactions, viz., oxidation of metal and reduction of either H^+ ions or dissolved oxygen. Mandula et al. investigated the corrosion of four valve metals, viz., Ti, Zr, Nb, and

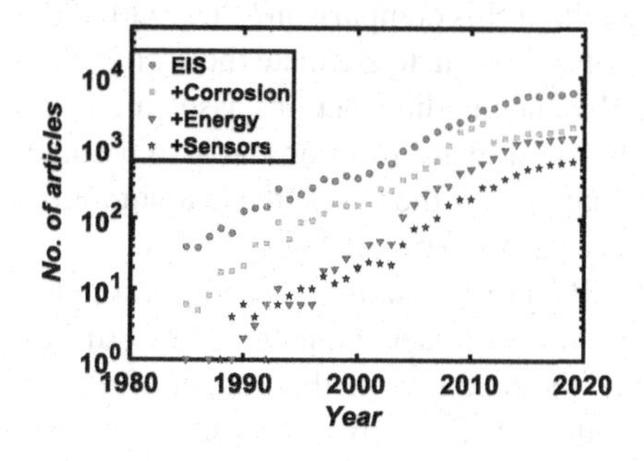

FIGURE 7.1 Number of journal articles containing EIS or "electrochemical impedance spectroscopy" in the main fields, viz., title, abstract, or keywords. Data sourced from Scopus®. The number of articles tagged with additional words, such as corrosion, sensors, and energy-related topics (battery, fuel cell, and supercapacitor), are also shown.

Ta in acidic fluoride media and used weight loss and electrochemical techniques (Rao et al. 2020). Although valve metals are generally very resistant to corrosion, they are attacked by acidic fluoride media. The experiments were conducted in hydrofluoric acid solutions, with and without the addition of NaF or H_2SO_4. HF is a weak acid, and HF solutions contain H^+, F^-, HF (in the molecular form), HF_2^- and $H_2F_3^-$. The concentration of each of these species can be calculated using the equilibrium constants of HF dissociation reactions. It is known that fluoride ions in neutral or alkali solutions do not attack the valve metals. The actual species that attack the valve metals were not known, and identifying them was one of the aims of the investigation.

When the nominal HF concentration (i.e., estimated concentration without accounting for HF dissociation) is increased, the concentration of each species also increases. When the predictors (here, the concentration of each species) are correlated, it becomes difficult to identify the species that are responsible for the attack on the valve metals. To break the predictor correlation, NaF or H_2SO_4 can be added; this would raise the concentration of F^- or H^+, respectively, without a proportionate increase in the concentration of other species. In Section 2.4.5, the EIS of Zr dissolution in 5 mM HF was described as an experimental example. Here, we consider the dissolution of Ta in acidic fluoride media and restrict our attention to the application of EIS. Weight loss experiments, performed at 30°C, were used to estimate the corrosion rate of Nb in several HF concentrations. Three electrochemical techniques, viz., potentiodynamic polarization (PDP), linear polarization (LP), and EIS, were used to characterize the dissolution. Tafel extrapolation of PDP was used to calculate the corrosion current density (i_{corr}) and the polarization resistance (R_p). EIS and LP were also used to estimate the R_p. The complex plane plots of EIS are shown in Figure 7.2a and b. First, EIS data were validated using linear KKT and then modeled using the equivalent circuit shown in Figure 7.2c. Since there are two capacitive loops and one inductive loop in all the data presented in Figure 7.2a and b, one of the Maxwell pairs in Figure 7.2c was permitted to hold negative values, and the model results matched the experimental data well. From the EEC best-fit values, R_{sol} and R_p values were recorded. From the R_p values estimated using PDP and LP, R_{sol} was subtracted to calculate the *true* R_p values and compared with those estimated from EIS. The comparison presented in Figure 7.3 shows that the match is reasonable, although not exact.

Now, what is the benefit of this comparison? PDP with Tafel extrapolation is commonly used in the field of corrosion, to estimate the corrosion rate. Tafel extrapolation assumes that the cathodic and anodic reactions associated with the metal dissolution process are simple electron transfer reactions. It also assumes that mass transfer is rapid. Of course, in a majority of the cases, these assumptions are not true, and the corrosion current density values estimated are suspect.

On the other hand, LP and EIS cannot be used to estimate i_{corr} and can only provide an estimate of R_p. The advantage is that neither of them makes any assumption about the complexity of the reactions or the impact of mass transfer on the results. Between the two techniques, EIS is superior to LP because even if the lowest measurement frequency is not sufficient to obtain an impedance with a negligible imaginary component, an EEC fit can be used to estimate R_p. In other words, even if the low-frequency data points in the complex plane plot do not fall on the abscissa, and the corresponding phase values are non-zero, as in Figure 7.2a and b, an appropriate

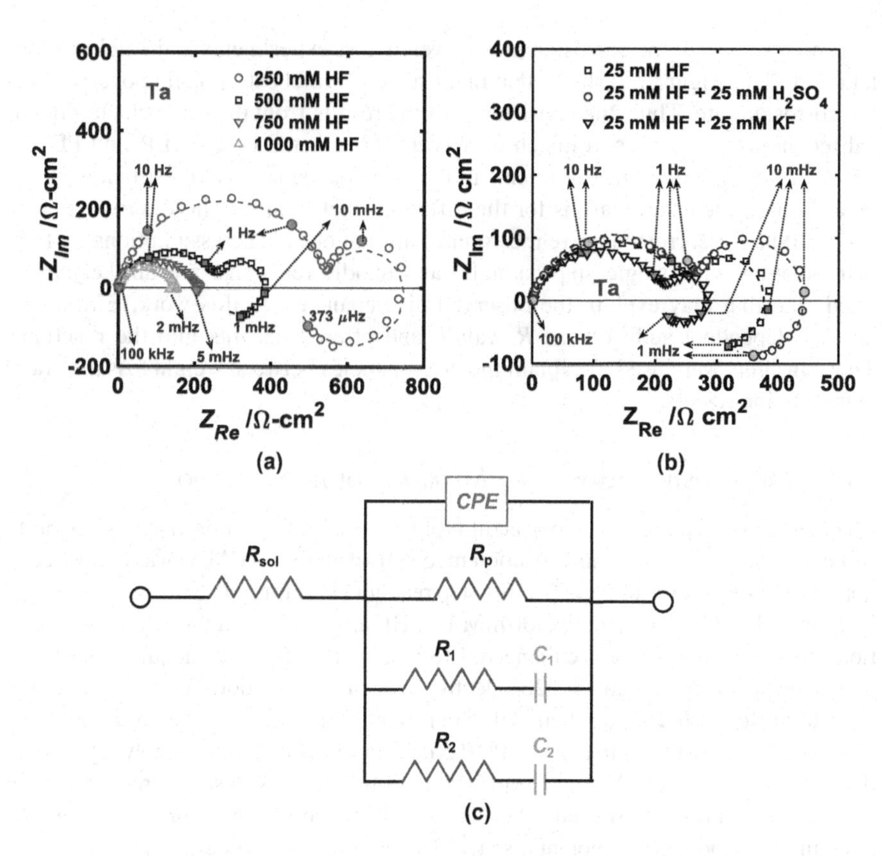

FIGURE 7.2 The complex plane plot of EIS of Ta dissolved in acidic fluoride media, with 1 M Na_2SO_4 as the supporting electrolyte, at the OCP. (a) Pure HF solutions (b) with H_2SO_4 or KF. (c) The equivalent electrical circuit used to model the data. (Adapted from Rao et al. (2020), under open source license CC BY.)

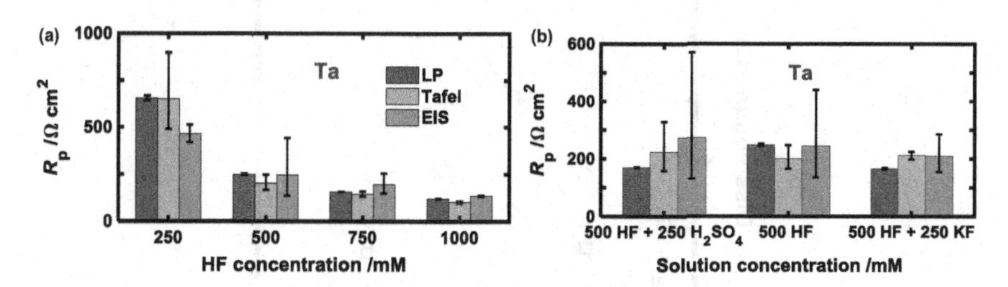

FIGURE 7.3 Polarization resistance values of Ta in acidic fluoride media, calculated from LP, Tafel extrapolation of PDP, and EIS. (a) Pure HF solutions (b) with H_2SO_4 or KF. (Adapted from Rao et al. (2020), under open source license CC BY.)

EEC model can be used to estimate R_p. EIS also shows that the reactions are not simple electron transfers, since multiple loops are seen in the complex plane plots. On the other hand, LP does not offer any insight into the reaction mechanism.

A comparison of the corrosion rate by weight loss experiments and Tafel extrapolation of PDP (Figure 7.4) shows that in most cases, Tafel extrapolation overpredicts the corrosion rate. Thus, the consistency of the results from the three electrochemical techniques, viz., a good match of R_p values estimated by EIS, LP, and PDP, is necessary but is not sufficient to ensure that the corrosion current estimates would be accurate. The exact reasons for the difference between weight loss experiments and PDP-based corrosion current estimates are not clear. The assumptions in Tafel extrapolation, viz., single step cathodic and anodic reactions, are unlikely to be valid, and that may explain the observed discrepancies. In this work, EIS served as an independent audit on the R_p values and offered insights into the reactions. Those insights were used to explain the discrepancies between weight loss and Tafel extrapolation results.

7.1.2 Mechanistic Analysis of a Metal Dissolution Reaction

The *mechanistic analysis* is a powerful tool to extract information from EIS data. In this example, we illustrate how reaction mechanism analysis (RMA) was employed to obtain a detailed description of a multi-step reaction. Fathima and Srinivasan reported PDP and EIS of Ti dissolved in 100 mM HF. HF attacks Ti, and under anodic conditions, the Ti oxidation rate is enhanced. EIS data of Ti in HF were acquired at several anodic dc potentials. A Parstat 2263 electrochemical workstation (Ametek, USA) was used to perform the electrochemical experiments. The working electrode potential was scanned at 10 mV s^{-1} to acquire PDP. EIS data were obtained by applying sinusoidal perturbations of 10 mV (*rms*) from 30 kHz to 100 mHz, with seven frequencies per decade, logarithmically spaced. The PDP results are shown in Figure 7.5, where the open markers indicate the potentials at which EIS data were acquired.

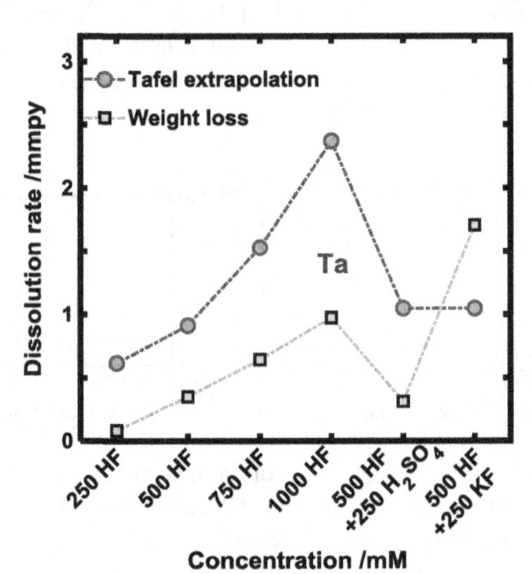

FIGURE 7.4 Ta dissolution rates as a function of solution composition in acidic fluoride media. (Adapted from Rao et al. (2020), under open source license CC BY.)

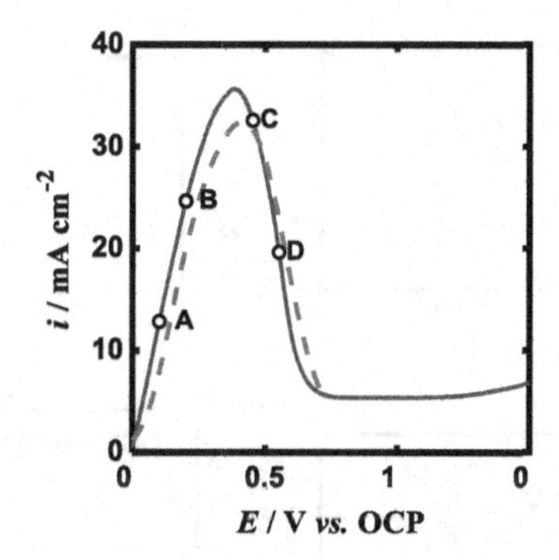

FIGURE 7.5 Potentiodynamic polarization of Ti in 100 mM HF and 1 M Na_2SO_4. The potential was scanned at 10 mV s^{-1}. The experimental data are shown as continuous lines, and the mechanistic model predictions are shown as dashed lines. (Adapted from Fasmin, Praveen, and Ramanathan (2015) under open source license CC BY.)

Figure 7.5 shows that, initially, the current increases with the potential, and after reaching a value of ~36 mA cm^{-2}, the current decreases with potential. The first region is termed as *active* to indicate active dissolution, while the second region is termed as *passive* to indicate that the surface is passivated. Although it is termed passive, it does not mean that the surface is completely passivated. Ti dissolution continues to occur, albeit at reduced rates. At very anodic potentials, the current increases slowly in the *transpassive region*. Here, a thick 3D film is present on the surface.

EIS data were acquired in two potentials in the active region (Figure 7.6a and b) and in two potentials in the passive region (Figure 7.6c and d). The complex plane plots of the EIS data are shown in Figure 7.6. The EIS data were first validated with KKT and modeled using a circuit with two Maxwell elements (Figure 7.2c). The double-layer was modeled with a constant phase element (CPE) instead of a capacitor. In Figure 7.6, the markers represent the experimental data, the continuous lines represent the EEC model fit, and the dashed lines represent the mechanistic model fit.

Figure 7.6a and b show that in the active region, three capacitive loops are seen. The high-frequency loop is a depressed semicircle and had to be modeled with a CPE. The mid and low-frequency loops do not show two well-separated capacitive loops, but fitting with equivalent circuits showed that two Maxwell pairs are required to model the data successfully. In the passive region, the mid and low-frequency impedance have a negative real component (Figure 7.6c and d). Since the slope of the current potential diagram (Figure 7.5) is negative in this region, this is along the expected lines. EEC fitting also showed that the low-frequency pattern could be captured only when one pair of Maxwell elements in Figure 7.2c were allowed to hold negative values, i.e., the behavior is inductive.

The EEC model results match the experimental results well. In contrast, RMA results can only be termed as semi-quantitative, i.e., the overall trends are matched, but the match is not quantitative. The mechanism proposed was

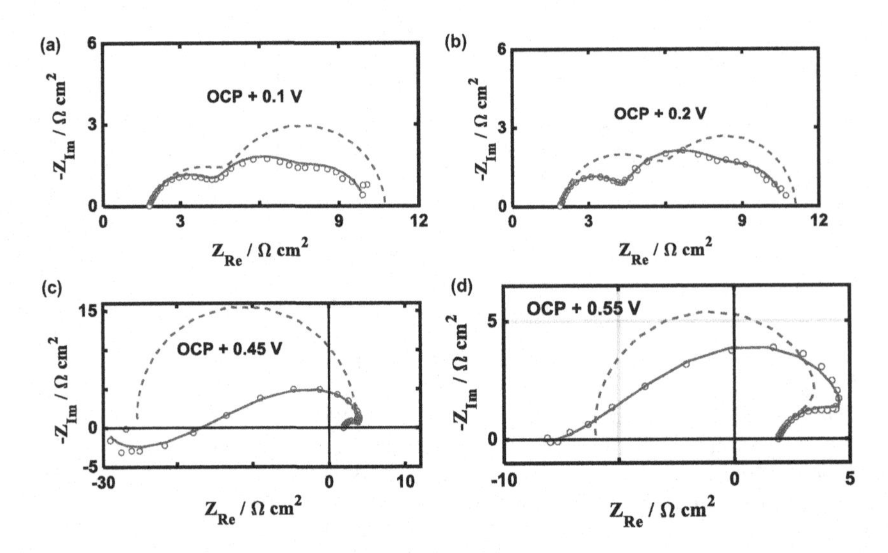

FIGURE 7.6 Complex plane plots of EIS data of Ti anodic dissolution in 100 mM HF and 1 M Na_2SO_4. In the active region, (a) OCP + 0.1 V and (b) OCP + 0.2 V. In the passive region, (c) OCP + 0.45 V and (d) OCP + 0.55 V. (Adapted from Fasmin, Praveen, and Ramanathan (2015) under open source license CC BY.)

$$Ti \underset{k_{-1}}{\overset{k_1}{\rightleftharpoons}} Ti^{3+}_{ads} + 3e^- \tag{7.1}$$

$$Ti^{3+}_{ads} \underset{k_{-2}}{\overset{k_2}{\rightleftharpoons}} Ti^{4+}_{ads} + e^- \tag{7.2}$$

$$Ti^{4+}_{ads} \xrightarrow{k_3} Ti^{4+}_{sol} \tag{7.3}$$

$$Ti^{3+}_{ads} \xrightarrow{k_4} Ti^{4+}_{sol} + e^- \tag{7.4}$$

This mechanism involves two intermediates, viz., Ti^{3+}_{ads} and Ti^{4+}_{ads}, and four steps. The reverse reactions in the third and fourth steps were neglected since the concentration of the dissolved species (Ti^{4+}_{sol}) in the solution was negligible. Attempts to model the reaction using fewer steps were not successful. The passivation in the PDP or the negative impedance at low frequencies in the passive region could not be captured when, for example, only the first three steps of the mechanism (viz., Eqs. 7.1–7.3) were employed. EEC fits were obtained using commercial software Zsimpwin, while the RMA fit was performed using a custom code written in MATLAB®. Now, although the RMA model fit is not as good as the EEC fit, a few points are worth noting. While the model fit of EEC is good, the model cannot be used to calculate PDP data and compare it with experimental results. The values of the model elements (Maxwell resistor and capacitors) cannot be easily linked to physical quantities or processes. Understanding the variation in the values of the Maxwell pair elements, as a function of dc potential bias, is even more challenging.

Based on the proposed mechanism, the fractional surface coverage values of the intermediate species were estimated, and the results are shown in Figure 7.7a. At small anodic overpotentials, the surface is mostly a bare metal. As the overpotential

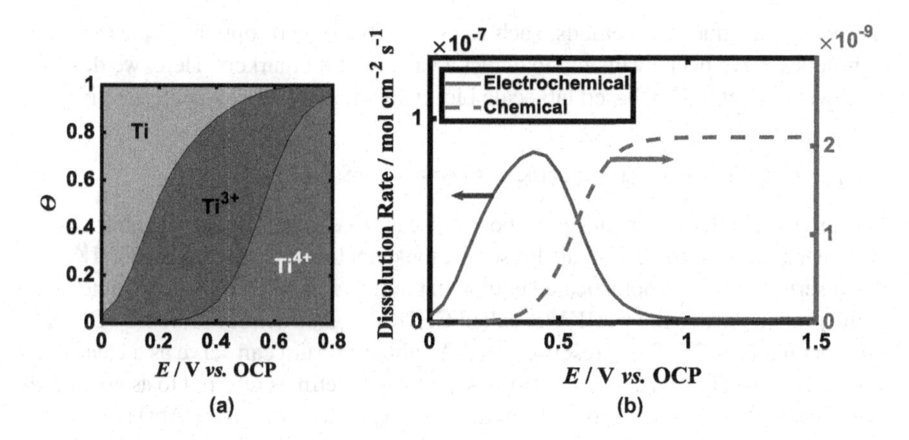

FIGURE 7.7 Anodic Ti dissolution in 100 mM HF and 1 M Na_2SO_4. (a) Fractional surface coverage of bare metal and intermediate species as a function of overpotential, estimated from the mechanism described in Eq. 7.1. (b) Chemical and electrochemical dissolution rates as a function of overpotential, as predicted by the mechanism. (Adapted from Fasmin, Praveen, and Ramanathan (2015) under open source license CC BY.)

increases, the surface is covered mostly with Ti_{ads}^{3+} intermediate species, probably in the form of a sub-oxide or fluoride or oxyfluoride. At large overpotentials, the surface is mostly covered with Ti_{ads}^{4+}, again, in the form of an oxide or fluoride or oxyfluoride. The third step in the mechanism is a chemical dissolution step since it does not involve electron transfer. The fourth step, on the other hand, is an electrochemical dissolution step. The model can be used to predict the dissolution rate by chemical and electrochemical dissolution steps, as shown in Figure 7.7b. The rate of chemical dissolution increases and saturates with overpotential, while that of the electrochemical dissolution shows a peak near 0.4 V anodic of the open circuit potential (OCP). The *rate constant* of the electrochemical step increases with potential. However, the *reactant* in the electrochemical step is Ti_{ads}^{3+}, and at large anodic overpotentials, the surface does not contain a significant amount of Ti_{ads}^{3+}, and hence the electrochemical dissolution rate is low. On the other hand, the rate constant of the chemical dissolution step is independent of the overpotential, but the fractional surface coverage of Ti_{ads}^{4+} increases and saturates with overpotential. Therefore, the chemical dissolution step also increases and saturates with overpotential. This example illustrates that although mechanistic modeling of EIS and PDP data can be more challenging and the RMA model fits of the EIS data may be poorer than the corresponding EEC model fits of the same EIS data, the insights offered by the mechanistic model is far better than those offered by the EEC model.

7.2 BIOSENSORS

Biosensors are devices that can be used to detect specific biomolecules, called biomarkers. Biosensors are visualized to be inexpensive, portable, and easy-to-use devices. The glucometer based on the electrochemical detection of blood sugar levels is a commercial example of biosensors. Biosensor development is an emerging field, and many transduction methods, such as electrochemical or optical or piezoelectric techniques, have been evaluated to detect a variety of biomarkers. Here, we describe the development of EIS-based lab-scale biosensors for two mosquito-borne diseases.

7.2.1 Detection of Chikungunya Protein Using EIS

Chikungunya (CHIK) is a mosquito-borne disease that affects a large number of people, particularly in tropical areas. Presently, the detection of CHIK virus (CHIKV) in blood serum requires sophisticated equipment and technically skilled personnel. One of the proteins inside the CHIKV is called CHIKV-non-structural protein 3 (CHIKV-nsP3 or simply nsP3). The presence of nsP3 in blood serum can serve as a clear diagnostic criterion of CHIK. CHIKV-nsP3 is a toxin, sometimes referred to as an *antigen* that triggers an immune response from the body. Specific *antibodies* (Abs) to this antigen were developed, and this CHIKV-nsP3-Ab binds only to nsP3 antigens. A biosensor was developed to generate a signal during the binding event (George et al. 2019).

First, we describe the principle behind an electrochemical biosensor to detect an antigen. An Au electrode was chosen as the substrate since Au is inert and biocompatible. The idea is to bind or immobilize the antibody on the Au electrode and then expose this surface to a *test sample* (e.g., blood serum or an equivalent liquid) that may contain the antigen. After that, a classical redox reaction such as that of a ferro/ferri redox couple would be performed on the electrode. If the *test sample* did not contain any antigen, then the impedance spectra would show a small charge-transfer resistance (R_t). If the nsP3 antigen were present in the *test sample*, then it would bind to the antibody. This antibody/antigen pair would hinder the redox reaction, and R_t would be higher. Thus, by monitoring R_t, it is possible to measure the antigen concentration in the test samples.

The CHIKV-nsP3-Ab does not bind strongly to the Au substrate. Therefore, a sequence of processes was carried out to immobilize the antibody on the sensor surface. Thiol molecules exhibit a good affinity to the Au surface. Hence, the Au surface was first exposed to a mixture of 11-mercapto-undecanoic acid and 6-mercapto-hexanol, dissolved in ethanol. This exposure leads to the formation of a self-assembled monolayer (SAM). Next, the surface was activated with *N*-hydroxy succinimide (NHS) and *N*-(3-dimethyl aminopropyl)-*N′*-ethyl carbodiimide hydrochloride (EDC). Finally, the surface was exposed to CHIKV-nsP3-Ab to immobilize the antibodies. A pictorial of the binding events is shown in Figure 7.8. The antibodies may not have covered the entire surface. In that case, if the sensor is exposed to a test solution containing antigens, some of the antigens would bind to the vacant sites (i.e., EDC/NHS-activated SAM). This binding would lead to an incorrect estimate of the antigen concentration. Hence, before exposing the sensor to test samples, the surface was exposed to a solution containing bovine serum albumin (BSA). BSA can easily bind to the vacant sites and block them. Such a detailed sequence of steps, or protocol, is necessary to obtain reproducible results.

After each step, the EIS data of the electrode in the ferro/ferri redox couple were acquired, and the results are shown as complex plane plots in Figure 7.9a. In the bare electrode, a semicircle followed by a 45° line is seen. When a SAM layer is formed, the mid and low-frequency impedance values increase significantly. The activation

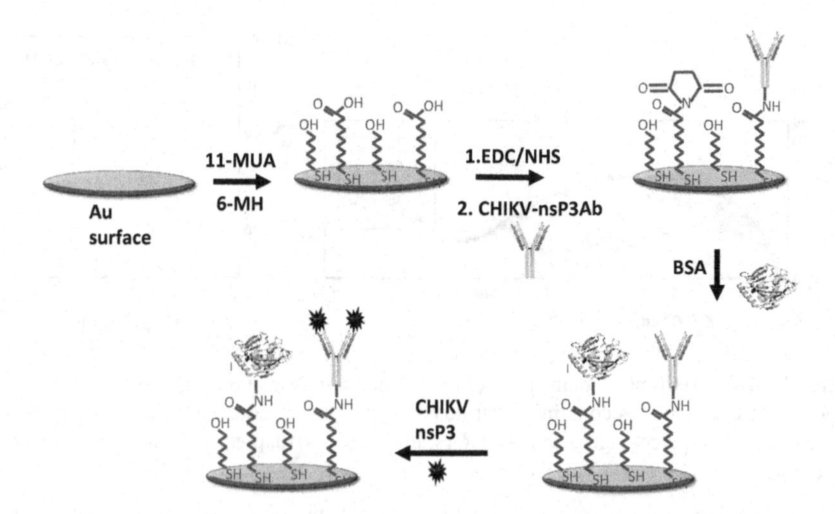

FIGURE 7.8 Pictorial representation of the fabrication of a CHIKV-nsP3 electrochemical biosensor and nsP3 detection.

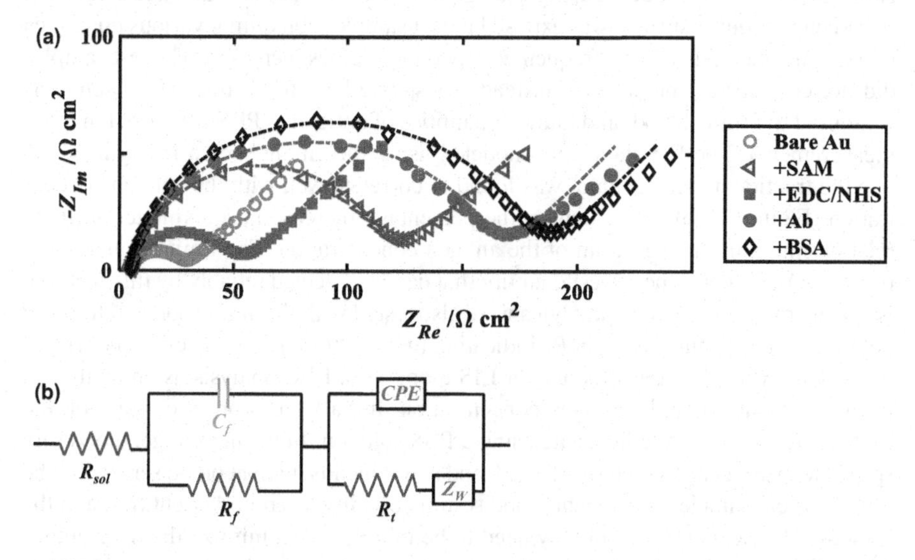

FIGURE 7.9 (a) Complex plane plots of impedance spectra acquired at various stages of CHIKV-nsP3 electrochemical biosensor fabrication. (b) Equivalent electrical circuit used to model the data, on electrodes with SAM onward. (Adapted from George et al. (2019) under open source license CC BY.)

process results in the formation of an NHS ester, and the impedance values decrease. This trend is similar to those reported in the literature, but the cause of the reduction in the mid and low-frequency impedance was not probed further. The addition of CHIKV-nsP3-Ab increased the mid and low-frequency impedance.

The bare electrode data can be modeled using a Randles circuit with semi-infinite Warburg impedance. The data with SAM and other modifications can be modeled

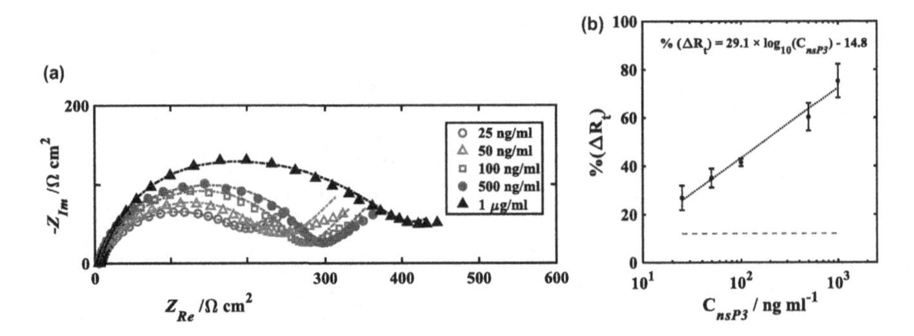

FIGURE 7.10 (a) Complex plane plots of impedance response of the fabricated biosensor, on exposure to test solutions containing various antigen concentrations. (b) The dose–response curve relating the percentage increase in R_t *vs.* $\log_{10}(C_{nsP3})$. (Adapted from George et al. (2019) under open source license CC BY.)

with a more complex circuit, shown in Figure 7.9b. The reaction is still considered as a simple electron transfer step, but the SAM and the other molecules are visualized as a film that can be modeled using a Voigt pair of a capacitor (C_F) and a resistor (R_F).

When the sensor surface was exposed to test samples containing various quantities of antigens, the mid and low-frequency impedance values increased. The test samples did not contain real blood sera; instead, phosphate-buffered saline (PBS) solutions were used to mimic blood, and known quantities of antigen in PBS were used in these measurements. The EIS data were modeled using the circuit shown in Figure 7.10a and the fractional increase in R_t was found to correlate well with the antigen concentration (Figure 7.10b). In particular, the percentage increase in R_t exhibited a linear relationship with the logarithm of the antigen concentration. The limit of quantification (LOQ), which is the lowest quantity that can be detected reliably by this method, is ~ 17 ng mL^{-1}. A few other analytes were also tested with this biosensor, and they did not yield a significant change in R_t, indicating that the biosensor is specific to CHIKV-nsP3. This work illustrates the use of EIS along with EEC analysis as an analytical technique to quantify the antigen concentration. Initially, as a proof of concept, the evaluation is done on the lab-scale using a PBS solution. In the next stage, blood sera spiked with antigen would be evaluated, and subsequently, patient blood sera would be tested. A real sample, i.e., patient blood serum contains several other entities, and the biosensor fabrication protocol may need to be modified to minimize the interference from all the entities other than the antigen in the blood serum.

7.2.2 DNA Sensing

Tuberculosis (TB) is a bacterial disease caused by *Mycobacterium tuberculosis* (MTB). Here again, the current detection methods use skilled personnel and expensive instruments. Teengam et al. (2018) developed an EIS-based biosensor to detect MTB DNA. By suitably modifying peptide nucleic acids (PNA), specific binding of DNA to a PNA can be ensured. This is somewhat equivalent to the specific binding between an antigen and a specific antibody. A conformationally constrained pyrrolidinyl PNA system, known as *acpcPNA*, was employed in this work.

The electrode fabrication and the biosensor design in this study were quite different compared to the previous example of CHIKV-nsP3 detection. Here, a Whatman filter paper was used as the substrate. First, it was coated with wax to create a hydrophobic surface. This coating ensures that when a small quantity of test sample is placed on the electrodes, the droplet will not spread. The *electrochemical cell* was essentially two pieces of paper that were held together with an adhesive tape (Figure 7.11). The electrodes were screen printed on the papers. Paper is essentially made of cellulose, and the probe acpcPNA molecules were attached to the cellulose substrate using the following procedure. On the other side of the working electrode, the paper was treated with a mixture of lithium chloride and sodium periodate to form aldehyde groups. Then the probe was attached to the aldehyde group with the help of sodium cyanoborohydride dissolved in dimethylformamide, as shown in Figure 7.11.

Impedance spectra were obtained in the 5 mM Ferro/Ferri redox couple, before and after attaching the probe acpcPNA molecules, and the results are shown in Figure 7.12. The spectra were modeled using a simple Randles circuit with a semi-infinite Warburg impedance. When the electrode was modified using acpcPNA, the mid and low-frequency impedance values decreased, and this was attributed to the better electron transfer property of acpcPNA, specifically to the electroactivity of guanine and adenine bases. The R_t of the bare electrode was about 34 kΩ, while that of the acpcPNA-modified electrode was about 19 kΩ.

Subsequently, a test solution containing the MTB DNA oligonucleotide molecules, called *targets or analytes*, was placed onto the working electrode, and the target molecules were allowed to bind with the probe molecules. The EIS data in the ferro/ferri redox couple were acquired again, and the results (Figure 7.12) show that the mid and low-frequency impedance increased significantly. The increase in R_t was correlated

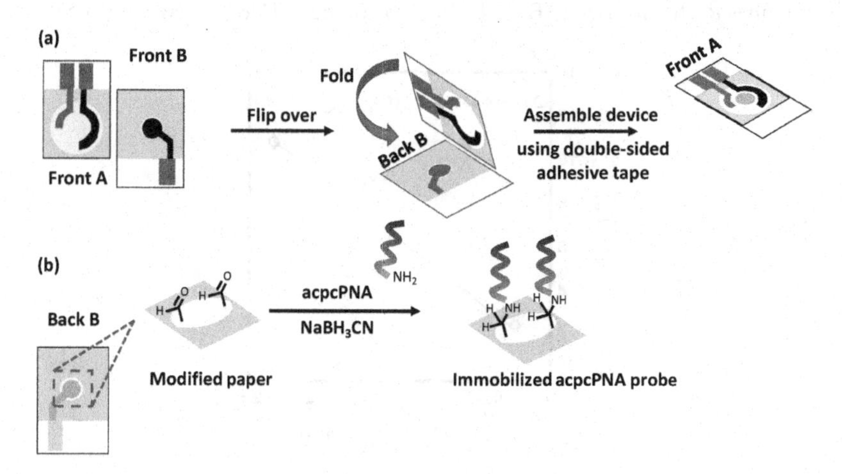

FIGURE 7.11 Pictorials illustrating (a) the design and operation of a paper-based electrochemical biosensor. (b) Covalent immobilization of acpcPNA on the working electrode. (Adapted from *Analytica Chimica Acta*, 1044, Prinjaporn Teengam, Weena Siangproh, Adisorn Tuantranont, Tirayut Vilaivan, Orawon Chailapakul, Charles S. Henry, Electrochemical impedance-based DNA sensor using pyrrolidinyl peptide nucleic acids for tuberculosis detection, pp. 102–109, Copyright (2018), with permission from Elsevier.)

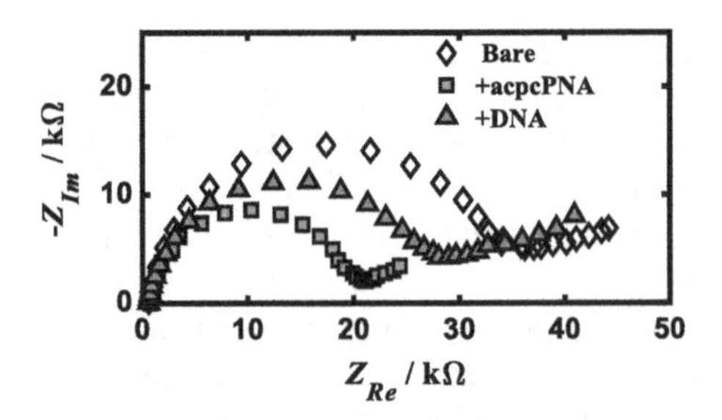

FIGURE 7.12 Complex plane plots of impedance spectra of a paper-based electrochemical biosensor to detect MTB DNA. (Adapted from *Analytica Chimica Acta*, 1044, Prinjaporn Teengam, Weena Siangproh, Adisorn Tuantranont, Tirayut Vilaivan, Orawon Chailapakul, Charles S. Henry, Electrochemical impedance-based DNA sensor using pyrrolidinyl peptide nucleic acids for tuberculosis detection, pp. 102–109, Copyright (2018), with permission from Elsevier.)

with the logarithm of the analyte concentration (Figure 7.13). The relationship was linear in the analyte concentration range of 2–200 nM. The LOQ was reported to be 3.69 nM. Further experiments showed that the probe is specific to the MTB DNA and that R_t did not significantly increase when it was exposed to other analytes. This example illustrates the use of EIS-based biosensors to quantify biomolecules such as DNA.

7.3 BATTERIES

Lithium-ion batteries (LIB) have a high energy density and they have been widely used in the past two decades in a variety of applications. The 2019 Nobel Prize in chemistry was awarded to the developers of LIB (Wrublewski 2020). Researchers employ a variety of techniques to characterize LIB, and EIS is one of them. Here, we consider a couple of

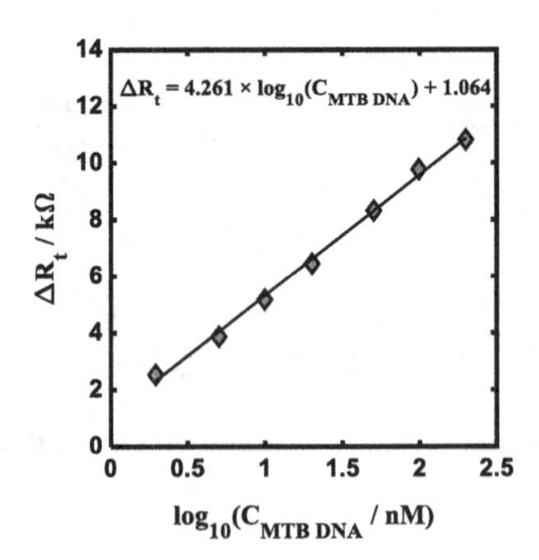

FIGURE 7.13 The dose–response curve, showing the increase in charge-transfer resistance as a function of the logarithm of MTB DNA concentration. (Adapted from *Analytica Chimica Acta*, 1044, Prinjaporn Teengam, Weena Siangproh, Adisorn Tuantranont, Tirayut Vilaivan, Orawon Chailapakul, Charles S. Henry, Electrochemical impedance-based DNA sensor using pyrrolidinyl peptide nucleic acids for tuberculosis detection, pp. 102–109, Copyright (2018), with permission from Elsevier.)

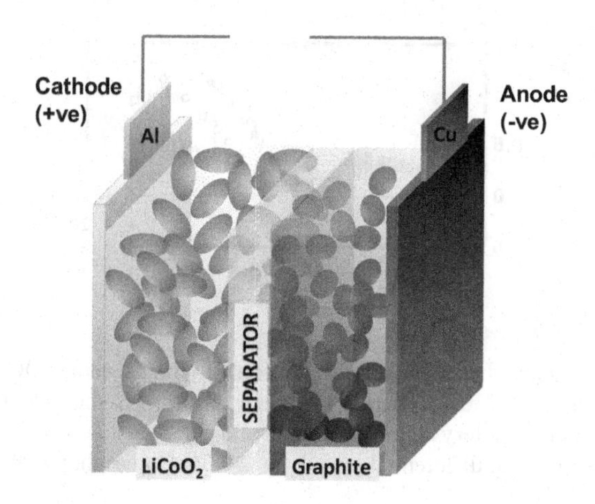

FIGURE 7.14 A pictorial of a section of a Li-ion battery.

examples from the literature. First, we introduce a few salient features of LIB and then proceed with the examples. A typical LIB consists of a graphite anode, a metal oxide, such as lithium-cobalt-oxide cathode, and a thin separator that keeps the cathode and anode apart (Figure 7.14). In battery terminology, the anode is the negative terminal, and the cathode is the positive terminal. The electrolyte is a mixture of organic solvents with a lithium salt. The anode materials are graphite particles attached to a Cu current collector plate using conducting binders. Likewise, the cathode materials are also in the form of particles attached to an aluminum current collector plate. When the battery is first charged, a layer called a *solid electrolyte interface* (SEI) forms next to the anode. It is not shown in the pictorial for simplicity. The reader can obtain more detailed information on LIB from suitable references (Wu 2015, Yuan, Liu, and Zhang 2011, Warner 2019). In this section, we first describe the use of EIS to measure the state of charge (SOC) and the cycle life aging of a LIB. In the second example, we describe the dependence of LIB EIS on the temperature and the insight obtained from the analysis of those EIS data.

7.3.1 Battery Status Evaluation

De Sutter et al. (2019) performed an extensive set of experiments to characterize the effect of aging and the SOC on LIB EIS. When the battery is fully charged, the SOC is 100%, and when it is fully discharged, the SOC is 0%. Li-ion cells with a nickel-manganese-cobalt-oxide cathode (NMC) and graphite anode were charged and discharged at specific rates. One full cycle of charging and discharging was denoted as a *full equivalent cycle* (FEC). After every 300 FEC, detailed characterization, including EIS, was performed. Several results were analyzed, and here, we limit our attention to the effect of SOC of a new battery and that of cycle life aging on EIS. A pseudo-galvanostatic measurement was employed, and the frequency range was from 10 kHz to 5 mHz. The dc current was set to zero to ensure that the SOC does not change during the measurement. The impedance spectra of a fresh battery, at various SOC, are presented as complex plane plots in Figure 7.15.

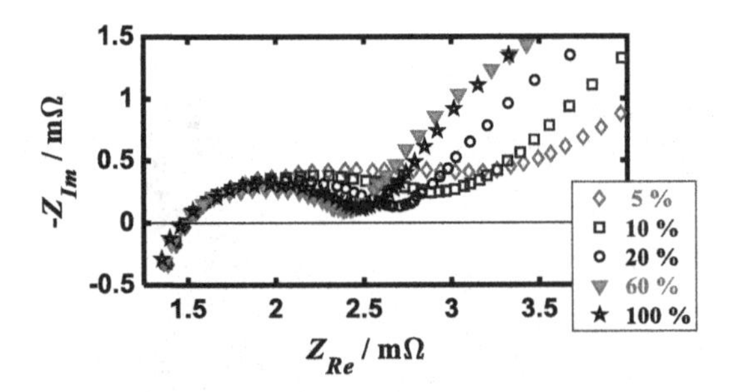

FIGURE 7.15 Complex plane plots of EIS of a fresh LIB, at several SOC values. (Adapted from *Electrochimica Acta*, 305, Lysander De Sutter, Yousef Firouz, Joris De Hoog, Noshin Omar, Joeri Van Mierlo, Battery aging assessment and parametric study of lithium-ion batteries by means of a fractional differential model, pp. 24–36, Copyright (2019), with permission from Elsevier.)

First, we note that the data points in the high-frequency region are in the fourth quadrant, i.e., Z_{Im} values are positive. Earlier (Section 2.4), we saw that in aqueous systems, the high-frequency EIS data might fall in the fourth quadrant. In LIB EIS, the high-frequency inductive behavior is usually assigned to the connecting cable. In the mid-frequency range, a depressed semicircle is seen, and at low frequencies, a Warburg-like impedance is seen.

We also note that the high-frequency data are more or less independent of the SOC, whereas the mid and low-frequency data strongly depend on the potential. The authors compared two models, one with the circuit shown in Figure 1.18a (Randles circuit with a capacitor to represent the double-layer and a resistor to represent the reaction) and another with a similar circuit, but with a CPE to represent the double-layer. Although the details of the model fit were not described, the authors may have employed only the mid-frequency range data for the optimization. The high and low-frequency data would require an inductor and a Warburg impedance, respectively, to adequately model the experimental results. First, the authors compared the results of two models and concluded that the circuit with a CPE offered a better description of the data. The values of the CPE pre-exponent (Q) and the charge transfer resistance (R_t) correlated well with the SOC.

The complex plane plots of EIS of batteries at various cycle life aging are shown in Figure 7.16. The high-frequency intercept of the abscissa depends strongly on the cycle life aging. As the cycle life aging increases, the ohmic resistance increases, presumably due to SEI growth. The increase is not exactly monotonic. This example shows that EIS with an appropriate circuit model can help in estimating the SOC and the cycle life aging of a Li-ion battery. Still, the correlation between the circuit parameters and the property is not always good, and there is scope for improvement.

7.3.2 Application of EIS in Battery Research

Momma et al. (2012) analyzed a commercial LIB with a graphite anode at several temperatures using EIS. The nominal capacity and potential of the battery were 0.8 Ah and 3.8 V respectively. The results, adapted from Momma et al. (2012), are shown as

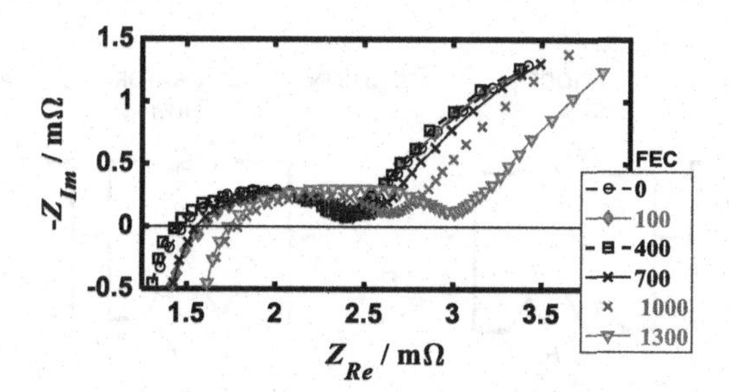

FIGURE 7.16 Complex plane plots of EIS of a LIB, at 50% SOC, and several cycle life aging. (Adapted from *Electrochimica Acta*, 305, Lysander De Sutter, Yousef Firouz, Joris De Hoog, Noshin Omar, Joeri Van Mierlo, Battery aging assessment and parametric study of lithium-ion batteries by means of a fractional differential model, pp. 24–36, Copyright (2019), with permission from Elsevier.)

complex plane plots in Figure 7.17. The frequency range employed was from 100 kHz to 10 mHz. Although it is common to use a pseudo-galvanostatic method in battery EIS measurements, here, the authors used 5 mV ac potential perturbation, i.e., pseudo-potentiostatic measurement. The dc potential of the electrode at 50% of the battery charge was ~3.8 V, and all the measurements were acquired when the battery was at a 50% charge. The temperature was varied from −20°C to + 20°C in steps of 5°C.

The high-frequency data were in the fourth quadrant, and here too, they were assigned to cable effects. As the temperature decreases, the mid and low-frequency impedance values increase significantly. The low-frequency data were not a 45° line, indicating either that the semi-infinite boundary layer diffusion effects do not manifest in this frequency range or that a finite Warburg impedance may describe the process. The authors modeled the data using the circuit shown in Figure 7.18.

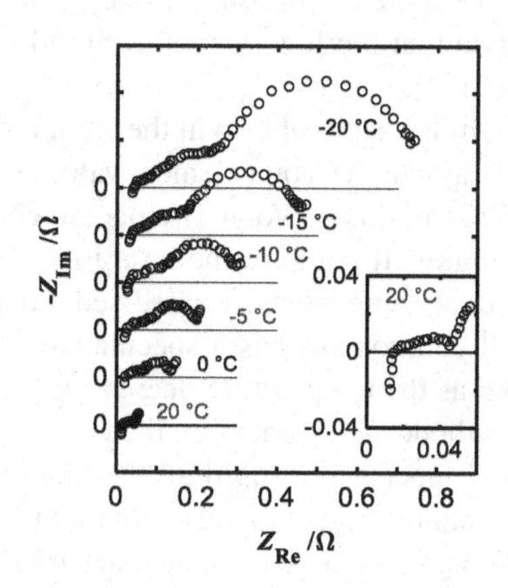

FIGURE 7.17 Complex plane plots of EIS of LIB at 50% SOC and various temperatures. Note that the origin is different for each plot. (Adapted from *Journal of Power Sources*, 216, L Toshiyuki Momma, Mariko Matsunaga, Daikichi Mukoyama, Tetsuya Osaka, Ac impedance analysis of lithium ion battery under temperature control, pp. 304–307, Copyright (2012), with permission from Elsevier.)

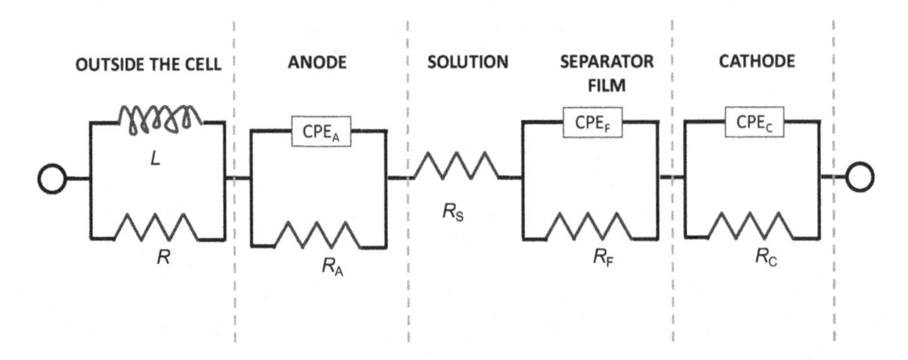

FIGURE 7.18 An equivalent electrical circuit used to model the data in Figure 7.17. (Adapted from *Journal of Power Sources*, 216, Toshiyuki Momma, Mariko Matsunaga, Daikichi Mukoyama, Tetsuya Osaka, Ac impedance analysis of lithium ion battery under temperature control, pp. 304–307, Copyright (2012), with permission from Elsevier.)

Here, the pair of elements R_1 and L are assigned to the cable and other elements 'outside' the battery. R_s represents the solution resistance. The anode, cathode, and separator film are modeled by the Voigt elements with the subscripts A, C, and F, respectively. In the Voigt elements, a CPE is used instead of a capacitor. At room temperature, the kinetics and diffusion are faster compared to the corresponding values at low temperatures. Hence, the impedance magnitude is less at higher temperatures. Since the kinetics and mass transfer effects manifest more clearly at mid and low frequencies, the decrease in impedance at higher temperatures is seen at low frequencies. The authors suggested that in the measured frequency range, mass transfer effects are not significant. When the battery was cycled between −20°C and +20°C, the impedance at 20°C did not change, and from this, we can conclude that the low temperature did not irreversibly alter the intrinsic properties of the LIB.

It is worth noting that while it is possible to fit the circuit model to the data, it is not possible to unambiguously assign the Voigt element values to the cathode, the anode, and the separator film. Since the three Voigt circuits are connected in series, swapping any two Voigt elements will not alter the total impedance. Unless the impedance spectra of the individual components are assessed independently, assigning the Voigt elements to the cell components has a speculative element in it. Regardless, the authors concluded that as the temperature decreased, the capacitance of the SEI decreased, while that of cathode and anode remained more or less a constant. On the other hand, a decrease in temperature caused all the three resistances, viz., R_A, R_C, and R_F, to increase by orders of magnitude. At room temperature, the resistance of SEI was extremely low and could not be estimated accurately. EIS measurements at low temperatures changed the time constants such that they are well separated, and multiple loops could be observed in the complex plane plots of impedance spectra. This example illustrates the utility of EIS in characterizing an electrochemical system, as well as the challenges in unambiguously assigning EEC element values to physical components or processes.

7.4 EXERCISE: APPLICATIONS – A FEW EXAMPLES

Q7.1 Choose an article that employs EIS for a specific application (e.g., corrosion, battery, fuel cells, biosensors, supercapacitors, and so on) and describe the impedance analysis. If data are available as Bode plots, perform equivalent circuit analysis to verify the circuit fit results.

Q7.2 In the published literature, identify equivalent circuits where swapping two elements or a group of elements will not alter the total impedance. In other words, identify the circuits where it is not possible to assign values to the elements unequivocally.

Q7.3 Select a few publications that provide a detailed reaction mechanism and a comparison of the experimental and model data. Verify the development of equations. Simulate the impedance spectra for the proposed set of kinetic parameters, and compare them with the experimental results.

8 Nonlinear EIS

We keep moving forward, opening new doors and doing new things, because we're curious and curiosity keeps leading us down new paths.

—**Walt Disney**

8.1 INTRODUCTION

During EIS measurement, only a small amplitude sine wave should be applied. Electrochemical systems are inherently nonlinear, and if we apply a large amplitude sine wave, then the response current will be periodic but not sinusoidal. Consider an irreversible simple electron transfer reaction shown in Eq. 8.1.

$$A \xrightarrow{k_1} B + e^- \tag{8.1}$$

The faradaic current response to a small and a large amplitude perturbation is shown in Figure 8.1

The potential is written as

$$E = E_{dc} + E_{ac0} \sin(\omega t) \tag{8.2}$$

while the current can be subjected to Fourier transform and written in Fourier series as

$$i = i_0 + i_1 \sin(\omega t + \phi_1) + i_2 \sin(2\omega t + \phi_2) + \cdots + i_n \sin(n\omega t + \phi_n) + \cdots \tag{8.3}$$

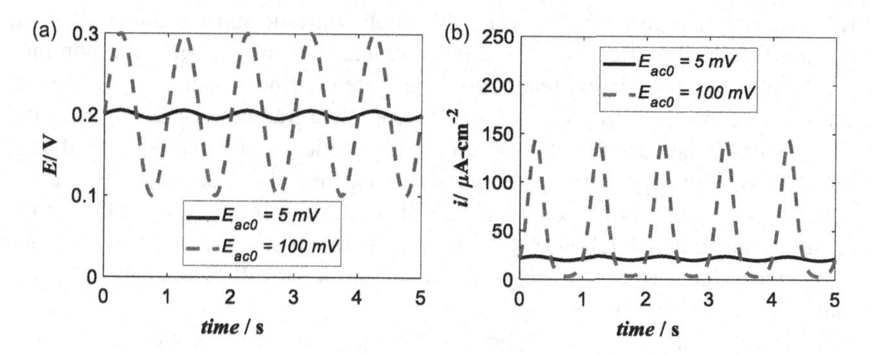

FIGURE 8.1 Applied potential and faradaic current response of an electrochemical reaction shown in Eq. 8.1. Here, $k_{10} = 10^{-6}$cm s^{-1}, $b_1 = 19\,V^{-1}$, $E_{dc} = 0.2\,V$, $E_{ac0} = 5$ or 100 mV, $[A] = 5$ mM, and $f = 1$ Hz. (a) Potential *vs.* time. (b) Faradaic current *vs.* time.

Here, the impedance magnitude is calculated by taking the ratio of E_{ac0} to the response at fundamental, i.e., i_1, and the impedance phase is $-\phi_1$. When E_{ac0} is small, the current response at higher harmonics, i.e., i_2, i_3,... would be negligible, but when E_{ac0} is large, higher harmonic should not be neglected. In this example, a perturbation amplitude of 5 mV is small, while 100 mV is large. In other cases, even a perturbation amplitude of 100 mV may be considered small. It depends on the system.

Strictly speaking, if the system exhibits nonlinearity, then the ratio of ac potential to ac current (E_{ac}/i_1) in the complex number form, also called the transfer function, cannot be called impedance. Only if the system is linear, causal, and stable, this ratio can be called impedance. Therefore, some researchers would consider 'nonlinear EIS' as an incorrect terminology. However, many in the electrochemical community call that ratio impedance anyway. Note that if one applies a large amplitude ac potential, the instrument will measure the vector ratio of potential to current and record the quantity as '*impedance*.' It is up to the user to verify if the system has been linear, causal, and stable. If nonlinear effects are present in the data and we recognize it, we call it nonlinear EIS or NLEIS data. If nonlinear effects are not present, or if nonlinear effects are present, but we fail to recognize it, we would refer to that as EIS data!

8.1.1 What Exactly Is NLEIS?

There are two aspects of NLEIS that we need to consider. In the first, we can apply a large ac potential at some frequency and measure the phase and magnitude of the current at that frequency (i.e., fundamental frequency) and use the ratio E_{ac}/i_1 (Victoria and Ramanathan 2011). In the second, we can measure the fundamental and besides, the higher harmonics of the current (Wong and MacFarlane 1995); in the latter case, we can either report the current magnitude and phase *vs.* frequency or report the ratio of $E_{ac}(\omega)/I_{ac}(n\omega)$ as '*impedance at higher harmonics*.' However, there is some difficulty associated with that, as described later. In early literature, NLEIS refers mostly to the ratio at the fundamental frequency and not the higher harmonics, but recently, emphasis on higher harmonics analysis is increasing.

Experimentally, measuring NLEIS at the fundamental frequency is trivial. In the FRA, one only has to increase the E_{ac0} (for potentiostatic studies, and I_{ac0} for galvanostatic studies) and collect the impedance data. In normal EIS, the nonlinear contributions are considered *undesirable*, but here nonlinear contributions are welcome and are expected to provide more insights than traditional EIS. On the other hand, to obtain the current values at higher harmonics, a more sophisticated setup is needed. We can apply an ac potential at a given frequency and collect the current at its second (or *n*th) harmonic, or we can collect the current and use FFT to extract the current at fundamental and other harmonics. In either case, it is better to apply a single pure sine wave since applying a multi-sine wave will complicate the signal, usually beyond redemption.

At fundamental frequency, nonlinear impedance can be defined as

$$Z_{NL} = \frac{E_{ac0}}{i_{ac}\big|_{\omega}} \tag{8.4}$$

for any arbitrary value of E_{ac0}, whereas the traditional linear impedance is defined as

$$Z_{\text{linear}} = \lim_{E_{ac0} \to 0} \frac{E_{ac0}}{i_{ac}|_{\omega}} \tag{8.5}$$

It is obvious that $\lim_{E_{ac0} \to 0} Z_{NL} = Z_{\text{linear}}$, a constant, and the unit of Z is Ω cm^2.

There is a difficulty in defining *impedance* at higher harmonics. One way (Wong and MacFarlane 1995) to define nonlinear impedance at higher harmonics is to write it (for example, second harmonic impedance) as

$$Z_{2\omega} = \frac{E_{ac0}}{i_{ac}|_{2\omega}} \tag{8.6}$$

The unit of $Z_{2\omega}$ as defined in Eq. 8.6 is Ω cm^2, but

$$\lim_{E_{ac0} \to 0} Z_{2\omega} = \infty \tag{8.7}$$

This is because at small and moderate values of E_{ac0},

$$i_{ac}|_{n\omega} \propto (E_{ac0})^n \tag{8.8}$$

An alternative (Smiechowski et al. 2011) is to define higher harmonic impedance (for example, second harmonic impedance) as

$$Z_{2\omega} = \frac{(E_{ac0})^2}{i_{ac}|_{2\omega}} \tag{8.9}$$

Now,

$\lim_{E_{ac0} \to 0} Z_{2\omega} =$ constant, but the unit of $Z_{2\omega}$ as defined in Eq. 8.9 is not Ω cm^2, but V Ω cm^2 and is not desirable. Hence, NLEIS data are often presented as current magnitude and phase, at various harmonics, as Bode plots.

While it is relatively easy to experimentally obtain NLEIS data at the fundamental frequency, analyzing it is more challenging, especially in the framework of reaction mechanism analysis. First, a few clarifications are in order. The electrolyte is assumed to be modeled correctly by a simple resistance, and therefore it will not exhibit a nonlinear effect. The double-layer capacitance is usually assumed to be an ideal capacitor or a constant phase element. In either case, it is considered a linear element. There is some evidence to suggest that the double-layer does exhibit nonlinearity (Xu and Riley 2011). Still, we will neglect it here and assume that the nonlinear component in the signal arises only from the faradaic reaction occurring at the electrode–electrolyte interface.

8.2 MATHEMATICAL BACKGROUND

NLEIS analysis requires familiarity with the Taylor series and Fourier series expansion of exponential of a sine wave, and a brief introduction is in order.

8.2.1 Taylor Series and Fourier Series

Consider the function $e^{a\sin\theta}$. It can be expanded in the Taylor series as

$$e^{a\sin\theta} = 1 + \frac{a\sin\theta}{1!} + \frac{a^2\sin^2\theta}{2!} + \frac{a^3\sin^3\theta}{3!} + \cdots \quad (8.10)$$

It is possible to rewrite the higher power terms as higher harmonics and rearrange, e.g.,

$$\sin^2\theta = \frac{1-\cos(2\theta)}{2} \quad (8.11)$$

$$\sin^3\theta = \frac{3\sin\theta - \sin 3\theta}{4} \quad (8.12)$$

and so on.

On the other hand, it is also possible to directly write $e^{a\sin\theta}$ in the Fourier series as

$$e^{a\sin\theta} = I_0(a) + 2\sum_{m=0}^{\infty} I_{2m+1}(a)\sin\{(2m+1)\theta\} + 2\sum_{m=1}^{\infty} I_{2m}(a)\cos\{2m\theta\} \quad (8.13)$$

(Abramowitz and Stegun 1972) where I is the modified Bessel function of the first kind and the subscripts 0, 1, and so on refer to the order of this function.

In 1715, **Brook Taylor** introduced the Taylor series. In 1807, **Jean-Baptiste Joseph Fourier** developed the Fourier series while working on the heat conduction problem. In 1888, **A. B. Basset** introduced *modified Bessel functions* in his treatise on hydrodynamics.

Note that $I_n(x)$ is one of the two solutions of

$$x^2\frac{d^2y}{dx^2} + x\frac{dy}{dx} = (x^2 + n^2)y = 0 \quad (8.14)$$

For integer values of 'n,' we can write

$$I_n(x) = \left(\frac{x}{2}\right)^n \sum_{k=0}^{\infty} \frac{\left(\frac{x}{2}\right)^{2k}}{k!(n+k)!} \quad (8.15)$$

The plots of $I_0(x)$, $I_1(x)$, and $I_2(x)$, are shown in Figure 8.2 so that the reader can understand the trends.

It can be shown that

$$\lim_{x\to 0} I_0(x) = 1 \tag{8.16}$$

$$\lim_{x\to 0} I_1(x) = \frac{x}{2} \tag{8.17}$$

$$\lim_{x\to 0} I_n(x) = \left(\frac{x}{2}\right)^n \frac{1}{n!} \tag{8.18}$$

for an integer 'n'.

8.2.2 NLEIS Analysis Methods

The methods to solve the NLEIS governing equations can be grouped into three categories. The current and mass balance equations can be written in the Taylor series (Darowicki 1994, 1995a, b, Diard, Le Gorrec, and Montella 1997d, 1998), and instead of truncating after the first derivative, we can truncate it at the second or third (or nth) derivative. Truncation will lead to approximation, but that is deemed acceptable. The algebra tends to be complex, but, at least in some cases, the current can be written (Darowicki 1994, 1995a, b, Diard, Le Gorrec, and Montella 1997d, 1998).

The second choice is to write the equations as they are and solve them numerically using computer programs (Victoria and Ramanathan 2011, Kaisare et al. 2011, Santhanam, Ramani, and Srinivasan 2012). Once the current is found as a function of time, FFT can be applied to extract the current at fundamental and other harmonics. A third possibility exists, i.e., write the current equation in the Fourier series, and the result can be used without any truncation error (Harrington 1997). The difficulty is in writing the Fourier series, and for most mechanisms, it is not possible to write it analytically, and numerical methods are found to be necessary anyway. In the case of second-order reactions such as a catalytic mechanism, it is even more difficult since handling nonlinear expressions is very challenging in the general matrix framework employed in (Harrington 1997).

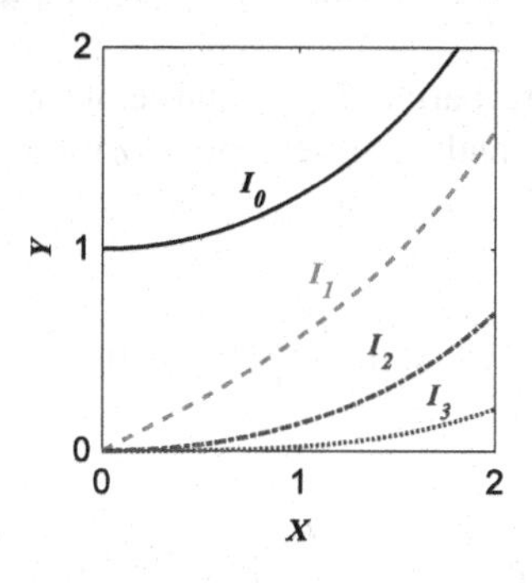

FIGURE 8.2 Plot of modified Bessel functions (I_n) near $x = 0$.

8.3 ESTIMATION OF NONLINEAR CHARGE-TRANSFER AND POLARIZATION RESISTANCES

Two important parameters that are often extracted from an impedance spectrum are charge-transfer resistance (R_t) and polarization resistance (R_p). They are defined (Diard, Le Gorrec, and Montella 1997a), respectively, as

$$R_t = \lim_{E_{ac0}\to 0}\lim_{\omega\to\infty} Z_F = \lim_{E_{ac0}\to 0}\lim_{\omega\to\infty}\frac{E_{ac0}}{i_F|_\omega}\left(\text{potentiostatic EIS}\right) \tag{8.19}$$

$$R_t = \lim_{i_{ac0}\to 0}\lim_{\omega\to\infty} Z_F = \lim_{i_{ac0}\to 0}\lim_{\omega\to\infty}\frac{E_F|_\omega}{i_{ac0}}\left(\text{galvanostatic EIS}\right) \tag{8.20}$$

and

$$R_p = \lim_{E_{ac0}\to 0}\lim_{\omega\to 0} Z_F = \lim_{E_{ac0}\to 0}\lim_{\omega\to 0}\frac{E_{ac0}}{i_F|_\omega}\left(\text{potentiostatic NLEIS}\right) \tag{8.21}$$

$$R_p = \lim_{i_{ac0}\to 0}\lim_{\omega\to 0} Z_F = \lim_{i_{ac0}\to 0}\lim_{\omega\to 0}\frac{E_F|_\omega}{i_{ac0}}\left(\text{galvanostatic NLEIS}\right) \tag{8.22}$$

However, this is not the only definition employed in the literature. A more rigorous definition of charge-transfer resistance is (Harrington 2015)

$$R_t = \left(\frac{\partial E}{\partial i}\right)_{C_i,\theta_i} \tag{8.23}$$

which reduces, in most cases, to the definition given above. Some authors consider R_t and R_p to be the same and use them interchangeably (He and Mansfeld 2009). Therefore, when reading the literature, it is important to note which definition is employed by the authors. In this book, we employ the definitions given in Eqs. 8.19–8.22.

When the potential perturbation (E_{ac0}) is large, we can define nonlinear charge-transfer resistance ($R_{t,\,NL}$) and nonlinear polarization resistance ($R_{p,\,NL}$) respectively as

$$R_{t,NL} = \lim_{\omega\to\infty} Z_F = \lim_{\omega\to\infty}\frac{E_{ac0}}{i_F|_\omega}\left(\text{potentiostatic NLEIS}\right) \tag{8.24}$$

$$R_{t,NL} = \lim_{\omega\to\infty} Z_F = \lim_{\omega\to\infty}\frac{E_F|_\omega}{i_{ac0}}\left(\text{galvanostatic NLEIS}\right) \tag{8.25}$$

and

$$R_{p,\mathrm{NL}} = \lim_{\omega\to 0} Z_F = \lim_{\omega\to 0} \frac{E_{\mathrm{ac}0}}{i_F|_\omega} (\text{potentiostatic NLEIS}) \tag{8.26}$$

$$R_{p,\mathrm{NL}} = \lim_{\omega\to 0} Z_F = \lim_{\omega\to 0} \frac{E_F|_\omega}{i_{\mathrm{ac}0}} (\text{galvanostatic NLEIS}) \tag{8.27}$$

For a given reaction mechanism, under the assumptions of negligible solution resistance, no film formation, and the Langmuir isotherm model for the adsorbed intermediates, one can develop the expressions for linear and nonlinear charge-transfer and polarization resistances (Diard, Le Gorrec, and Montella 1997a, b, c).

8.4 CALCULATION OF THE NLEIS RESPONSE OF ELECTROCHEMICAL REACTIONS

8.4.1 Simple Electron Transfer Reaction

At first, we illustrate the Taylor series approach to calculate the NLEIS response of a simple electron transfer reaction. Consider

$$\mathrm{Fe}^{2+} \underset{k_{-1}}{\overset{k_1}{\rightleftarrows}} \mathrm{Fe}^{3+} + \mathrm{e}^- \tag{8.28}$$

The faradaic current is given by

$$i_F = F\left(k_{10}\mathrm{e}^{b_1 E}\left[\mathrm{Fe}^{2+}\right] - k_{-10}\mathrm{e}^{b_{-1}E}\left[\mathrm{Fe}^{3+}\right]\right) \tag{8.29}$$

Under the application of an ac potential E_{ac} along with a dc potential E_{dc}, this becomes

$$i_F = F\left(k_{1\mathrm{dc}}\mathrm{e}^{b_1 E_{\mathrm{ac}}}\left[\mathrm{Fe}^{2+}\right] - k_{-1\mathrm{dc}}\mathrm{e}^{b_{-1}E_{\mathrm{ac}}}\left[\mathrm{Fe}^{3+}\right]\right) \tag{8.30}$$

By expanding the term $\mathrm{e}^{b_1 E_{\mathrm{ac}}}$ and $\mathrm{e}^{b_{-1}E_{\mathrm{ac}}}$ in the Taylor series, Eq. 8.30 can be written as

$$i_F = F\left(k_{1\text{-dc}}[\mathrm{Fe}^{2+}]\left\{1 + b_1 E_{\mathrm{ac}} + \frac{b_1^2 E_{\mathrm{ac}}^2}{2!} + \cdots\right\} - k_{-1\text{-dc}}\left[\mathrm{Fe}^{3+}\right]\left\{1 + b_{-1}E_{\mathrm{ac}} + \frac{b_{-1}^2 E_{\mathrm{ac}}^2}{2!} + \cdots\right\}\right) \tag{8.31}$$

The Taylor series can be truncated after the cubic term (for example). Then, one can substitute $E_{\mathrm{ac}} = E_{\mathrm{ac}0} \sin(\omega t)$ and use the identities Eqs. 8.11 and 8.12, to write the faradaic current as

$$i_F = i_{F\text{-dc}} + \left(i_{F\text{-ac}}|_\omega\right)\sin\left(\omega t + \phi_1\right) + \left(i_{F\text{-ac}}|_{2\omega}\right)\sin\left(2\omega t + \phi_2\right) + \left(i_{F\text{-ac}}|_{3\omega}\right)\sin\left(3\omega t + \phi_3\right) \tag{8.32}$$

where

$$i_{F\text{-dc}} = F\left(k_{1\text{-dc}}\left[\mathrm{Fe}^{2+}\right]\left\{1+\frac{b_1^2E_{\mathrm{ac}0}^2}{4}\right\}-k_{-1\text{-dc}}\left[\mathrm{Fe}^{3+}\right]\left\{1+\frac{b_{-1}^2E_{\mathrm{ac}0}^2}{4}\right\}\right) \tag{8.33}$$

$$\left(i_{F\text{-ac}}\big|_{\omega}\right) = F\left(k_{1\text{-dc}}\left[\mathrm{Fe}^{2+}\right]\left\{b_1E_{\mathrm{ac}0}+\frac{1}{8}b_1^3E_{\mathrm{ac}0}^3\right\}-k_{-1\text{-dc}}\left[\mathrm{Fe}^{3+}\right]\left\{b_{-1}E_{\mathrm{ac}0}+\frac{1}{8}b_{-1}^3E_{\mathrm{ac}0}^3\right\}\right) \tag{8.34}$$

$$\left(i_{F\text{-ac}}\big|_{2\omega}\right) = F\left(k_{1\text{-dc}}\left[\mathrm{Fe}^{2+}\right]\left\{\frac{-b_1^2E_{\mathrm{ac}0}^2}{4}\right\}+k_{-1\text{-dc}}\left[\mathrm{Fe}^{3+}\right]\left\{\frac{b_{-1}^2E_{\mathrm{ac}0}^2}{4}\right\}\right) \tag{8.35}$$

$$\left(i_{F\text{-ac}}\big|_{3\omega}\right) = F\left(k_{1\text{-dc}}\left[\mathrm{Fe}^{2+}\right]\left\{\frac{-b_1^3E_{\mathrm{ac}0}^3}{24}\right\}+k_{-1\text{-dc}}\left[\mathrm{Fe}^{3+}\right]\left\{\frac{b_{-1}^3E_{\mathrm{ac}0}^3}{24}\right\}\right) \tag{8.36}$$

We thus arrive at the following conclusion. At small and moderate values of $E_{\mathrm{ac}0}$,

$$i_{F\text{-ac}}\big|_{n\omega} = \lambda_0\left(E_{\mathrm{ac}0}\right)^n \tag{8.37}$$

At large values of $E_{\mathrm{ac}0}$,

$$i_{F\text{-ac}}\big|_{n\omega} = \left(\left(E_{\mathrm{ac}0}\right)^n\sum_{m=0}^{\infty}\lambda_i\left(E_{\mathrm{ac}0}\right)^{2m}\right) \tag{8.38}$$

Note that λ_i is a constant in Eq. 8.38.

Next, we illustrate the Fourier series approach.

Substituting Eq. 8.13 in Eq. 8.30, we can write the faradaic current as

$$i_F = F\left\{\begin{array}{l} +\left(k_{1\mathrm{dc}}\left[\mathrm{Fe}^{2+}\right]\right)\left\{\begin{array}{l}\left[I_0\left(b_1E_{\mathrm{ac}0}\right)+2\sum_{m=0}^{\infty}I_{2m+1}\left(b_1E_{\mathrm{ac}0}\right)\sin\left(\left[2m+1\right]\omega t\right)\right. \\ \left.+2\sum_{m=1}^{\infty}I_{2m}\left(b_1E_{\mathrm{ac}0}\right)\cos\left(2m\omega t\right)\right]\end{array}\right\} \\ -\left(k_{-1\mathrm{dc}}\left[\mathrm{Fe}^{3+}\right]\right)\left\{\begin{array}{l}\left[I_0\left(b_{-1}E_{\mathrm{ac}0}\right)+2\sum_{m=0}^{\infty}I_{2m+1}\left(b_{-1}E_{\mathrm{ac}0}\right)\sin\left(\left[2m+1\right]\omega t\right)\right. \\ \left.+2\sum_{m=1}^{\infty}I_{2m}\left(b_{-1}E_{\mathrm{ac}0}\right)\cos\left(2m\omega t\right)\right]\end{array}\right\}\end{array}\right\} \tag{8.39}$$

Then the correct expression for the dc component of the faradaic current is

$$i_{F\text{-dc}} = F\left\{\left(k_{1\text{dc}}\left[\text{Fe}^{2+}\right]\right)I_0\left(b_1E_{\text{ac}0}\right) - \left(k_{-1\text{dc}}\left[\text{Fe}^{3+}\right]\right)I_0\left(b_{-1}E_{\text{ac}0}\right)\right\} \tag{8.40}$$

Similarly, the time-varying component of the faradaic current is

$$i_{F\text{-ac}} = 2F\left\{\begin{array}{l}\left(k_{1\text{dc}}\left[\text{Fe}^{2+}\right]\right)\left\{\begin{bmatrix}\sum\limits_{m=0}^{\infty} I_{2m+1}\left(b_1E_{\text{ac}0}\right)\sin\left(\left[2m+1\right]\omega t\right)\\ +\sum\limits_{m=1}^{\infty} I_{2m}\left(b_1E_{\text{ac}0}\right)\cos\left(2m\omega t\right)\end{bmatrix}\right\}\\ -\left(k_{-1\text{dc}}\left[\text{Fe}^{3+}\right]\right)\left\{\begin{bmatrix}\sum\limits_{m=0}^{\infty} I_{2m+1}\left(b_{-1}E_{\text{ac}0}\right)\sin\left(\left[2m+1\right]\omega t\right)\\ +2\sum\limits_{m=1}^{\infty} I_{2m}\left(b_{-1}E_{\text{ac}0}\right)\cos\left(2m\omega t\right)\end{bmatrix}\right\}\end{array}\right\} \tag{8.41}$$

At the fundamental frequency (i.e., corresponding to ω), the current is

$$i_{F\text{-ac}}\Big|_{\omega} = 2F\left\{\begin{array}{l}\left(k_{1\text{dc}}\left[\text{Fe}^{2+}\right]\right)I_1\left(b_1E_{\text{ac}0}\right)\\ -\left(k_{-1\text{dc}}\left[\text{Fe}^{3+}\right]\right)I_1\left(b_{-1}E_{\text{ac}0}\right)\end{array}\right\}\sin\left(\omega t\right) \tag{8.42}$$

The faradaic impedance can then be written as

$$Z_{F,\text{NL}} = \frac{E_{\text{ac}}}{i_{F\text{-ac}}\big|_{\omega}} = \frac{E_{\text{ac}0}}{2F\left\{\left(k_{1\text{dc}}\left[\text{Fe}^{2+}\right]\right)I_1\left(b_1E_{\text{ac}0}\right) - \left(k_{-1\text{dc}}\left[\text{Fe}^{3+}\right]\right)I_1\left(b_{-1}E_{\text{ac}0}\right)\right\}} \tag{8.43}$$

Using Eq. 8.17, we can verify that

$$\lim_{E_{\text{ac}0}\to 0} Z_{F,\text{NL}} = \frac{1}{F\left\{\left(b_1k_{1\text{-dc}}\left[\text{Fe}^{2+}\right]\right) - \left(b_{-1}k_{-1\text{-dc}}\left[\text{Fe}^{3+}\right]\right)\right\}} = Z_F \tag{8.44}$$

For an exemplary system given by Eq. 8.28, the NLEIS response is calculated, and the nonlinear impedance at the fundamental frequency is presented as a complex plane plot in Figure 8.3. The nonlinear faradaic impedance of this particular reaction is independent of the applied perturbation frequency but depends on the value of E_{dc} and $E_{\text{ac}0}$, as seen in Eq. 8.24. At a given E_{dc}, it can be represented by a resistor whose resistance depends on $E_{\text{ac}0}$. When the value of $E_{\text{ac}0}$ is small, then the value of the resistance is independent of $E_{\text{ac}0}$. In this particular example, $R_{t,\text{NL}}$ and $R_{p,\text{NL}}$ are

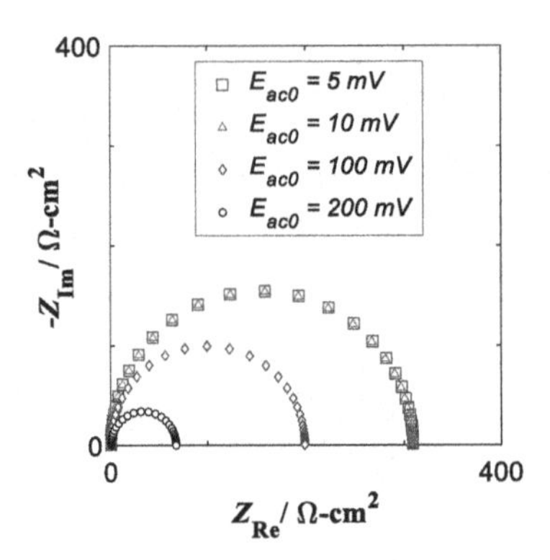

FIGURE 8.3 Complex plane plots of a nonlinear impedance response of a simple electron transfer reaction given in Eq. 8.28. $k_{10} = k_{-10} = 10^{-3}$cm s^{-1}, $b_1 = 19.47\,\text{V}^{-1}$, $b_{-1} = -19.47\,\text{V}^{-1}$, $[Fe^{2+}] = [Fe^{3+}] = 5\,\text{mM}$, $E_{dc} = 0.3\,\text{V}$ *vs.* OCP, and $C_{dl} = 20$ μF cm^{-2}. (Adapted from *Electrochimica Acta*, 56, S. Noyel Victoria,S. Ramanathan, Effect of potential drifts and ac amplitude on the electrochemical impedance spectra, 2606–2615, Copyright (2011), with permission from Elsevier.)

identical and Figure 8.3 shows that $R_{t,\,\text{NL}}$ decreases with E_{ac0}. For this reaction, the magnitude of the nth harmonic ($i_{n\omega}$) is independent of frequency, and the phase is 0° for odd harmonic and 90° for even harmonic.

Next, we illustrate the numerical method. The faradaic current is given as

$$i_F = F\left(k_{1dc}e^{b_1E_{ac}}\left[Fe^{2+}\right] - k_{-1dc}e^{b_{-1}E_{ac}}\left[Fe^{3+}\right]\right) \tag{8.45}$$

Now, let us synthesize the sine wave (i.e., create a table of time t and corresponding potential $E_{ac0}\sin(\omega t)$). At any time t, we can calculate the value of i_F using the above expression. For various time values, we will have corresponding current values. This current vector should be subjected to FFT, so that the dc, fundamental, and higher harmonic components are numerically calculated. Let us denote the magnitude at the fundamental frequency as i_1 and the phase as ψ. Normally, we choose the phase of the applied potential to be zero, i.e., in the expression $E_{ac} = E_{ac0}\sin(\omega t + \phi)$, we choose $\phi = 0$. Hence, the impedance is given by

$$Z = \frac{E_{ac0}}{i_1}e^{-j\psi} \tag{8.46}$$

If we employ a frequency of f (e.g., 1 Hz) for the applied sine wave, normally, it is desirable to use 16 or 32 points within this period to simulate the sine wave, i.e., the sampling frequency should be $2^n f$, where n is a positive integer since it makes it simpler for the FFT algorithm to handle the data.

8.4.2 Reaction with an Adsorbed Intermediate

Consider the two-step reaction analyzed in Section 5.1.2 by assuming that the ac potential amplitude is small. The reactions are

$$\mathrm{M} \xrightarrow{k_1} \mathrm{M}^+_{\mathrm{ads}} + \mathrm{e}^-$$
$$\mathrm{M}^+_{\mathrm{ads}} \xrightarrow{k_2} \mathrm{M}^+_{\mathrm{sol}} \tag{8.47}$$

The mass balance equation is

$$\Gamma \frac{\mathrm{d}\theta}{\mathrm{d}t} = k_1(1-\theta) - k_2\theta \tag{8.48}$$

where Γ is the total number of surface sites per unit area and θ is the fractional surface coverage of the adsorbed intermediate.

The *charge balance* equation, giving the faradaic current density, is

$$i_F = Fk_1(1-\theta) \tag{8.49}$$

When a dc potential (E_{dc}) is applied, the surface coverage θ_{ss} is given by

$$\theta_{ss} = \frac{k_{1\text{-dc}}}{k_{1\text{-dc}} + k_2} \tag{8.50}$$

Initially, only a *dc* potential E_{dc} is applied. Then, at time $t = 0$, an *ac* potential E_{ac} is superimposed on the *dc* potential.

The mass balance equation can be rearranged as

$$\frac{\mathrm{d}\theta}{\mathrm{d}t} = \frac{1}{\Gamma}\left[k_{1\text{-dc}}\mathrm{e}^{b_1 E_{ac0}\sin(\omega t)}(1-\theta) - k_2\theta\right] \tag{8.51}$$

with the initial condition

$$t = 0 : \theta = \theta_{ss} \tag{8.52}$$

The rate constants will change, and the surface coverage will also change. Equation 8.51 must be integrated to obtain θ as a function of time, with the initial condition at $t = 0$, $\theta = \theta_{ss}$. Note that the rate constants are a function of potential, and the potential is a function of time ($E = E_{dc} + E_{ac0}\sin(\omega t)$). Although the mass balance equation is a linear ordinary differential equation, we are not aware of an analytical solution to Eq. 8.51 and hence used numerical methods.

The integration results for a given frequency and amplitude are shown in Figure 8.4. The kinetic parameters used in the simulation are given in the figure caption. Before applying the sinusoidal potential, the steady-state surface coverage was 0.83. However, after superimposing the ac potential, Figure 8.4a bottom left inset shows that the average value of the fractional surface coverage (θ_{av}, averaged over one cycle time) does not remain at θ_{ss}. Instead, it drifts and stabilizes after some time (50 ms) to a value of 0.92, as shown in Figure 8.4b. Even when the value of E_{ac0} is small, the values of θ_{av} and θ_{ss} are different (Kaisare et al. 2011).

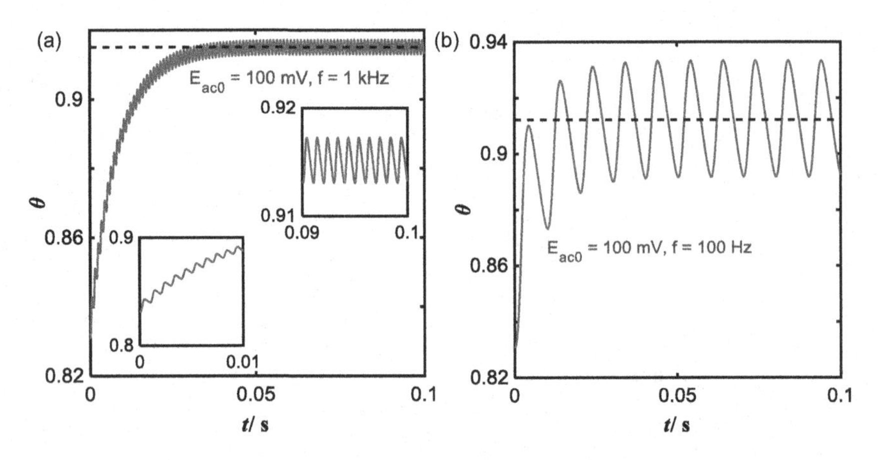

FIGURE 8.4 Fractional surface coverage as a function of time for an electrochemical reaction described by Eq. 8.47. The parameters are $E_{dc} = 0.2\,V$ *vs.* OCP, $k_{10} = 10^{-9}\,mol\ cm^{-2}s^{-1}$, $b_1 = 19.47\,V^{-1}$, $k_{20} = 10^{-8}\,mol\ cm^{-2}s^{-1}$, $\Gamma = 10^{-9}\,mol\ cm^{-2}$, and $C_{dl} = 10^{-5}\,F\ cm^{-2}$. (a) $f = 1\,kHz$ and $E_{ac0} = 100\,mV$. The bottom left inset shows the expanded view of oscillations in the beginning, and the top right inset shows the expanded view of steady periodic oscillations obtained after some time. (b) $f = 100\,Hz$ and $E_{ac0} = 100\,mV$.

It is possible to estimate the average value of fractional surface coverage during oscillation (θ_{av}) using a heuristic formula (Kaisare et al. 2011).

$$\theta_{av} \simeq \frac{k_{1\text{-dc}} I_0\left(b_1 E_{ac0}\right)}{\left(k_{1\text{-dc}} I_0\left(b_1 E_{ac0}\right) + k_2\right)} \tag{8.53}$$

Essentially, in the expression used to calculate θ_{ss}, the electrochemical rate constants (k_i) must be multiplied by the correction factor $I_0(b_i E_{ac0})$. After sufficient wait time, steady periodic oscillations of fractional surface coverage are seen (Figure 8.4a, top right inset). It is clear that data must be acquired only after a sufficient wait time and analyzed. The wait time necessary to achieve steady periodic results depends on the kinetic parameters, double-layer capacitance, and solution resistance (Santhanam, Ramani, and Srinivasan 2012), but does not vary strongly with frequency (Kaisare et al. 2011). A comparison of Figure 8.4a and b shows that the amplitude of θ oscillations increases with frequency, but θ_{av} does not vary with frequency. It is worth noting that these insights are obtained only from the numerical solution results.

After numerical integration, the fractional surface coverage (θ), which is available as a function of time, is substituted in Eq. 8.49, to calculate the faradaic current as a function of time. The current is then subjected to FFT to obtain the magnitude (i_1) and the phase (ψ) of the current at the fundamental frequency. From these values, the impedance or admittance can easily be calculated. This is repeated for all the frequency values needed. Note that FFT results also give higher harmonic current (phase and magnitude).

Compared to the linearization methods described in Chapter 5, the main advantage of this method is that there is no restriction of *small amplitude* for the applied potential. This works equally well for small or large amplitude perturbations. As long as the numerical computations are executed correctly and accurately, this method would yield the correct results. It must be noted that the computational time to generate NLEIS data by this method is high. As of 2020, on a modern desktop computer, the computational time of the code for linearized equations (which are applicable only for small E_{ac0}) is less than a second, but the computational time of the corresponding nonlinear version by numerical methods can take several minutes or even hours.

The extent of the change in the average value of θ depends on the amplitude of alternating potential (E_{ac0}) employed (Kaisare et al. 2011). However, at a given E_{ac0}, it does not depend strongly on the frequency of the ac potential (Kaisare et al. 2011). The time it takes to stabilize does not depend strongly on the frequency or the E_{ac0}. Thus, for a given system characteristic (kinetic parameters and double-layer capacitance), at a fixed E_{dc}, the duration needed to obtain steady periodic results is more or less fixed for all perturbations (i.e., for any E_{ac0} value and any frequency).

This means when a perturbation is applied, the current ensuing immediately after the application of this perturbation should not be used to calculate the impedance. When a periodic perturbation cycle begins, sufficient wait time must be given to obtain a steady periodic response, before the data are acquired. This is true for any type of perturbation, such as single sine, multi-sine, or triangular waves (which are used in cyclic voltammetry). Figure 8.5a shows the NLEIS response at the fundamental frequency, presented as impedance in complex plane plots. Figure 8.5b and c show the expanded view at mid and low frequencies, respectively, to enable estimation of $R_{t,\,NL}$ and $R_{p,\,NL}$, respectively. In this case, $R_{t,\,NL}$ increases with E_{ac0}, while $R_{p,\,NL}$ initially decreases and later increases with E_{ac0}. In general, $R_{t,\,NL}$ and $R_{p,\,NL}$ may increase or decrease with E_{ac0}, although in our experience, we find that for many systems, $R_{t,\,NL}$ decreases with E_{ac0}.

It is possible to estimate $R_{t,\,NL}$, again using a heuristic formula (Fasmin and Srinivasan 2015). For the reaction described by Eq. 8.47, charge-transfer resistance (R_t) under small-amplitude perturbations is given by

$$(R_t)^{-1} = F\left(b_1 k_{1dc}\left(1-\theta_{ss}\right)\right) \tag{8.54}$$

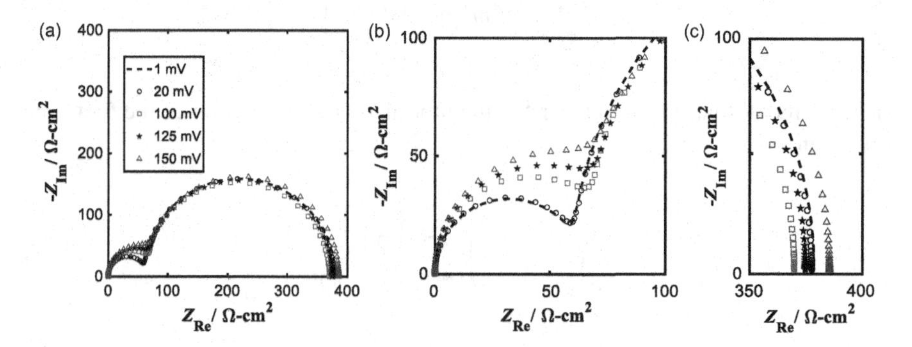

FIGURE 8.5 (a) Complex plane plots of nonlinear EIS data of the reaction described in Eq. 8.47. Parameters employed are the same as those in Figure 8.4. (b) High-frequency data and (c) Low-frequency data, expanded for clarity. Note that R_t increases with E_{ac0}, and R_p initially decreases with E_{ac0} but increases later.

Under large amplitude perturbations, $R_{t,\,\mathrm{NL}}$ is given by

$$\left(R_{t,\mathrm{NL}}\right)^{-1} = F\left(\frac{2I_1\left(b_1E_{\mathrm{ac}0}\right)}{E_{\mathrm{ac}0}}k_{1\mathrm{dc}}\left(1-\theta_{\mathrm{av}}\right)\right) \tag{8.55}$$

Essentially, one can obtain an expression $R_{t,\,\mathrm{NL}}$ from the expression for R_t, by substituting θ_{av} for θ_{ss} and "$2 \times I_1(b_iE_{\mathrm{ac}0})/E_{\mathrm{ac}0}$" for b_i. This method appears to work for other mechanisms as well (Fasmin and Srinivasan 2015).

Diard, Le Gorrec, and Montella (1997a) showed that when the current density (i) and potential (E) are related by an arbitrary function

$$i = f(E) \tag{8.56}$$

the nonlinear polarization resistance $R_{p,\mathrm{NL}}$ can be written as

$$\left(R_{\mathrm{sol}} + R_{p,\mathrm{NL}}\right)^{-1} = \sum_{k=0}^{\infty} b_k f^{(2k+1)} E_{\mathrm{ac}0}^{(2k+1)} \tag{8.57}$$

where

$$b_0 = 1, b_k = \frac{b_{k-1}}{4k(k+1)}, \text{for } k \geq 1 \tag{8.58}$$

$$\text{Note that } f^{(2k+1)} = \frac{\mathrm{d}^{(2k+1)}f}{\mathrm{d}E^{(2k+1)}} \tag{8.59}$$

An equivalent representation is

$$\left(R_{\mathrm{sol}} + R_{p,\mathrm{NL}}\right) = \frac{1}{\sum_{m=0}^{\infty}\left[2^{2m}m!(m+1)!\right]^{-1} \times \left(\frac{\mathrm{d}^{2m+1}i}{\mathrm{d}E^{2m+1}}\right)E_{\mathrm{ac}0}^{2m}} \tag{8.60}$$

Here we recall that, for an integer 'n,' the modified Bessel function of the first kind is written as

$$I_n(z) = \left(\frac{1}{2}z\right)^n \sum_{k=0}^{\infty} \frac{\left(\frac{1}{2}z\right)^{2k}}{k!(n+k)!} \tag{8.61}$$

Using the operator notation

$$\left(D_E\right) = \frac{\mathrm{d}}{\mathrm{d}E} \tag{8.62}$$

$$\text{such that}\left(D_E\right)^n = \frac{\mathrm{d}^n}{\mathrm{d}E^n}, \tag{8.63}$$

one can write (Fasmin and Srinivasan 2017)

$$\left(R_{\text{sol}} + R_{p,\text{NL}}\right)^{-1} = \frac{2}{E_{\text{ac}0}}\left[I_1\left(E_{\text{ac}0}D_E\right)\right](i) = \frac{2}{E_{\text{ac}0}}\left[I_1\left(E_{\text{ac}0}D_E\right)\right]\left(f(E)\right) \tag{8.64}$$

In general, for odd 'n,' the current corresponding to the nth harmonic can be written as

$$i\big|_{n\omega} = 2(-1)^{\left(\frac{n-1}{2}\right)}\sin(n\theta)\times\left\{\left[I_n\left(E_{\text{ac}0}D\right)\right](i)\right\} \tag{8.65}$$

For even 'n,' the current corresponding to the nth harmonic can be written as

$$i\big|_{n\omega} = 2(-1)^{\frac{n}{2}}\cos(n\theta)\times\left\{\left[I_n\left(E_{\text{ac}0}D\right)\right](i)\right\} \tag{8.66}$$

The fractional surface coverage can be written as

$$\theta = \theta\big|_{\text{SS}} + \frac{E_{\text{ac}}}{1!}\left(\frac{\mathrm{d}\theta}{\mathrm{d}E}\right) + \frac{\left(E_{\text{ac}}\right)^2}{2!}\left(\frac{\mathrm{d}^2\theta}{\mathrm{d}E^2}\right) + \frac{\left(E_{\text{ac}}\right)^3}{3!}\left(\frac{\mathrm{d}^3\theta}{\mathrm{d}E^3}\right) + \cdots \tag{8.67}$$

The rate constant of the electrochemical step can be written as

$$k_i = k_{i\text{dc}} + \frac{E_{\text{ac}}}{1!}b_i k_{i\text{dc}} + \frac{\left(E_{\text{ac}}\right)^2}{2!}b_i^2 k_{i\text{dc}} + \frac{\left(E_{\text{ac}}\right)^3}{3!}b_i^3 k_{i\text{dc}} + \cdots \tag{8.68}$$

For the reaction described by Eq. 8.47, substituting Eqs. 8.67 and 8.68 in Eq. 8.48, and rearranging the terms, we get

$$\frac{\mathrm{d}\theta}{\mathrm{d}E} = \frac{b_1 k_{1\text{dc}}\left(1-\theta_{\text{ss}}\right)}{k_{1\text{dc}} + k_2 + j\omega\Gamma} \tag{8.69}$$

The higher-order derivatives are

$$\frac{\mathrm{d}^2\theta}{\mathrm{d}E^2} = \frac{2!}{\left(k_{1\text{dc}} + k_2\right)}\left[\frac{b_1^2 k_{1\text{dc}}}{2!} - \left\{\frac{\left(b_1 k_{1\text{dc}}\right)}{1!}\frac{\mathrm{d}\theta}{\mathrm{d}E}\frac{1}{1!} + \frac{\left(b_1^2 k_{1\text{dc}}\right)}{2!}\theta_{\text{ss}}\frac{1}{0!}\right\}\right] \tag{8.70}$$

$$\frac{\mathrm{d}^3\theta}{\mathrm{d}E^3} = \frac{3!}{\left(k_{1\text{dc}} + k_2\right)}\left[\frac{b_1^3 k_{1\text{dc}}}{3!} - \left\{\frac{\left(b_1 k_{1\text{dc}}\right)}{1!}\frac{\mathrm{d}^2\theta}{\mathrm{d}E^2}\frac{1}{2!} + \frac{\left(b_1^2 k_{1\text{dc}}\right)}{2!}\frac{\mathrm{d}\theta}{\mathrm{d}E}\frac{1}{1!} + \frac{\left(b_1^3 k_{1\text{dc}}\right)}{3!}\theta_{\text{SS}}\frac{1}{0!}\right\}\right] \tag{8.71}$$

and so on. In general,

$$\frac{d^n\theta}{dE^n} = \frac{n!}{(k_{1dc}+k_2)}\left[\frac{(b_1^n)k_{1dc}}{n!} - \sum_{m=1}^{n}\left\{\frac{(b_1^m k_{1dc})}{m!}\frac{d^{(n-m)}\theta}{dE^{(n-m)}}\frac{1}{(n-m)!}\right\}\right] \text{ for } n>0 \tag{8.72}$$

When the angular frequency $\omega \to 0$, the expression for the first-order derivative is simplified as

$$\frac{d\theta}{dE} = \frac{(b_1 - b_2)k_{1dc}(1-\theta_{ss})}{k_{1dc}+k_{2dc}} \tag{8.73}$$

It is worth noting that the higher-order derivatives depend on ω only through the first-order derivative. Similarly, the current can be expanded, and we can show that for any value of 'n' ($n > 0$)

$$\frac{d^n i_F}{dE^n} = F\left[b_1^n k_{1dc}(1-\theta_{ss}) - \frac{k_{1dc}}{n!}\sum_{m=0}^{n-1}\left\{\begin{pmatrix} n \\ m \end{pmatrix}(b_1)^m \frac{d^{(n-m)}\theta}{dE^{(n-m)}}\right\}\right] \tag{8.74}$$

Using Eq. 8.64, we can evaluate the $R_{p,\mathrm{NL}}$ as a function of E_{ac0}, and the results are presented as 'analytical' in Figure 8.6. The series in Eq. 8.64 was truncated to 50 terms, and R_{sol} was assumed to be zero. Also, the impedance at 1 mHz was calculated, and the magnitude was presented as 'numerical' in Figure 8.6, and it is seen that both methods predict $R_{p,\mathrm{NL}}$ equally well.

It is worth noting that if impedance data were acquired immediately after the perturbation is applied, then the resulting spectra differ from the one acquired after stabilization of θ_{av}, and examples are shown in Figure 8.6. In the case of small E_{ac0}, the difference is negligible (Figure 8.7a), whereas, when E_{ac0} is large, the difference is significant (Figure 8.7b). If the data were acquired before steady periodic currents

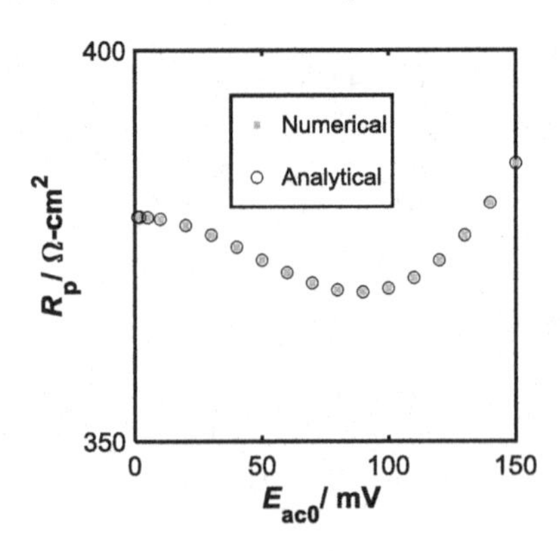

FIGURE 8.6 Polarization resistance, from the magnitude of EIS at 10^{-3} Hz, and analytical expression given in Eq. 8.64, truncated at 100 terms. The reaction mechanism is given in Eq. 8.47. $E_{dc} = 0.2$ V *vs.* OCP, $k_{10} = 10^{-9}$ mol cm^{-2}s^{-1}, $b_1 = 19.47$ V^{-1}, $k_2 = 10^{-8}$ mol cm^{-2}s^{-1}, and $\Gamma = 10^{-9}$ mol cm^{-2}. Note that the results by both methods overlap and are practically indistinguishable.

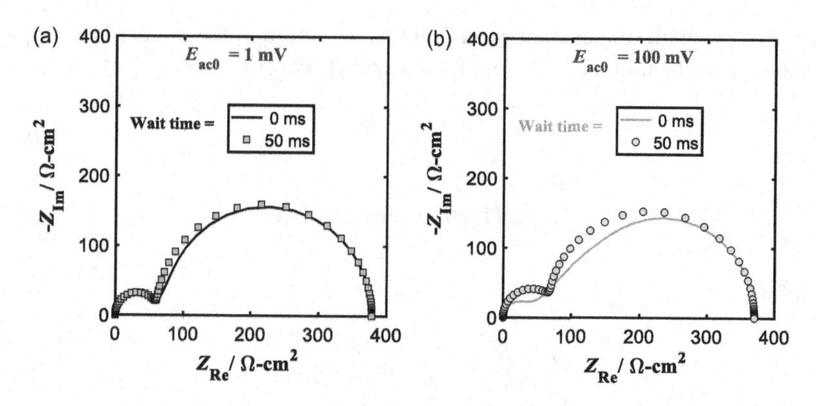

FIGURE 8.7 Complex plane plots of nonlinear impedance spectra obtained immediately after application of perturbation, and after steady periodic response (i.e., after 50 ms) (a) $E_{ac0} = 1$ mV and (b) $E_{ac0} = 100$ mV. The parameters are the same as those employed in Figure 8.4.

manifest, the resulting spectra would not be KKT compliant (Kaisare et al. 2011), since the stability criterion is violated.

Nonlinear EIS of a three-step mechanism with two adsorbed intermediates and a catalytic mechanism have been simulated (Kaisare et al. 2011, Santhanam, Ramani, and Srinivasan 2012, Fasmin and Srinivasan 2015). The conclusions drawn from the above simulations were found to apply to all those cases as well.

8.5 SIMULATION NLEIS UNDER GALVANOSTATIC CONDITIONS

So far, we have assumed that the data are acquired under potentiostatic mode. The experiment can also be conducted under galvanostatic mode. When small-amplitude perturbations are applied, both modes will yield the same result. However, when large amplitude perturbations are employed, it is not possible to make one to one comparison of the results of potentiostatic and galvanostatic NLEIS, as shown below.

Under galvanostatic mode, an ac current of fixed amplitude and a given frequency will be superimposed on a dc current, and the potential across the electrode is measured. From this, the impedance value at that frequency is calculated. This is repeated at different frequencies, and thus, the impedance spectrum is acquired. The analysis for galvanostatic mode for a reaction with an adsorbed intermediate is presented below.

Consider the two-step reaction analyzed in the previous section.

$$\begin{aligned} &M \xrightarrow{k_1} M^+_{ads} + e^- \\ &M^+_{ads} \xrightarrow{k_2} M^+_{sol} \end{aligned} \tag{8.75}$$

Now, instead of applying an ac potential on top of the dc potential, we apply an ac current on top of the dc current. Then the total current is

$$i = i_{dc} + i_{ac} = i_{dc} + i_{ac0}\sin(\omega t) \tag{8.76}$$

If R_{sol} is very small, the current essentially passes through the double-layer capacitance ($i_{C_{dl}}$) and the faradaic reaction (i_F) in parallel. We can write

$$i_{C_{dl}} = C_{dl} \frac{dE}{dt} \quad (8.77)$$

$$i_F = Fk_1(1-\theta) \text{ where } k_1 = k_{10}\, e^{b_1 E} \quad (8.78)$$

Therefore,

$$C_{dl} \frac{dE}{dt} + Fk_1(1-\theta) = i_{dc} + i_{ac0} \sin(\omega t), \quad (8.79)$$

This equation, along with mass balance, Eq. 8.48, can be rearranged as

$$\frac{dE}{dt} = \frac{\left(i_{dc} + i_{ac0} \sin(\omega t)\right) - Fk_{10}e^{b_1 E}(1-\theta)}{C_{dl}} \quad (8.80)$$

$$\frac{d\theta}{dt} = \frac{1}{\Gamma}\left(k_{10}e^{b_1 E}(1-\theta) - k_2\theta\right) \quad (8.81)$$

These two equations must be solved simultaneously to obtain the potential E and fractional surface coverage θ.

The initial conditions are

$$E\big|_{t=0} = E_{dc} \text{ and } \theta\big|_{t=0} = \theta_{ss} \quad (8.82)$$

To find the initial conditions, Eqs. 8.80 and 8.81 should be set to zero, and the two nonlinear algebraic equations must be solved simultaneously. In general, this will require numerical methods, but in this particular case, an analytical solution can be written as

$$\theta_{ss} = \left(\frac{i_{dc}}{F}\right)\frac{1}{k_2} \quad (8.83)$$

$$E_{dc} = \frac{1}{b_1} In\left(\frac{i_{dc}}{Fk_{10}(1-\theta_{ss})}\right) \quad (8.84)$$

The potential can be expressed in the Fourier series as

$$E = E_0 + E_1 \sin(\omega t + \psi_1) + E_2 \sin(2\omega t + \psi_2) + \cdots + E_n \sin(n\omega t + \psi_n) + \cdots \quad (8.85)$$

The notation must be carefully marked. Under potentiostatic mode, the alternating potential is given by $E_{ac} = E_{ac0}\sin(\omega t)$. The magnitude of E_{ac} is given by E_{ac0} and is fixed by the user. The current response is written as

$$i = i_0 + i_1 \sin(\omega t + \phi_1) + i_2 \sin(2\omega t + \phi_2) + \cdots, \quad (8.86)$$

The magnitude of i_{ac} is calculated by subjecting the measuring current to FFT and is written as i_1.

Under small signal conditions, Eq. 8.86 could be written as

$$i = i_{dc} + i_{ac0} \sin(\omega t + \phi_1) \tag{8.87}$$

but under large signal conditions, we limit the usage to i_1, i_2, and so on in the potentiostatic mode.

In the galvanostatic mode, the ac current chosen by the user is written as $i_{ac} = i_{ac0} \sin(\omega t)$.

The potential measured is given by Eq. 8.85. Under small-signal conditions, one could write E_1 as E_{ac0}, but under large-signal conditions, we limit the usage to E_1, E_2, and so on in the galvanostatic mode.

Equations 8.80 and 8.81 are solved for a particular set of parameters, and then the ratio of E_1 to i_{ac0} is recorded as nonlinear impedance magnitude. The phase of the nonlinear impedance is given by ψ_1. The remaining terms in Eq. 8.85 would be recorded as higher harmonics.

The complex plane plots of NLEIS response simulated under galvanostatic conditions, for a small i_{ac0} and a large i_{ac0}, are presented in Figure 8.8. Recall that when a sinusoidal potential is superimposed on a dc potential, the response takes some time to stabilize, and that steady periodic response is obtained only after some time. Likewise, under galvanostatic conditions, too, we need to wait for some time to record a steady periodic response. The results in Figure 8.8 are obtained after a sufficient wait time.

The relationship between potential and current is $E = iZ$, and, in general, the magnitude and phase of Z vary with frequency. Under potentiostatic NLEIS measurement, E_{ac0} is fixed, and the magnitude and phase of i_1 will vary with frequency. On the other hand, under galvanostatic NLEIS measurement, i_{ac0} is fixed, and E_1 will vary with frequency. Thus, it is not possible to identify a current perturbation amplitude (i_{ac0}) corresponding to a potential perturbation amplitude (E_{ac0}) for NLEIS measurement.

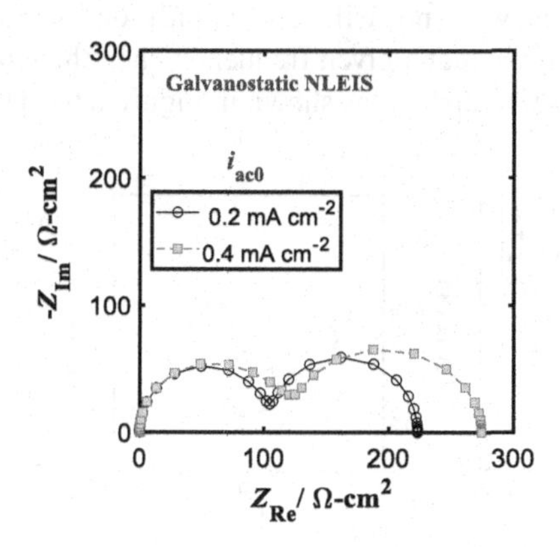

FIGURE 8.8 Complex plane plots of impedance spectra, simulated under galvanostatic conditions. The reaction mechanism is given in Eq. 8.75. $i_{dc} = 0.5\,mA\ cm^{-2}$, $k_{10} = 10^{-9}$mol cm^{-2}s^{-1}, $b_1 = 19.47\,V^{-1}$, $k_{20} = 10^{-8}$mol cm^{-2}s^{-1}, $\Gamma = 10^{-9}$mol cm^{-2}, and $C_{dl} = 10^{-5}$ F cm^{-2}.

At this stage, we note that a numerical solution appears to be the only viable method for simulating galvanostatic NLEIS response. While it is possible to write faradaic current density as an explicit function of potential, for most reactions, it is not possible to write potential as an explicit function of current density. Therefore, in most cases, the Taylor series expansion or Fourier series expansion methods cannot be employed to predict the NLEIS response of electrochemical systems under galvanostatic conditions.

8.6 SIMULATION OF INSTABILITY IN ELECTROCHEMICAL SYSTEMS

One of the assumptions in EIS data analysis is that the experimental system is stable. When the applied perturbation is removed, the system will come back to the original state. Often, it is found that the experimental system characteristics change over time; e.g., the open-circuit potential of a corroding metal may change over time, due to the change in surface roughness, or temperature, or in the active species concentration in solution. All efforts must be taken to ensure that the electrochemical system is operated under stable conditions for EIS measurements; however, it is not always possible to do so, and one can use numerical methods to predict the EIS response of an unstable system, as illustrated here.

Consider an electrochemical system, where the open-circuit potential (OCP) initially changes rapidly and later changes slowly. This can be modeled using an exponential change.

$$E_{\Delta} = E_{\text{change}}\left(1 - e^{-bt}\right) \tag{8.88}$$

Here, E_{change} is the maximum change and it can be positive or negative. In contrast, the parameter 'b' indicates how fast or slow the change is and is restricted to positive values only. Usually, when a dc bias is applied, the equipment will apply the bias against a reference electrode. Then the potential bias, measured vs. OCP, will drift over time. If there were no drift, and if only one sinusoidal perturbation is applied for EIS measurement at a given frequency, then the temporal form of potential wave perturbations will appear, as shown in Figure 8.9a. The drift itself is shown

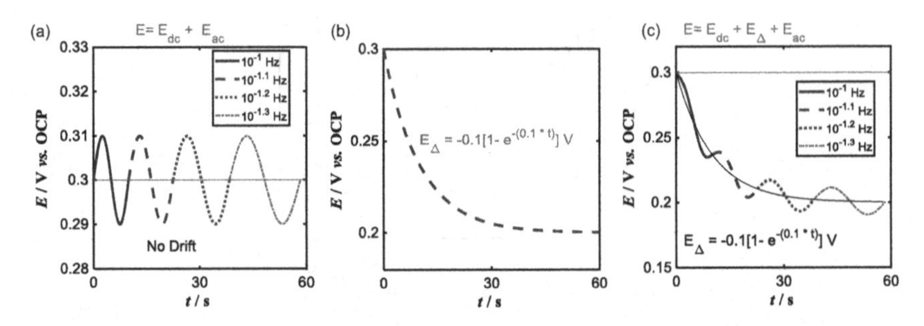

FIGURE 8.9 Temporal profile of (a) sinusoidal potential waves at various frequencies, applied in sequence when there is no potential drift in the system, (b) potential drift alone, and (c) sinusoidal potential waves at various frequencies, applied in sequence, during the potential drift.

in Figure 8.9b. When a sequence of sinusoidal potential waves of various frequencies is superimposed on the dc potential, the temporal profile of the potential *vs.* OCP will appear, as shown in Figure 8.9c.

As an example, the mechanism described in Eqs. 8.89 and 8.90 is analyzed.

$$\mathrm{M} \xrightarrow{k_1} \mathrm{M}^{+}_{\mathrm{ads}} + \mathrm{e}^{-} \tag{8.89}$$

$$\mathrm{M}^{+}_{\mathrm{ads}} \xrightarrow{k_2} \mathrm{M}^{2+}_{\mathrm{sol}} + \mathrm{e}^{-} \tag{8.90}$$

Note that the second reaction is also an electrochemical reaction. The mass balance equation is the same as Eq. 8.48

$$\Gamma \frac{\mathrm{d}\theta}{\mathrm{d}t} = k_{10}\mathrm{e}^{b_1 E}(1-\theta) - k_2\theta \tag{8.91}$$

whereas the faradaic current is given by

$$i_F = F\left(k_1(1-\theta) + k_2\theta\right) \tag{8.92}$$

At each frequency, the starting dc bias is different from the original dc bias (E_{dc}), and correspondingly, the starting fractional surface coverage is also different from the original fractional surface coverage (θ_{ss}). During the sinusoidal perturbation, the potential drift is combined with the sinusoidal potential and Eq. 8.91 is integrated to obtain θ as a function of time. It must be noted that since the potential drifts, the rate constants k_1 and k_2 also drift, and the current can be calculated by substituting θ and k's as a function of time, in Eq. 8.92. The resulting current array can be subjected to FFT to calculate the response at fundamental, and hence the impedance.

When there is no drift, the complex plane of the impedance spectrum shows two loops (Figure 8.10a) for the set of kinetic parameters given in the figure caption. At low frequencies, an inductive loop is observed while at high frequencies, a capacitive loop is seen. If the OCP increases with time (i.e., E_Δ is negative), then the resulting spectrum is shown as a complex plane plot in Figure 8.10b. At low frequencies,

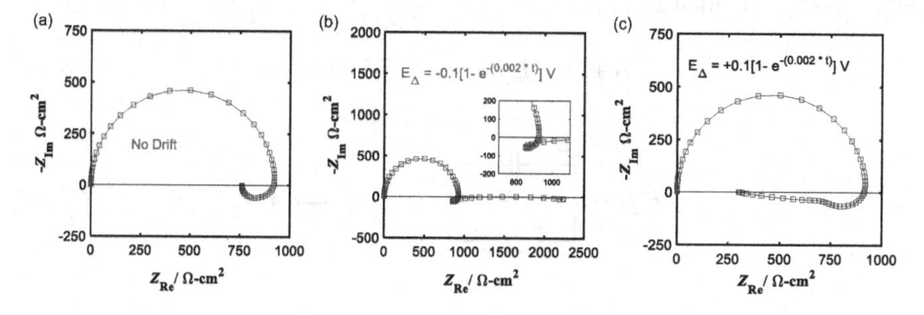

FIGURE 8.10 Complex plane plot of impedance spectra of a reaction system described in Eqs. 8.89 and 8.90, (a) without drift, (b) with OCP increasing with time, (c) with OCP decreasing with time. Starting dc bias $E_{dc} = 0.3$ V *vs.* OCP. $E_{ac0} = 10$ mV, $k_{10} = 10^{-11}$ mol cm^{-2}s^{-1}, $b_1 = 15$ V^{-1}, $k_{20} = 10^{-10}$mol cm^{-2}s^{-1}, $b_2 = 5$ V^{-1}, $\Gamma = 3 \times 10^{-9}$mol cm^{-2}, and $C_{dl} = 10^{-5}$ F cm^{-2}.

another loop-like structure manifests. The data appear like a greatly depressed capacitive loop. If not analyzed correctly, this could lead to incorrect interpretations, e.g., the underlying mechanism could be thought of as a reaction with at least two adsorbed intermediates.

Figure 8.10c shows the complex plane plot of the impedance spectrum obtained when the OCP decreases with time and E_Δ is positive. Now, the low-frequency data appear as a weak inductive loop, i.e., in the fourth quadrant of the plot of $-Z_{\mathrm{Im}}$ *vs.* Z_{Re}. If the data were analyzed assuming that the systems were stable, the drifts would lead to incorrect estimation of kinetic parameters and polarization resistance.

At high frequencies, the contribution of the faradaic current to the total impedance is small, and only double-layer capacitance determines the net impedance. At low frequencies, the contribution of the double-layer charging current to the total current is negligible, and the faradaic current determines the net impedance. Usually, during EIS measurements, high-frequency data would be acquired first, and low-frequency data would be acquired later. By the time low-frequency data acquisition begins, the system would have drifted close to the final condition, and a further change in potential is relatively small. Yet, the duration of data acquisition at low frequencies is more, and hence it has a significant effect on the low-frequency impedance data. The numerical methodology can be applied to simulate the effect of instability on impedance spectra of other reactions (Victoria and Ramanathan 2011), as well as harmonics (Pachimatla et al. 2019).

8.7 INCORPORATION OF SOLUTION RESISTANCE EFFECTS IN NLEIS SIMULATIONS

When solution resistance is not negligible, we should include a resistor to model the electrochemical system (Figure 8.11). Under galvanostatic mode, incorporation of solution resistance effects in NLEIS simulations is straightforward. The potential across the electrode–electrolyte interface can be simulated without considering the solution resistance, and the impedance of the interface (a capacitor in parallel with the faradaic process) should be added to the potential across the solution, to obtain the total potential change as a function of time.

Under potentiostatic conditions, the governing equations of the circuit shown in Figure 8.11 are written as

$$E_1 + E_2 = E = E_{\mathrm{dc}} + E_{\mathrm{ac0}} \sin(\omega t) \tag{8.93}$$

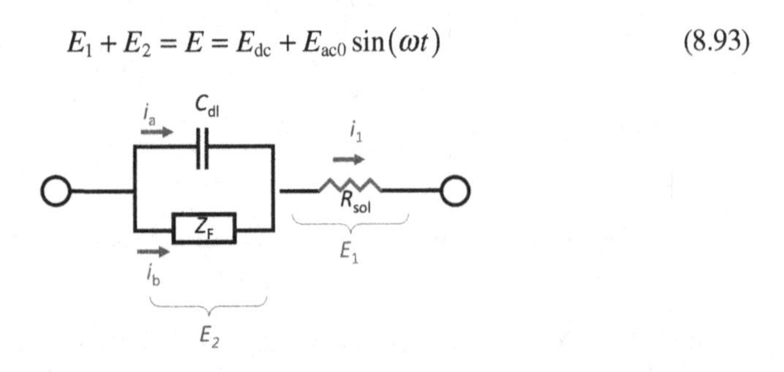

FIGURE 8.11 Equivalent circuit of an electrochemical system, with an electrode–electrolyte interface and significant solution resistance.

$$i_a + i_b = i_1 \tag{8.94}$$

$$i_a = C_{\text{dl}} \frac{\text{d}E_2}{\text{d}t} \tag{8.95}$$

$$i_1 = \frac{E_1}{R_{\text{sol}}} \tag{8.96}$$

$$i_b = \frac{E_2}{Z_F} \tag{8.97}$$

Note that the faradaic impedance Z_F relates E_2 and i_b. For a simple irreversible electron transfer reaction, the faradaic current would be written as $i_b = Fk_{10}e^{b_1 E_2}$.

Rearranging Eqs. 8.93–8.97, for an irreversible simple electron transfer reaction, we get

$$C_{\text{dl}} \frac{\text{d}E_2}{\text{d}t} + Fk_{10}e^{b_1 E_2} = \frac{(E - E_2)}{R_{\text{sol}}} \tag{8.98}$$

The initial condition can be obtained by setting Eq. 8.98 to zero and solving for E_2 to get $E_{2\text{dc}}$. Integration of Eq. 8.98 and substitution in Eqs. 8.93 and 8.96 would yield the current (i_1), from which the impedance can be calculated. The results show that if solution resistance is not zero, then, even for a simple electron transfer reaction, the response stabilizes only after some time. Hence, some wait time is necessary before data acquisition (Santhanam, Ramani, and Srinivasan 2012). The time it takes to stabilize depends on the value of R_{sol}, C_{dl}, and kinetic parameters. If spectra are acquired before stabilization, the result will be flagged as non-compliant by KKT.

In general, when the solution resistance is large, and the applied potential is sinusoidal, the potential across the electrode–electrolyte interface is not sinusoidal. It can be viewed as distorted sinusoidal, and it has higher harmonic components. The extent of distortion depends on the degree of non-linearity of the faradaic process and the applied sine wave amplitude and frequency. A part of the potential drop occurs at the electrode–electrolyte interface, and the rest occurs across the solution. What fraction of the applied potential drop occurs at the electrode–electrolyte interface depends on the applied frequency and the system variables such as the kinetic parameters, solution resistance, and the applied dc bias (Figure 8.12).

The impedance spectra, calculated for two different E_{ac0} values and a few R_{sol} values, are shown as complex plane plots in Figure 8.13. Note that the abscissa shows (Z_{Re}-R_{sol}) instead of Z_{Re}, for ease of comparison.

If the effect of solution resistance is to only shift the complex plane plot to the right side in the complex plane plot, then all the spectra in Figure 8.13a would be identical. In this particular case, the charge-transfer resistance (R_t) and the polarization resistance (R_p) are the same, since the faradaic impedance is independent of the perturbation frequency. Figure 8.13a clearly shows that when R_{sol} increases, the 'measured' or 'apparent' polarization resistance (R_p) increases. When R_{sol} is of the same order of magnitude as the measured R_p, the effect is large. A comparison of Figure 8.13a and b

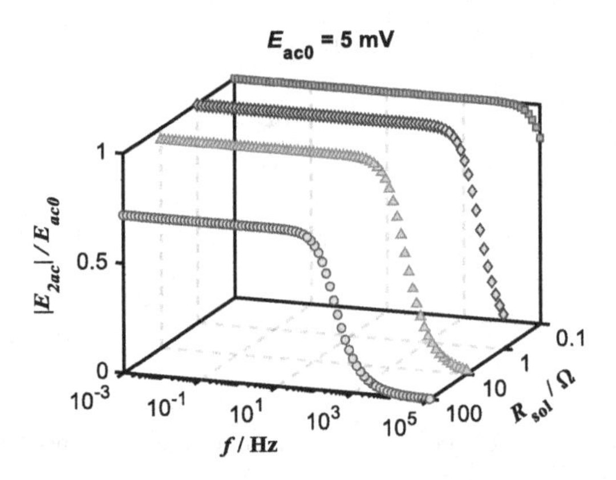

FIGURE 8.12 Fraction of ac perturbation that occurs across the electrode–electrolyte interface, for various values of R_{sol}. A simple irreversible electron transfer reaction is considered. E_{dc} =0.4 V *vs.* OCP, $k_{10} = 10^{-12}$mol cm^{-2}s^{-1}, $b_1 = 20$ V^{-1}, and $C_{dl} = 10^{-5}$ F cm^{-2}. (Adapted with permission from Springer Nature Customer Service Centre GmbH, Springer, *Journal of Solid State Electrochemistry*, Numerical investigations of solution resistance effects on nonlinear electrochemical impedance spectra, Sruthi Santhanam, Vimala Ramani and Ramanathan Srinivasan, Copyright (2011).)

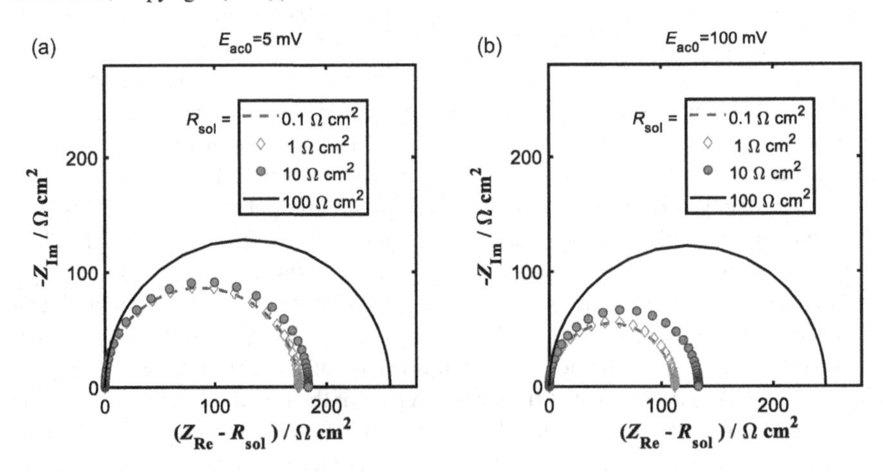

FIGURE 8.13 Complex plane plots of the impedance spectra for the electron transfer reaction with E_{ac0} = (a) 5 mV and (b) 100 mV. The parameter values are k_{10} =10^{-12}mol cm^{-2}s^{-1}, $b_1 = 20$ V^{-1}, and $C_{dl} = 10$ μF cm^{-2}. R_{sol} was varied from 0.1 to 100 Ω cm^2. Note that in the abscissa, the R_{sol} is subtracted from Z_{Re} to enable a clear comparison between the spectra. (Adapted with permission from Springer Nature Customer Service Centre GmbH, Springer, *Journal of Solid State Electrochemistry*, Numerical investigations of solution resistance effects on nonlinear electrochemical impedance spectra, Sruthi Santhanam, Vimala Ramani and Ramanathan Srinivasan, Copyright (2011).)

shows that when R_{sol} is small, then an increase in E_{ac0} from 5 to 100 mV results in a decrease in R_p. On the other hand, when R_{sol} is large (100 Ω cm^2), the 'apparent' R_p does not change to a large extent; this would suggest that the system response is linear when the perturbation amplitude is varied from 5 to 100 mV. However, the fact is that the system response is distorted by large solution resistance, and the distorted spectra happen to be very similar, thus leading to an incorrect conclusion.

It is possible to predict the effect of R_{sol} on more complex reactions. Consider the reaction shown in Eq. 8.99

$$\begin{aligned} &M \xrightarrow{k_1} M^+_{ads} + e^- \\ &M^+_{ads} \xrightarrow{k_2} M^+_{sol} \end{aligned} \tag{8.99}$$

The mass balance equation is

$$\Gamma \frac{d\theta}{dt} = k_1(1-\theta) - k_2\theta \tag{8.100}$$

The total current balance is

$$C_{dl}\frac{dE_2}{dt} + \left[Fk_{10}e^{b_1E_2}(1-\theta)\right] = \frac{E-E_2}{R_{sol}} \tag{8.101}$$

where E_2 is the potential drop across the electrode–electrolyte interface. We assume that mass transfer is rapid.

The steady-state values of E_2 and θ can be obtained by setting the above two equations to zero. The resulting nonlinear algebraic equations (Eqs. 8.102 and 8.103) can be solved simultaneously using numerical methods.

$$\theta_{ss} = \frac{k_{1\text{-dc}}}{k_{1\text{-dc}} + k_2} \tag{8.102}$$

$$\frac{E_{dc} - E_{2dc}}{R_{sol}} = Fk_{1\text{-dc}}(1-\theta) \tag{8.103}$$

When a sinusoidal potential (E_{ac}) is applied, Eqs. 8.100 and 8.101 have to be integrated numerically, with the initial conditions: at $t = 0$, $E_2 = E_{2dc}$, and $\theta = \theta_{ss}$. Initially, the average fractional surface coverage and average E_2 will drift, and after some time, steady periodic oscillations will ensue. The resulting potential (E_2) as a function of time can be used to calculate E_1 using Eq. 8.93, and subsequently, the current (i_1) using Eq. 8.96. The current should be subjected to Fourier transform to obtain the magnitude and phase at fundamental (as well as at higher harmonics, if required), and the impedance can be calculated as the vector ratio of applied potential (E_{ac}) to the resulting current at fundamental.

The results of such a simulation, for a given set of kinetic parameters and dc bias, are presented as complex plane plots in Figure 8.14, for a small and large value of R_{sol}. Note that the abscissa shows (Z_{Re}-R_{sol}) rather than Z_{Re}. When the solution resistance is small, increasing the perturbation amplitude from 5 to 200 mV causes a slight

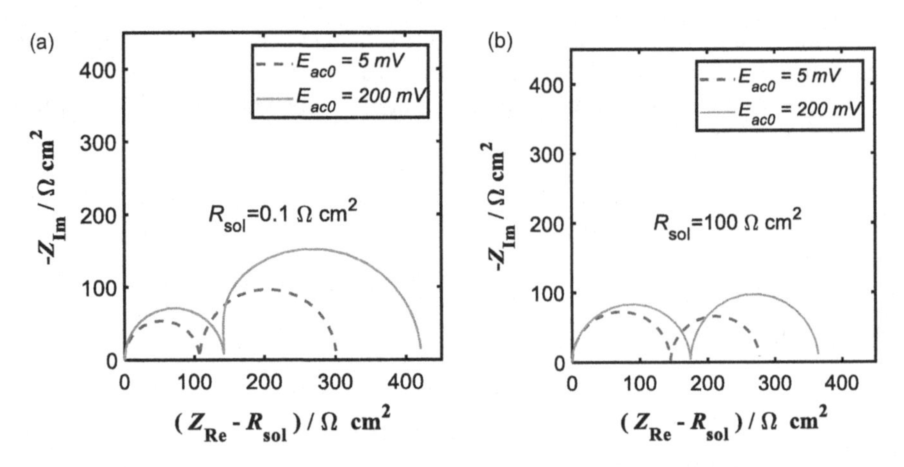

FIGURE 8.14 Solution resistance effect. A complex plane plot of impedance spectra corresponding to the mechanism shown in Eq. 8.99. $k_{10} = 10^{-11}$ mol cm^{-2}s^{-1}, $b_1 = 15$ V^{-1}, $k_2 = 10^{-8}$ mol cm^{-2}s^{-1}, $\Gamma = 10^{-7}$ mol cm^{-2}, $E_{dc} = 0.5$ V *vs.* OCP, and $C_{dl} = 10$ μF cm^{-2}. R_{sol} = (a) 0.1 Ω cm² and (b) 100 Ω cm². Note that in the abscissa, the R_{sol} is subtracted from Z_{Re} to enable clear comparison between the spectra. (Adapted with permission from Springer Nature Customer Service Centre GmbH, Springer, *Journal of Solid State Electrochemistry*, Numerical investigations of solution resistance effects on nonlinear electrochemical impedance spectra, Sruthi Santhanam, Vimala Ramani, and Ramanathan Srinivasan, Copyright (2011).)

increase in R_t and a large increase in R_p, as seen in Figure 8.14a. The second loops, i.e., at mid and low frequencies, appear as a semicircle at 5 mV and as a distorted semicircle at 200 mV, due to nonlinear effects. On the other hand, when the solution resistance is large (100 Ω cm²), a similar increase in E_{ac0} causes a slight increase in R_t and a moderate increase in R_p. The nonlinear effect (i.e., the difference between the results of 5 and 200 mV perturbations) is relatively less at higher R_{sol} because the actual potential at the interface is reduced. The effects of E_{ac0} on R_t and R_p are strongly influenced by the R_{sol} value and cannot be predicted *a priori* and has to be calculated by solving the actual governing equations, usually by numerical methods. This methodology can be easily extended to other mechanisms (Santhanam, Ramani, and Srinivasan 2012).

8.8 FRUMKIN ISOTHERM – SIMULATION OF THE IMPEDANCE RESPONSE

In most of the reports employing complex reaction mechanisms, it is assumed that the Langmuir model can describe the adsorption isotherm of the intermediate species. When the intermediate species are charged, the Frumkin model offers a better description of the physical phenomenon than the Langmuir model, but then it is also a more complex model; with the Frumkin model, calculating the polarization current itself would require the use of numerical methods. The governing equations can be solved numerically to simulate the impedance response, as illustrated here.

Consider the following two-step mechanism with an adsorbed intermediate,

$$\begin{aligned} &M \xrightarrow{k_1} M^{+}_{ads} + e^{-} \\ &M^{+}_{ads} \xrightarrow{k_2} M^{+}_{sol} \end{aligned} \tag{8.104}$$

The mass balance equation is

$$\Gamma \frac{d\theta}{dt} = k_1(1-\theta) - k_2\theta \tag{8.105}$$

and the charge balance equation is

$$i_F = Fk_1(1-\theta) \tag{8.106}$$

For this illustration, we assume that the solution resistance is negligible and that mass transfer is rapid. It is easy to incorporate solution resistance effects, as described in Section 8.7, although incorporating mass transfer resistance is a lot more challenging.

When a reversible step describes the adsorption of a species, the Frumkin model proposes that the equilibrium constant is a function of the fractional surface coverage (Gileadi 2011). Specifically, the standard Gibbs energy of adsorption is assumed to be linearly dependent on the surface coverage. The isotherm model is given by

$$\frac{\theta}{1-\theta} e^{\left(\frac{r\theta}{RT}\right)} = \frac{\theta}{1-\theta} e^{(g\theta)} = K_{eq0} e^{\left(\frac{FE}{RT}\right)} \tag{8.107}$$

where r is the parameter describing the rate of change of Gibbs energy with fractional surface coverage and g is the dimensionless Frumkin parameter given by $g = \frac{r}{RT}$. When the adsorbed species are charged, the interactions will be repulsive, and r will be a positive number, with typical values between 20 and 60 kJ mol^{-1} (Gileadi 2011).

Although the Frumkin model defines the form of the equilibrium constant and not the individual rate constants, the reaction rate constants are assumed to depend on the surface coverage in the same way that the equilibrium constant depends on the surface coverage, i.e.,

$$k_1 = k_{10} e^{b_1 E} e^{\beta_1 g\theta} \tag{8.108}$$

where β_1 is the coverage parameter and is a negative number since it corresponds to the reaction producing the intermediate species. Now, even though the mechanism described in Eq. 8.104 employs only irreversible steps, the rate constant of the first step can be described by Eq. 8.108 and the rate constant of the second step can be written as

$$k_2 = k_{20} e^{\beta_2 g\theta} \tag{8.109}$$

β_2 is the coverage parameter and is a positive number since it corresponds to a reaction consuming the intermediate species. Note that the second step does not involve electron transfer, and hence k_2 is independent of potential.

The steady-state values of the surface coverage are obtained by substituting Eqs. 8.108 and 8.109 in Eq. 8.105 and setting it to zero, i.e.,

$$k_{10}e^{b_1 E_{dc}} e^{\beta_1 g \theta_{ss}} (1-\theta_{ss}) - k_{20}e^{\beta_2 g \theta_{ss}} \theta_{ss} = 0 \quad (8.110)$$

For a given E_{dc}, Eq. 8.110 can be solved using numerical methods to obtain θ_{ss}. When an ac potential is superimposed on a dc potential, Eq. 8.105 becomes

$$\frac{d\theta}{dt} = \frac{1}{\Gamma}\left[k_{10}e^{b_1 (E_{dc}+E_{ac0}\sin(\omega t))} e^{\beta_1 g \theta} (1-\theta) - k_{20}e^{\beta_2 g \theta} \theta\right] \quad (8.111)$$

Equation 8.111 should be numerically integrated, with the initial condition, at $t = 0$, $\theta = \theta_{ss}$. The solution of this equation will be the time-dependent surface coverage $\theta(t)$, and an example is shown in Figure 8.15a. Initially, the average fractional surface coverage drifts and later steady periodic results are obtained. After that, the oscillations in the surface coverage as a function of time $\theta(t)$ can be recorded. This series and the time-dependent reaction rate constant $k_{10}e^{b_1 (E_{dc}+E_{ac0}\sin(\omega t))} e^{\beta_1 g \theta}$ are substituted in Eq. 8.106 to obtain the transient current. The current $i(t)$ is subjected to FFT to obtain the response at fundamental and higher harmonics, and hence the impedance.

It is possible to linearize Eqs. 8.111 and 8.106 to obtain an analytical solution for $\frac{d\theta}{dE}$ and Z_F, as shown in Section 5.1.9, but the results will be valid only for small E_{ac0} values. The numerical methods illustrated here can be used for small as well

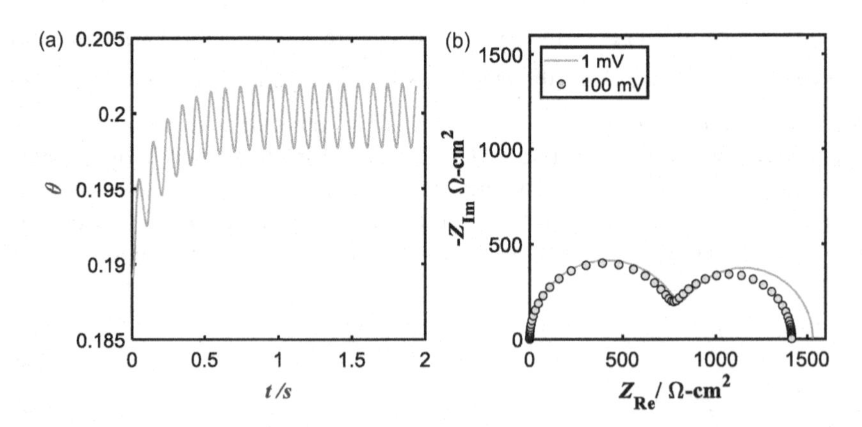

FIGURE 8.15 (a) Fractional surface coverage as a function of time for an electrochemical reaction described by Eq. 8.104. The parameter values are $k_{10} = 5 \times 10^{-10}$mol cm^{-2}s^{-1}, $b_1 = 15\,V^{-1}$, $k_2 = 10^{-10}$mol cm^{-2}s^{-1}, $\beta_1 = -1$, $\beta_2 = 1$, g = 20, $\Gamma = 10^{-8}$mol cm^{-2}, $C_{dl} = 10$ µF cm^{-2}, $E_{dc} = 0.3$ V *vs.* OCP, and $E_{ac0} = 100$ mV. (b) Complex plane plots of impedance spectra at two perturbation amplitudes. Results for a different set of parameters are shown in Victoria and Ramanathan (2011).

as large perturbation amplitudes, and information on fundamental as well as higher harmonics can be obtained.

Numerical methods have dramatically expanded the scope of the physical phenomena that can be simulated. Both time-domain response and frequency-domain response can be obtained by simulation, offering new and interesting insights. While the computational effort, compared to that for simulation of a linearized version of EIS, is relatively more, numerical methods have freed us from the limitations such as small amplitude perturbation, Langmuir adsorption isotherm, stable system, zero solution resistance, and potentiostatic mode. Now, the EIS response of electrodes undergoing reactions with adsorbed intermediates obeying more realistic isotherms such as the Frumkin or Temkin model can be handled. Large amplitude perturbations, solution resistance effects, and instabilities such as drift in potentials or concentrations of adsorbed species or dissolved species can be incorporated into the model. The response under multi-sine perturbations and EIS under galvanostatic mode can be simulated easily and accurately.

The main challenge at this point is the large time complexity (i.e., amount of computer time) required to perform the numerical calculations. In the year 2020, while an impedance spectrum using analytical expressions can be calculated in a few milliseconds in modern desktop computers, numerical simulations take several minutes to produce the same result. This is a serious shortcoming of numerical simulation of NLEIS, particularly when simulated results have to be compared with experimental data, and optimization of the parameters is required. With further improvement in computational speed and algorithms, it is hoped that this difficulty can be overcome and that the full potential of the numerical simulations can be realized.

8.9 EXERCISE – NONLINEAR EIS

- Q8.1 Plot $e^{a \sin\theta}$ and the Taylor series expansion (up to four terms) for θ varying from 0 to 4π, with $a = 3$. Explore the effect of varying 'a.'
- Q8.2 Expand sin(x) in the Taylor series of 'x' and truncate after three terms. Compare the values of sin(x) with those obtained from the truncated series, for x varying from 0 to 2π. Repeat this for a series truncated after five terms. List your observations.
- Q8.3 Plot $I_0(x)$ and $0.2 \times e^x$, for x varying from 0.01 to 30 and compare. Similarly, plot $I_1(x)$ and compare with $0.2 \times (e^x - 1)$. The ordinate scale should be logarithmic.
- Q8.4 From the definition of $I_n(x)$ given in Eq. 8.15, derive Eqs. 8.16–8.18.
- Q8.5 Derive Eq. 8.38.
- Q8.6 Generate the plots shown in Figure 8.4, at a few other frequencies and amplitudes. Explore the effect of changing kinetic parameters, especially Γ, on the stabilization time.
- Q8.7 Derive Eq. 8.60 from Eqs. 8.57–8.59.
- Q8.8 Derive Eq. 8.72 from Eqs. 8.48, 8.67, and 8.68.
- Q8.9 Derive Eq. 8.74.
- Q8.10 Using Eqs. 8.53 and 8.55, calculate $R_{t,\,\mathrm{NL}}$ with parameters shown in Figure 8.4.

Q8.11 Using Eq. 8.64, calculate $R_{p,\,\mathrm{NL}}$ shown in Figure 8.6. Change the kinetic parameter values and the dc bias and explore their effect on $R_{p,\,\mathrm{NL}}$.

Q8.12 For the mechanism given in Eq. 8.1, write the governing equation and initial condition for the potential, under galvanostatic conditions. An analytical solution is possible if the perturbation and response amplitudes are very small.

Q8.13 For the set of kinetic parameters given in Figure 8.15, find θ_{ss} as a function of E_{dc} from 0 to 0.5 V. Explore the effect of changing the parameter 'g.'

Appendix 1

Complex Numbers Refresher

1. $j=\sqrt{-1}$ is the imaginary number. Normally, the imaginary number is denoted by i, but since the current density is also denoted by i, we use j to denote the imaginary number. MATLAB® accepts both i and j as the imaginary numbers.
2. A complex number has a real part 'x' and an imaginary part 'y' and can be written as $z=x+jy$ in the Cartesian coordinate system.
3. A complex number can also be written as $z=re^{j\theta}$ in the polar coordinate system.
4. The transformation between the two coordinate systems is given by $r=\sqrt{x^2+y^2}$, $\theta=\tan^{-1}\left(y/x\right)$, and $x=r\cos\theta$ and $y=r\sin\theta$.
5. The complex conjugate of $z=x+jy$ is given by $\overline{z}=x-jy$.
6. De Moivre's theorem: $e^{j\theta}=\cos\theta+j\sin\theta$.
7. $e^z=e^{x+jy}=e^x\left(\cos y+j\sin y\right)$.

Appendix 2
Differential Equations Refresher

1. A differential equation containing only one independent variable is an *ordinary differential equation* (*ode*). An example of an *ode* is $\frac{dE}{dt} + P_1(E) = Q_1(t)$, where t is the independent variable, E is the dependent variable, and P_1 and Q_1 are algebraic functions of E and t, respectively. The symbol d denotes the ordinary derivative. Note that the number of *dependent* variables in an *ode* can be more than one, as described below in item no. 5.
2. If more than one independent variable is present, it is a *partial differential equation* (*pde*). An example of a *pde* is $\frac{\partial^2 E}{\partial x^2} + \frac{\partial E}{\partial t} + P_2(E) = Q_2(t)$. P_2 and Q_2 are algebraic functions of E and t, respectively. The symbol ∂ denotes the partial derivative.
3. An *ode* is *linear* if the solutions of the *ode* can be added together to form another valid solution. An example of a linear *ode* is $\frac{dE}{dt} + K_1 E = Q_3(t)$, where K_1 is a constant and Q_3 is a function of t. An example of a *nonlinear ode* is $\frac{dE}{dt} + K_1 E^2 = Q_3(t)$.
4. If time t is the independent variable and the values of the dependent variables are known at a particular time t_0, then the problem is an *initial value problem* (IVP). An example of IVP is $\frac{dE}{dt} + K_1 E = K_2$ and $E|_{t=0} = K_3$. Here, K_1, K_2, and K_3 are constants.
5. The case where many differential equations have to be solved simultaneously is known as a *system of differential equations*. An example of a *system of odes* is $\frac{d\theta}{dt} = P_1(E,\theta,t)$ and $\frac{dE}{dt} = P_2(E,\theta,t)$, with initial conditions $E|_{t=0} = K_1$ and $\theta|_{t=0} = K_2$. Here, θ and E are the dependent variables.
6. A *system of (linear or nonlinear) odes* can be solved by numerical integration, by converting the differential equations to difference equations and changing the independent variable in small steps. Usually (but not always), if the solution varies rapidly in a region, the *step size* of the independent variable has to be small. If the solution varies slowly, then the step size can be large. Modern mathematical software such as MATLAB® have solvers to handle this seamlessly.

7. In numerical integration, an *ode* is called *stiff* if the solution varies slowly, and yet, the step size of the independent variable has to be small to obtain the correct solution. Certain numerical methods are more suited for handling stiff equations, and other methods are more suited for handling nonstiff equations. An example of MATLAB® implementation of a stiff solver is *ode15s* and that of a nonstiff solver is *ode45*.

Appendix 3
Multi-sine Waves

In FRAs, it is also possible to apply many sine waves (with different frequencies, amplitudes, and phases) simultaneously and measure and resolve the current using Fourier transform. Fourier transform of a periodic signal allows us to resolve the signal into its Fourier series. This is sometimes called as a 'multi-sine' technique. It is not possible to use a multi-sine technique with the lock-in amplifier equipment. The advantage of simultaneously applying many waves is that the total run time of the experiment can be reduced. The disadvantage is that the contribution from 'noise' to the data is higher compared to a single sine case. If the amplitude (E_{ac0}) is increased to a large level, then the nonlinear effects will corrupt the data.

Before one chooses the multi-sine technique in the hope of obtaining EIS data in a short time, one should know what exactly is involved in employing the multi-sine technique. Here we list a few key points, without providing detailed examples.

If we want to apply many sine waves simultaneously, we must choose the frequency, amplitude, and phase of each wave. The frequencies are usually chosen such that they are either 'odd harmonics' or 'log spaced.' Let the frequency of the first wave (base) be f_1. In the case of odd harmonics, the remaining waves are chosen to be the odd harmonics of the first. The period of the combined multi-sine wave will be the same as that of the first wave. When odd harmonics are used, the amplitude of individual waves has to be kept low. Otherwise, nonlinear responses at the lower frequencies will corrupt the results at higher frequencies. However, if the amplitude is reduced too much, the *signal-to-noise* ratio will become poor, and the results will be noisy.

If two or more sine waves, which are not harmonics of the fundamental wave, are added, the resulting wave has a period that is longer than the period of the fundamental. The implication of this is that if the multi-sine wave consists of many nonharmonic waves, then the resulting wave will have a long period, and hence the experiment will need a long data acquisition time. This negates the advantage of using a multi-sine wave.

A parameter called 'crest factor' is important in the synthesis of a multi-sine wave. It is the ratio of peak amplitude to *rms* value. A large crest factor indicates that the peak value is very high, and the nonlinear effects are likely to manifest. To an extent, the phase values of the input sine waves can be adjusted such that the crest factor is low. One option is to employ 'random values' for the phases.

When a multi-sine wave is subjected to FFT, we would like to get the correct amplitude and phase at the respective frequencies and zero amplitude at other frequencies. However, under certain conditions, the calculated amplitude and phase values at the respective frequencies can be incorrect, and the results may also show nonzero amplitude and phase at other frequencies. This is called a spectral leakage. This can arise if (a) the input multi-sine wave is not exactly periodic, which would be because the input waves are not harmonics, or (b) the input waves are not repeated at the right time. The spectral leakage can be reduced (but cannot be eliminated) by modifying the multi-sine wave using a technique called 'windowing.'

Appendix 4
Experiments and Analysis – Few Hints

What is the best way to conduct EIS experiments, obtain good quality data, and analyze it? The only correct answer is 'it depends on the system,' but that is not a satisfactory answer to most questioners. At the risk of simplifying too much, here are a few suggestions.

1. If possible, conduct the experiments in a Faraday cage. This is a fancy name for a metal box in which the electrochemical cell must be kept. The box would normally have a hole to allow the wires to connect the cell to the instrument. The front door is usually hinged but can also be of a sliding type. If a metal box is not available, a wooden box that is covered with a metal mesh will do. This reduces electrical interference from outside.
2. Choose the reference electrode position carefully. It has to be close to the working electrode but not too close. The distance between the working electrode and the reference electrode must be ~two times the diameter of the working electrode.
3. The connecting wires must be electrically shielded. Some equipment manufacturers provide leads that are shielded most of the way. Otherwise, make the shield yourself by covering the wires with metal mesh and by grounding the mesh.
4. Conduct experiments with a single sine wave rather than with a multi-sine wave.
5. Potentiostatic mode is what you should start with. However, if you work in fuel cells or batteries, the galvanostatic mode is usually the norm.
6. Repeat each experiment at least twice and check the spread of the data. If the data is not repeatable, it is not worth anything. Debug the experimental setup to obtain repeatable data.
7. To understand the mechanism of an electrochemical reaction, conduct the experiments for at least a few *dc* potentials – the more, the better in terms of the confidence in your analysis. Also, remember that the more *dc* potentials you use, the more difficult it is to fit the RMA equations to the data. The EEC model fit is done one potential at a time anyway, and hence this difficulty does not arise.
   i. If E_{ac0} is the amplitude used for the experiments, the dc potentials must be such that they differ by at least $2E_{ac0}$. That is to say, if $E_{ac0} = 20\,\text{mV}$, then do not acquire one set of data at $E_{dc} = 0.1\,\text{V}$ and another set at $E_{dc} = 0.11\,\text{V}$. The second E_{dc} must be at least 0.14 V.

   ii. The above recommendation applies to experiments conducted in the potentiostatic mode. For galvanostatic experiments, use the same idea to choose the i_{dc} values.
8. Validate the data using KKT or equivalent methods such as the measurement model analysis or linear KKT.
   i. If data is repeatable but not KKT compliant, then you have either a well-defined, repeatable drift or nonlinear effects.
   ii. Reduce the E_{ac0} and check if the new EIS data is different from the original data. If it is, then nonlinear effects are present; reduce the E_{ac0} further until it has no impact on the EIS data. Then subject the final data to KKT. If the new data is KKT compliant, then nonlinear effects are present in the original data, which could make it KKT noncompliant. On the other hand, if the new data acquired with the lowest E_{ac0} (with an acceptable signal-to-noise ratio) is not KKT compliant, then drift effects are present. They have to be eliminated, and data must be acquired again.
9. First, use equivalent electrical circuits to fit the data. The number of capacitors or inductors you need to fit the data will give you an idea of the type of reaction mechanism and the minimum number of adsorbed species you need to employ.
10. *RMA with Linearized Equations:*
   i. First, fit only the polarization data and use those values as an initial guess for fitting the EIS data.
   ii. While fitting the EIS data, initially, use only data corresponding to 1 dc potential and polarization data and assess if the fit is good. A constrained optimization algorithm can be used, i.e., fit the model to EIS data, with the constraint that the polarization data is within a few % (or a few mA cm^{-2} as appropriate).
   iii. If the fit with EIS data at one potential (and polarization constraint) is good, then add more EIS data (i.e., data at other E_{dc}, one by one) and optimize the parameters. Optimizing the parameters by analyzing all data sets at one go does not work well unless your initial guess of the parameter set is very good.
11. *Analysis with Nonlinear Equations:* It is too time-consuming, and as of May 2020, we are not aware of any work on parameter optimization performed with nonlinear EIS data yet. Hopefully, in a few more years, computers will become faster (and/or parallel computing will be cheaper), and better algorithms may be available so that nonlinear EIS data can be analyzed to extract the kinetic parameters.

Appendix 5
Manufacturers and Suppliers of EIS Equipment

An incomplete list of suppliers of EIS equipment in 2020, in no particular order, is given below. In electrical and electronic engineering, where two terminal connections are common, frequency response analyzers of a wide frequency range (typically a few kHz to a few GHz) are common. However, most electrochemical experiments involve three terminal connections (three-electrode cells), and useful information is often found in the low-frequency range (a few Hz to a few μHz). The suppliers who focus on electrochemical impedance spectra equipment are listed here. The electronics are getting cheaper and more sophisticated every day, and it is likely that, in the future, a greater number of manufacturers would offer instruments to measure EIS.

1. Solartron, UK, now a part of AMETEK, USA
2. Princeton Applied Research (PAR), USA, now a part of AMETEK, USA
3. BioLogic, France
4. Gamry, USA
5. CH Instruments, USA
6. Metrohm, Switzerland
7. eDAQ, Australia
8. BAS Inc., USA
9. IVIUM, Netherlands
10. Gill AC, UK
11. Zahner, Germany
12. Stanford Research System (SRS), USA
13. OrigaLys, France
14. Zivelab, Korea
15. Corrtest Instruments, China
16. Digi-Ivy, USA
17. Kanopy Techno Solutions, India
18. Keithley/Tektronix, USA
19. Japan Analytical Industry Co. Ltd, Japan
20. Bioanalytical Systems Inc., USA
21. WonATech Co Ltd., Korea
22. MicruX Technologies, Spain
23. Bank Elektronik-Intelligent Controls GmbH, Germany

References

Abramowitz, M., and I. A. Stegun. 1972. *Handbook of mathematical functions: With formulas, graphs, and mathematical tables*. New York: Dover Publications.

Agarwal, P., O. D. Crisalle, M. E. Orazem, and L. H. Garcia-Rubio. 1995. Application of measurement models to impedance spectroscopy: II. Determination of the stochastic contribution to the error structure. *Journal of the Electrochemical Society*. 142:4149–4158.

Agarwal, P., M. E. Orazem, and L. H. Garcia-Rubio. 1992. Measurement models for electrochemical impedance spectroscopy: I. Demonstration of applicability. *Journal of the Electrochemical Society*. 139:1917–1927.

Agarwal, P., M. E. Orazem, and L. H. Garcia-Rubio. 1995. Application of measurement models to impedance spectroscopy: III. Evaluation of consistency with the Kramers-Kronig relations. *Journal of the Electrochemical Society*. 142:4159–4168.

Albery, W. J. 1975. *Electrode kinetics*. Oxford: Oxford University Press.

Bard, A. J., and L. R. Faulkner. 1980. *Electrochemical methods: Fundamentals and applications*. New York: Wiley.

Bard, A. J., and M. V. Mirkin. 2012. *Scanning electrochemical microscopy*. Boca Raton: CRC Press.

Bockris, J. O. M., A. K. Reddy, and M. Gamboa-Aldeco. 2002. *Modern electrochemistry. Fundamentals of electrodics*. New York: Kluwer Academic Publishers.

Bojinov, M. 1997a. The ability of a surface charge approach to describe barrier film growth on tungsten in acidic solutions. *Electrochimica Acta*. 42:3489–3498.

Bojinov, M. 1997b. Modelling the formation and growth of anodic passive films on metals in concentrated acid solutions. *Journal of Solid State Electrochemistry*. 1:161–171.

Bojinov, M., I. Betova, and R. Raicheff. 1996. A model for the transpassivity of molybdenum in acidic sulphate solutions based on ac impedance measurements. *Electrochimica Acta*. 41:1173–1179.

Bojinov, M., S. Cattarin, M. Musiani, and B. Tribollet. 2003. Evidence of coupling between film growth and metal dissolution in passivation processes. *Electrochimica Acta*. 48:4107–4117.

Bojinov, M. S., V. I. Karastoyanov, and B. T. Tzvetkov. 2010. Barrier layer growth and nanopore initiation during anodic oxidation of tungsten and niobium. *ECS Transactions*. 25:89–104.

Boukamp, B. A. 1995. A linear Kronig-Kramers transform test for immittance data validation. *Journal of the Electrochemical Society*. 142:1885–1894.

Boukamp, B. A. 2004. Electrochemical impedance spectroscopy in solid state ionics: Recent advances. *Solid State Ionics*. 169:65–73.

Boukamp, B. A. 2008. Complex models in Electrochemical Impedance Spectroscopy. In *8th Symposium on Electrochemical Impedance Spectroscopy*. Trest, Czech.

Breugelmans, T., J. Lataire, T. Muselle, E. Tourwé, R. Pintelon, and A. Hubin. 2012. Odd random phase multisine electrochemical impedance spectroscopy to quantify a nonstationary behavior: Theory and validation by calculating an instantaneous impedance value. *Electrochimica Acta*. 76:375–382.

Breugelmans, T., E. Tourwé, J.-B. Jorcin, A. Alvarez-Pampliega, B. Geboes, H. Terryn, and A. Hubin. 2010. Odd random phase multisine EIS for organic coating analysis. *Progress in Organic Coatings*. 69:215–218.

Breugelmans, T., E. Tourwé, Y. Van Ingelgem, J. Wielant, T. Hauffman, R. Hausbrand, R. Pintelon, and A. Hubin. 2010. Odd random phase multisine EIS as a detection method for the onset of corrosion of coated steel. *Electrochemistry Communications*. 12:2–5.

Brug, G., A. Van Den Eeden, M. Sluyters-Rehbach, and J. Sluyters. 1984. The analysis of electrode impedances complicated by the presence of a constant phase element. *Journal of Electroanalytical Chemistry and Interfacial Electrochemistry*. 176:275–295.

Buttry, D. A., and M. D. Ward. 1992. Measurement of interfacial processes at electrode surfaces with the electrochemical quartz crystal microbalance. *Chemical Reviews*. 92:1355–1379.

Cattarin, S., M. Musiani, and B. Tribollet. 2002. Nb electrodissolution in acid fluoride medium steady-state and impedance investigations. *Journal of the Electrochemical Society*. 149:B457–B464.

Chao, C., L. Lin, and D. Macdonald. 1981. A point defect model for anodic passive films: I. Film growth kinetics. *Journal of the Electrochemical Society*. 128:1187–1194.

Chao, C., L. Lin, and D. Macdonald. 1982. A point defect model for anodic passive films: III. Impedance response. *Journal of the Electrochemical Society*. 129:1874–1879.

Compton, R. G., and C. E. Banks. 2011. *Understanding voltammetry*. London: World Scientific.

Cordoba-Torres, P., T. Mesquita, O. Devos, B. Tribollet, V. Roche, and R. Nogueira. 2012. On the intrinsic coupling between constant-phase element parameters α and Q in electrochemical impedance spectroscopy. *Electrochimica Acta*. 72:172–178.

Darowicki, K. 1994. Fundamental-harmonic impedance of first-order electrode reactions. *Electrochimica Acta*. 39:2757–2762.

Darowicki, K. 1995a. The amplitude analysis of impedance spectra. *Electrochimica Acta*. 40:439–445.

Darowicki, K. 1995b. Corrosion rate measurements by non-linear electrochemical impedance spectroscopy. *Corrosion Science*. 37:913–925.

De Sutter, L., Y. Firouz, J. De Hoog, N. Omar, and J. Van Mierlo. 2019. Battery aging assessment and parametric study of lithium-ion batteries by means of a fractional differential model. *Electrochimica Acta*. 305:24–36.

Diard, J.-P., B. Le Gorrec, and C. Montella. 1997a. Deviation from the polarization resistance due to non-linearity I. Theoretical formulation. *Journal of Electroanalytical Chemistry*. 432:27–39.

Diard, J.-P., B. Le Gorrec, and C. Montella. 1997b. Deviation of the polarization resistance due to non-linearity II. Application to electrochemical reactions. *Journal of Electroanalytical Chemistry*. 432:41–52.

Diard, J.-P., B. Le Gorrec, and C. Montella. 1997c. Deviation of the polarization resistance due to non-linearity III. Polarization resistance determination from non-linear impedance measurements. *Journal of Electroanalytical Chemistry*. 432:53–62.

Diard, J.-P., B. Le Gorrec, and C. Montella. 1997d. Non-linear impedance for a two-step electrode reaction with an intermediate adsorbed species. *Electrochimica Acta*. 42:1053–1072.

Diard, J.-P., B. Le Gorrec, and C. Montella. 1998. Corrosion rate measurements by non-linear electrochemical impedance spectroscopy. Comments on the paper by K. Darowicki, Corros. Sci. 37, 913 (1995). *Corrosion Science*. 40:495–508.

Diard, J., B. Le Gorrec, and C. Montella. 2020. *Handbook of electrochemical impedance spectroscopy*. Available from https://dx.doi.org/10.13140/RG.2.2.27472.33288

Engelhardt, G. R., R. P. Case, and D. D. Macdonald. 2016. Electrochemical impedance spectroscopy optimization on passive metals. *Journal of the Electrochemical Society*. 163:C470–C476.

Esteban, J. M., and M. E. Orazem. 1991. On the application of the Kramers-Kronig relations to evaluate the consistency of electrochemical impedance data. *Journal of the Electrochemical Society*. 138:67–76.

Fasmin, F., B. Praveen, and S. Ramanathan. 2015. A kinetic model for the anodic dissolution of Ti in HF in the active and passive regions. *Journal of the Electrochemical Society*. 162:H604–H610.

Fasmin, F., and S. Ramanathan. 2015. Effect of CO poisoning of PEM fuel cell anode on impedance spectra-simulations. *ECS Transactions*. 66:1–14.

Fasmin, F., and R. Srinivasan. 2015. Detection of nonlinearities in electrochemical impedance spectra by Kramers–Kronig transforms. *Journal of Solid State Electrochemistry*. 19:1833–1847.

Fasmin, F., and R. Srinivasan. 2017. Nonlinear electrochemical impedance spectroscopy. *Journal of the Electrochemical Society*. 164:H443–H455.

Fletcher, S. 1994. Tables of degenerate electrical networks for use in the equivalent-circuit analysis of electrochemical systems. *Journal of the Electrochemical Society*. 141:1823–1826.

Franceschetti, D. R., and J. R. Macdonald. 1977. Electrode kinetics, equivalent circuits, and system characterization: Small-signal conditions. *Journal of Electroanalytical Chemistry and Interfacial Electrochemistry*. 82:271–301.

Gabrielli, C., M. Keddam, and H. Takenouti. 1993. Kramers Kronig transformation in relation to the interface regulating device. Paper read at *Electrochemical Impedance: Analysis and Interpretation* (STP 1188), West Conshohocken, PA.

George, A., M. Amrutha, P. Srivastava, V. Sai, S. Sunil, and R. Srinivasan. 2019. Label-free detection of chikungunya non-structural protein 3 using electrochemical impedance spectroscopy. *Journal of the Electrochemical Society*. 166:B1356.

Gileadi, E. 2011. *Physical electrochemistry: Fundamentals, techniques, and applications*. Weinheim: Wiley-VCH.

Haili, C. 1987. *The corrosion of iron rotating hemispheres in 1 M sulfuric acid: An electrochemical impedance study*. Berkeley: University of California.

Harrington, D. A. 1996. Electrochemical impedance of multistep mechanisms: Mechanisms with diffusing species. *Journal of Electroanalytical Chemistry*. 403:11–24.

Harrington, D. A. 1997. Theory of electrochemical impedance of surface reactions: Second-harmonic and large-amplitude response. *Canadian Journal of Chemistry*. 75:1508–1517.

Harrington, D. A. 2015. The rate-determining step in electrochemical impedance spectroscopy. *Journal of Electroanalytical Chemistry*. 737:30–36.

He, Z., and F. Mansfeld. 2009. Exploring the use of electrochemical impedance spectroscopy (EIS) in microbial fuel cell studies. *Energy and Environmental Science*. 2:215–219.

Hirschorn, B., M. E. Orazem, B. Tribollet, V. Vivier, I. Frateur, and M. Musiani. 2010a. Constant-phase-element behavior caused by resistivity distributions in films II. Applications. *Journal of the Electrochemical Society*. 157:C458–C463.

Hirschorn, B., M. E. Orazem, B. Tribollet, V. Vivier, I. Frateur, and M. Musiani. 2010b. Determination of effective capacitance and film thickness from constant-phase-element parameters. *Electrochimica Acta*. 55:6218–6227.

Hsu, C., and F. Mansfeld. 2001. Technical note: Concerning the conversion of the constant phase element parameter Y0 into a capacitance. *Corrosion*. 57:747–748.

Hurvich, C. M., and C.-L. Tsai. 1989. Regression and time series model selection in small samples. *Biometrika*. 76:297–307.

Kaisare, N. S., V. Ramani, K. Pushpavanam, and S. Ramanathan. 2011. An analysis of drifts and nonlinearities in electrochemical impedance spectra. *Electrochimica Acta*. 56:7467–7475.

Kendig, M., and F. Mansfeld. 1983. Corrosion rates from impedance measurements: An improved approach for rapid automatic analysis. *Corrosion*. 39:466–467.

Kerner, Z., and T. Pajkossy. 1998. Impedance of rough capacitive electrodes: The role of surface disorder. *Journal of Electroanalytical Chemistry*. 448:139–142.

Kerner, Z., and T. Pajkossy. 2000. On the origin of capacitance dispersion of rough electrodes. *Electrochimica Acta*. 46:207–211.

Kong, D.-S. 2010. Anion-incorporation model proposed for interpreting the interfacial physical origin of the faradaic pseudocapacitance observed on anodized valve metals–With anodized titanium in fluoride-containing perchloric acid as an example. *Langmuir.* 26:4880–4891.

Koster, D., G. Du, A. Battistel, and F. La Mantia. 2017. Dynamic impedance spectroscopy using dynamic multi-frequency analysis–A theoretical and experimental investigation. *Electrochimica Acta.* 246:553–563.

Lasia, A. 2002. Electrochemical impedance spectroscopy and its applications. In: Conway B.E., Bockris J.O., & White R.E. (eds), *Modern aspects of electrochemistry*, vol. 32, 143–248. Boston, MA: Springer.

Lasia, A. 2014. *Electrochemical impedance spectroscopy and its applications.* New York: Springer.

Lin, L., C. Chao, and D. Macdonald. 1981. A point defect model for anodic passive films II. Chemical breakdown and pit initiation. *Journal of the Electrochemical Society.* 128:1194–1198.

Macdonald, D. D. 1992. The point defect model for the passive state. *Journal of the Electrochemical Society.* 139:3434–3449.

Macdonald, D. D. 1999. Passivity–The key to our metals-based civilization. *Pure and Applied Chemistry.* 71:951–978.

Macdonald, D. D. 2011. The history of the point defect model for the passive state: A brief review of film growth aspects. *Electrochimica Acta.* 56:1761–1772.

Macdonald, D. D., and M. Urquidi-Macdonald. 1985. Application of Kramers-Kronig transforms in the analysis of electrochemical systems I. Polarization resistance. *Journal of the Electrochemical Society.* 132:2316–2319.

Macdonald, J. R. 1997. Possible universalities in the ac frequency response of dispersed, disordered materials. *Journal of Non-Crystalline Solids.* 210:70–86.

MacDonald, M. A., and H. A. Andreas. 2014. Method for equivalent circuit determination for electrochemical impedance spectroscopy data of protein adsorption on solid surfaces. *Electrochimica Acta.* 129:290–299.

Mandula, T. R., and R. Srinivasan. 2017. Electrochemical impedance spectroscopic studies on niobium anodic dissolution in HF. *Journal of Solid State Electrochemistry.* 21:3155–3167.

Molina Concha, M. B., M. Chatenet, C. Montella, and J. P. Diard. 2013. A Faradaic impedance study of E-EAR reaction. *Journal of Electroanalytical Chemistry.* 696:24–37.

Momma, T., M. Matsunaga, D. Mukoyama, and T. Osaka. 2012. Ac impedance analysis of lithium ion battery under temperature control. *Journal of Power Sources.* 216:304–307.

Moré, J. J. 1978. The Levenberg-Marquardt algorithm: Implementation and theory. In: Watson G.A. (ed), Numerical Analysis. Lecture Notes in Mathematics, vol. 630, 105–116. Berlin, Heidelberg: Springer. https://doi.org/10.1007/BFb0067700

Olsson, D. M., and L. S. Nelson. 1975. The Nelder-Mead simplex procedure for function minimization. *Technometrics.* 17:45–51.

Orazem, M., J. Esteban, and O. Moghissi. 1991. Practical applications of the Kramers-Kronig relations. *Corrosion.* 47:248–259.

Orazem, M. E., and B. Tribollet. 2011. *Electrochemical impedance spectroscopy.* Vol. 48. New Hoboken, NJ: John Wiley & Sons.

Pachimatla, R. 2020. *Mechanistic analysis of electrochemical reactions using nonlinear electrochemical impedance spectroscopy.* Chennai: Indian Institute of Technology.

Pachimatla, R., M. Thomas, O. C. Safeer Rahman, and R. Srinivasan. 2019. Analysis of instabilities in electrochemical systems using nonlinear electrochemical impedance spectroscopy. *Journal of the Electrochemical Society.* 166:H304–H312.

Pajkossy, T. 1997. Capacitance dispersion on solid electrodes: Anion adsorption studies on gold single crystal electrodes. *Solid State Ionics.* 94:123–129.

Pajkossy, T. 2005. Impedance spectroscopy at interfaces of metals and aqueous solutions—Surface roughness, CPE and related issues. *Solid State Ionics*. 176:1997–2003.

Pajkossy, T., T. Wandlowski, and D. M. Kolb. 1996. Impedance aspects of anion adsorption on gold single crystal electrodes. *Journal of Electroanalytical Chemistry*. 414:209–220.

Popkirov, G. 1996. Fast time-resolved electrochemical impedance spectroscopy for investigations under nonstationary conditions. *Electrochimica Acta*. 41:1023–1027.

Posada, D., and T. R. Buckley. 2004. Model selection and model averaging in phylogenetics: Advantages of Akaike information criterion and Bayesian approaches over likelihood ratio tests. *Systematic Biology*. 53:793–808.

Prasad, Y. N., V. V. Kumar, and S. Ramanathan. 2009. Electrochemical impedance spectroscopic studies of copper dissolution in arginine–hydrogen peroxide solutions. *Journal of Solid State Electrochemistry*. 13:1351–1359.

Rao, T. M., R. P. Meethal, M. Amrutha, and R. Srinivasan. 2020. Studies on group IV and V valve metal corrosion in acidic fluoride media. *Journal of the Electrochemical Society*. 167:081505.

Sadkowski, A. 1999. Small signal (local) analysis of electrocatalytic reaction. Pole-zero approach. *Journal of Electroanalytical Chemistry*. 465:119–128.

Sadkowski, A. 2004. On benefits of the zero-pole representation of electrochemical impedance spectroscopy data close to discontinuity point. *Polish Journal of Chemistry*. 78:1245–1253.

Santhanam, S., V. Ramani, and R. Srinivasan. 2012. Numerical investigations of solution resistance effects on nonlinear electrochemical impedance spectra. *Journal of Solid State Electrochemistry*. 16:1019–1032.

Shukla, P. K., M. E. Orazem, and O. D. Crisalle. 2004. Validation of the measurement model concept for error structure identification. *Electrochimica Acta*. 49:2881–2889.

Smiechowski, M. F., V. F. Lvovich, S. Srikanthan, and R. L. Silverstein. 2011. Non-linear impedance characterization of blood cells-derived microparticle biomarkers suspensions. *Electrochimica Acta*. 56:7763–7771.

Springer, T. E., T. Rockward, T. A. Zawodzinski, and S. Gottesfeld. 2001. Model for polymer electrolyte fuel cell operation on reformate feed: Effects of CO, H_2 dilution, and high fuel utilization. *Journal of the Electrochemical Society*. 148:A11.

Srinivasan, R., V. Ramani, and S. Santhanam. 2013. Multi-sine EIS-drift, non linearity and solution resistance effects. *ECS Transactions*. 45:37–50.

Stoynov, Z. 1992. Rotating Fourier transform—New mathematical basis for non-stationary impedance analysis. *Electrochimica Acta*. 37:2357–2359.

Stoynov, Z. 1993. Nonstationary impedance spectroscopy. *Electrochimica Acta*. 38:1919–1922.

Teengam, P., W. Siangproh, A. Tuantranont, T. Vilaivan, O. Chailapakul, and C. S. Henry. 2018. Electrochemical impedance-based DNA sensor using pyrrolidinyl peptide nucleic acids for tuberculosis detection. *Analytica Chimica Acta*. 1044:102–109.

Urquidi-Macdonald, M., S. Real, and D. D. Macdonald. 1986. Application of Kramers-Kronig transforms in the analysis of electrochemical impedance data II. Transformations in the complex plane. *Journal of the Electrochemical Society*. 133:2018–2024.

Urquidi-Macdonald, M., S. Real, and D. D. Macdonald. 1990. Applications of Kramers Kronig transforms in the analysis of electrochemical impedance data—III. Stability and linearity. *Electrochimica Acta*. 35:1559–1566.

Van Ingelgem, Y., T. Breugelmans, R. Pintelon, and A. Hubin. 2011. The detection of non-linearities using odd random phase multisine EIS and its application in corrosion investigations. Boston, MA: The Electrochemcial Society. https://doi.org/10.1149/ma2011-02/21/1671

Venkatesh, R. P., and S. Ramanathan. 2010a. Electrochemical characterization of Cu dissolution and chemical mechanical polishing in ammonium hydroxide–hydrogen peroxide based slurries. *Journal of Applied Electrochemistry*. 40:767–776.

Venkatesh, R. P., and S. Ramanathan. 2010b. Electrochemical impedance spectroscopic studies of copper dissolution in glycine–hydrogen peroxide solutions. *Journal of Solid State Electrochemistry*. 14:2057–2064.

Victoria, S. N., and S. Ramanathan. 2011. Effect of potential drifts and ac amplitude on the electrochemical impedance spectra. *Electrochimica Acta*. 56:2606–2615.

Warner, J. T. 2019. *Lithium-ion battery chemistries: A primer.* Amsterdam: Elsevier.

Wong, D. K., and D. R. MacFarlane. 1995. Harmonic impedance spectroscopy. Theory and experimental results for reversible and quasi-reversible redox systems. *Journal of Physical Chemistry*. 99:2134–2142.

Wrublewski, D. T. 2020. Analysis for science librarians of the 2019 Nobel Prize in chemistry: Lithium-ion batteries. *Science & Technology Libraries*. 39:51–67.

Wu, Y. 2015. *Lithium-ion batteries: Fundamentals and applications*. Boca Raton: CRC Press.

Xu, N., and J. Riley. 2011. Nonlinear analysis of a classical system: The double-layer capacitor. *Electrochemistry Communications*. 13:1077–1081.

Xu, X., A. Makaraviciute, S. Kumar, C. Wen, M. Sjödin, E. Abdurakhmanov, U. H. Danielson, L. Nyholm, and Z. Zhang. 2019. Structural changes of mercaptohexanol self-assembled monolayers on gold and their influence on impedimetric aptamer sensors. *Analytical Chemistry*. 91:14697–14704.

Yuan, X., H. Liu, and J. Zhang. 2011. *Lithium-ion batteries: Advanced materials and technologies*. Boca Raton: CRC Press.

Ziino, E., S. Marnoto, and J. M. Halpern. 2020. Investigation to minimize electrochemical impedance spectroscopy drift. *ECS Transactions*. 97:737.

Index

Note: *Italic* page numbers refer to figures.

Printed in the United States
by Baker & Taylor Publisher Services